Integrated Manufacturing Systems Engineering

IFIP – The International Federation for Information Processing

IFIP was founded in 1960 under the auspices of UNESCO, following the First World Computer Congress held in Paris the previous year. An umbrella organization for societies working in information processing, IFIP's aim is two-fold: to support information processing within its member countries and to encourage technology transfer to developing nations. As its mission statement clearly states,

> IFIP's mission is to be the leading, truly international, apolitical organization which encourages and assists in the development, exploitation and application of information technology for the benefit of all people.

IFIP is a non-profitmaking organization, run almost solely by 2500 volunteers. It operates through a number of technical committees, which organize events and publications. IFIP's events range from an international congress to local seminars, but the most important are:

- the IFIP World Computer Congress, held every second year;
- open conferences;
- working conferences.

The flagship event is the IFIP World Computer Congress, at which both invited and contributed papers are presented. Contributed papers are rigorously refereed and the rejection rate is high.

As with the Congress, participation in the open conferences is open to all and papers may be invited or submitted. Again, submitted papers are stringently refereed.

The working conferences are structured differently. They are usually run by a working group and attendance is small and by invitation only. Their purpose is to create an atmosphere conducive to innovation and development. Refereeing is less rigorous and papers are subjected to extensive group discussion.

Publications arising from IFIP events vary. The papers presented at the IFIP World Computer Congress and at open conferences are published as conference proceedings, while the results of the working conferences are often published as collections of selected and edited papers.

Any national society whose primary activity is in information may apply to become a full member of IFIP, although full membership is restricted to one society per country. Full members are entitled to vote at the annual General Assembly, National societies preferring a less committed involvement may apply for associate or corresponding membership. Associate members enjoy the same benefits as full members, but without voting rights. Corresponding members are not represented in IFIP bodies. Affiliated membership is open to non-national societies, and individual and honorary membership schemes are also offered.

Integrated Manufacturing Systems Engineering

Edited by

Pierre Ladet

Laboratoire d'Automatique de Grenoble
Grenoble
France

and

François Vernadat

INRIA-Lorraine
Villers-Nancy
France

Published by Chapman & Hall on behalf of the
International Federation for Information Processing (IFIP)

CHAPMAN & HALL

London · Glasgow · Weinheim · New York · Tokyo · Melbourne · Madras

Published by Chapman & Hall, 2–6 Boundary Row, London SE1 8HN, UK

Chapman & Hall, 2–6 Boundary Row, London SE1 8HN, UK

Blackie Academic & Professional, Wester Cleddens Road, Bishopbriggs, Glasgow G64 2NZ, UK

Chapman & Hall GmbH, Pappelallee 3, 69469 Weinheim, Germany

Chapman & Hall USA, 115 Fifth Avenue, New York, NY 10003, USA

Chapman & Hall Japan, ITP-Japan, Kyowa Building, 3F, 2-2-1 Hirakawacho, Chiyoda-ku, Tokyo 102, Japan

Chapman & Hall Australia, 102 Dodds Street, South Melbourne, Victoria 3205, Australia

Chapman & Hall India, R. Seshadri, 32 Second Main Road, CIT East, Madras 600 035, India

First edition 1995

Printed in Great Britain by T J Press, Padstow, Cornwall

ISBN 0 412 72680 7

A catalogue record for this book is available from the British Library

∞ Printed on permanent acid-free text paper, manufactured in accordance with ANSI/NISO Z39.48-1992 and ANSI/NISO Z39.48-1984 (Permanence of Paper).

CONTENTS

Foreword

At a time when most industrial companies must lean their management and manufacturing operations, reengineer their business processes and integrate their applications and information systems to be more competitive on a global market, many enterprises face a crucial need for advanced systems engineering tools and methods. Existing and/or new manufacturing systems must be formally modelled, analysed, specified and prototyped before they can be put into operation. This happens at various levels of details and from different angles or viewpoints. Then, business operations can be integrated and operated on the basis of these models.

It is therefore necessary to provide the manufacturing industry with adequate tools and techniques for modelling, analysis, specification and integration of modern manufacturing systems.

Considering this need and the growing interest of the European research community in this field, it was decided in 1993 to organise a European Workshop on Integrated Manufacturing Systems Engineering (IMSE'94) to be held in Grenoble, 12-14 December 1994. This has been made possible with the support of INRIA, and especially INRIA Rhône-Alpes which has accepted to organise the Workshop, in association with EU DG III, IFAC Technical Committee on Manufacturing Modelling, Management and Control, IFIP Working Group WG 5.7 on Production Management, Groupement Scientifique Interdisciplinaire pour la Productique (GSIP) du CNRS and Ecole Nationale Supérieure de Génie Industriel (ENSGI). 20 papers have been selected from the 57 papers of the Workshop proceedings to make this book.

The aim of the book is to present advanced topics on manufacturing systems modelling, specification and integration as a contribution to an emerging discipline called *Enterprise Engineering*. It is intended to be a companion book to the book entitled *Enterprise Modeling and Integration Principles*, also by Chapman & Hall (to be published in 1996).

The book is organised into eight parts. First, an introduction on issues and challenges for integrated manufacturing systems engineering is given. Then, Prof. J. Browne et al. provide a discussion on the emerging concept of the Extended Enterprise, breaking down organisational boundaries and defined as inter-enterprise networking. The third part is devoted to Enterprise Engineering as a new engineering discipline for defining, modelling, analysing and simulating the processes of the manufacturing enterprise in compliance with the CIMOSA framework. The fourth part presents specific methods and tools for business process modelling and analysis while the fifth part concentrates on system specification and formal methods to model more precisely the manufacturing system, and eventually execute the specification. The sixth part presents some detailed manufacturing system analysis methods based on Petri nets. The seventh part deals with manufacturing system coordination and integration. Especially, principles of integration platforms are discussed and illustrated with AMBAS and CCE-CNMA. Finally, the eighth part addresses pre-normative and standardisation issues on enterprise modelling and integration, including a paper on the work by the IFAC/IFIP Task Force on Architectures for Enterprise Integration and a paper on the European Standardisation Committee (CEN) activities.

The editors would like to sincerely thank all authors and anonymous referees for their contribution to this book.

Prof. P. Ladet and Dr. F. Vernadat
Grenoble, June 1995

PART ONE

Introduction

1

The dimensions of Integrated Manufacturing Systems Engineering

P. Ladet[1], *F. Vernadat*[2]
[1]*Laboratoire d'Automatique de Grenoble (LAG), Institut de la Production Industrielle, ENSGI-INPG, 46 avenue Félix Viallet, F-38031 Grenoble, France*
Phone: +33 76 57 48 32; Fax: +33 76 57 43 17; e-mail: joyaux@imag.fr
[2]*INRIA Rhône-Alpes, 46 avenue Félix Viallet, F-38031 Grenoble, France*
Phone: +33 76 57 47 77; Fax: +33 76 57 47 54;
e-mail: Francois.Vernadat@inria.fr

Abstract

Modern, integrated manufacturing systems need to be engineered in a systematic way like any other complex dynamic systems. Due to the extreme complexity and interdisciplinarity nature of manufacturing system design, analysis, reengineering and continuous improvement, and due to the trend for internetworking of enterprises, a new discipline called Enterprise Engineering is emerging. Different aspects or dimensions of Enterprise Engineering in the context of integrated manufacturing systems engineering are reviewed in the paper.

Keywords

Enterprise Engineering, integrated manufacturing systems, interdisciplinarity aspects, integration, system life cycle, complexity

1. INTRODUCTION

Manufacturing enterprises play an essential role in industrialised countries both in terms of employment and revenues. Central to these are manufacturing systems, which have become significantly complex systems to design and to control. They can be defined as socio-economic discrete event dynamic systems.

Nowadays, the economic environment of most manufacturing enterprises is drastically changing. The economy of scale is being replaced by an economy of scope. Customisation, i.e. product adaptation to specific customer requirements, is driving the demand. Globalisation, i.e. the necessity to be present on world-wide markets, implies a timely deployment strategy in terms of product manufacturing and distribution, sometimes forcing strategic alliances with partner companies. Fierce competition with emerging countries forces industrialised countries to produce at lower cost, with higher quality and in shorter delays. Finally, the consequence of widespread automation and the need to gain on productivity are pushing manufacturing enterprises to lean their management and manufacturing operations, therefore employing less and less people.

Proper design or reengineering of their manufacturing systems is now a must for most industrial companies to face international competition. Modern manufacturing systems must be:

- flexible/agile
- reactive
- integrated and
- cost-effective

Manufacturing systems, like any other type of complex systems, need to be designed and engineered in a systematic way by means of structured approaches relying on sound principles and supported by efficient tools and methods. We call such an emerging approach *Enterprise Engineering*.

This book only provides a first step in this direction and deals with:

- the concept of the Extended Enterprise
- some Enterprise Engineering approaches under development
- techniques for business process modelling and reengineering
- formal approaches for (integrated) manufacturing systems specification
- Petri net techniques for modelling and analysing the physical part of the manufacturing system as well as production planning aspects such as process planning
- techniques for manufacturing system coordination and integration (both at the plant level and at the enterprise level), and
- pre-normative and standardisation issues in the areas of enterprise modelling and integration

2. ENTERPRISE ENGINEERING

Enterprise Engineering is the art of designing, implementing, maintaining and continuously improving enterprise systems (processes and components) so that the enterprise can fulfil its mission according to its business objectives.

Enterprise Engineering embraces under one term several engineering disciplines such as industrial engineering, systems engineering, information systems engineering and production engineering. It also relies on techniques from management sciences, organisation sciences, applied mathematics (especially, simulation and Operations Research) or human resource management.

Traditionally, these various disciplines have been applied separately and in ad hoc ways when designing or reengineering some of the enterprise business processes or any part of the enterprise. The overall process remains essentially sequential (performed step by step, "throwing the project over the wall" from one engineering domain to the next one) and technical barriers, mainly due to protected "islands of competence", are typical attitudes encountered in manufacturing systems engineering departments. Organisation and human resource management aspects remain clearly separated from technical aspects such as functional design, information system design, manufacturing plant layout design or computer network system design. Today, most companies realise that management, organisational and technical issues are closely inter-related, especially in the context of integrated systems.

There is a need to break down these barriers and develop structured methodologies to be supported by models and computer tools to quickly and efficiently engineer/reengineer modern manufacturing systems. Such methodologies should help and guide system designers to:

- improve integration of enterprise components to increase enterprise productivity and efficiency on the basis of enhanced communication, cooperation and coordination of enterprise operations. This is the aim of CIM (Computer-Integrated Manufacturing);
- simplify and lean management and manufacturing procedures to gain on costs and delays;

- parallelise work and operate according to the Just-In-Time (JIT) philosophy to reduce time-to-market and reduce backlogs and work-in-process inventories; and
- capitalise on previous knowledge and know-how to learn from past experience and build the enterprise memory, as required by Continuous Process Improvement (CPI).

Methodologies supporting Enterprise Engineering must lead to solutions which help the enterprise to compete on quality, costs and delays (QCD) and to face management of change because of the fierce competition on markets in a rapidly changing world.

Several aspects or dimensions characterising Enterprise Engineering must be considered. In this paper, we focus on:

- the dimension of interdisciplinarity
- the dimension of system life cycle
- the dimension of integration
- the dimension of complexity

3. THE DIMENSION OF INTERDISCIPLINARITY (OR VIEWS)

The complexity of manufacturing systems, as systems created by humans for their own needs, has progressively increased to the point of being comparable to the complexity of natural systems. There are two reasons for this:

- their management and control functions are themselves becoming more and more complex as well as their integration needs; and
- the role of human operators in the control loop, recently rehabilitated after an era of over automation, brings a part of uncertainty into the control process. This is a dimension which is not easy to formalise.

The control of production systems in general, and of manufacturing systems in particular, relies on a representative modelling of these systems as well as the control and management policies to be applied to them. However, each function and each control level, due to its nature and the concepts handled, requires a priori a different modelling tool. For instance, a scheduling function, a machine control function, resource sharing and synchronisation, flow modelling on the shop floor or human operator integration, all call for different modelling and analysis tools.

An interdisciplinary approach for Enterprise Engineering can therefore be defined on a three-layer basis, corresponding to three levels of granularity in system modelling:

1. The first layer concerns the elementary level: A given function, a view of a manufacturing process, a system component are often themselves very complex. Their analysis assumes the availability of methods and modelling tools able to take into account their detailed specificities and intricacies to a level as close to reality as possible. This regularly leads to advanced enhancement of existing tools as well as to the definition of new tools, usually derived from previous ones. Advances in Petri nets provide a good example (see the Manufacturing System Analysis section of this book for more details).

2. The second layer concerns the level of tool cooperation: The tools used to represent a given function or a given mechanism must be able to exchange information, to be synchronised or to cooperate with one another as well as to take into account the function or mechanism environment, i.e. the other functional elements of the system. This is the problem of structured representation of an application perceived from different approaches and disciplinarity angles at different levels of abstraction, each one relying on a different perception of the system modelled in terms of a given tool. The problem is to be able to

interface these tools by the definition of consistent information systems or even a unique information system to support collaborative decision making and engineering thinking.

3. The third layer concerns the global level: It consists in developing an all-embracing modelling environment able to represent the set of all facets of a manufacturing system or more precisely to federate tools which, although remaining different and specialised to the aspects they have been developed for, have been designed in a perspective for integrated analysis and design. This is an area still under development and it is a long way far from complete achievement. Relevant papers related to this area can be found in the Enterprise Engineering section of this book.

For each of these layers and throughout the system development life cycle, an interdisciplinary approach must be used for Enterprise Engineering in general, and for integrated manufacturing systems engineering in particular. This concerns engineering sciences on one hand and management sciences and human sciences on the other hand.

Too often, the design and reengineering of manufacturing systems is primarly considered as a matter of industrial engineering and application domain engineering (such as mechanical engineering, electrical engineering, or food industry engineering for instance), i.e. as a technical problem. In fact, it is also, and sometimes most of all, an organisational problem, a management problem, a human problem and an economic problem. In the case of integrated systems, it also becomes a problem for computer scientists and information system designers.

The challenge in Enterprise Engineering is to develop a framework supported by methodologies and tools to provide these different areas of expertise or viewswith

- a way to understand the "language" or viewpoint of other disciplines
- a way to indicate where and how each one fits in the framework
- a way to federate the viewpoints and expertises of each of them

so that they can operate in synergy to develop/reengineer enterprise systems faster, better and at a reasonable cost to meet specified business objectives.

4 THE DIMENSION OF SYSTEM LIFE CYCLE

The system life cycle covers the set of activities going from enterprise system inception to engineering, implementation, operation, improvement and finally system dismantlement.

In addition to the interdisciplinarity dimension, it is now widely understood that this process is not strictly sequential (it does not follow a "waterfall" approach) but involves a number of iterations, redoings and revisions at all levels. Furthermore, emerging principles of Concurrent Engineering (CE) principles can be applied to the design and engineering of enterprise systems (i.e. for an entire plant or just a subsystem of it). Moreover, since there is always something changing within the enterprise or in its environment, operational and management systems must be regularly analysed to be improved or corrected. These are the areas of strategic management on the administrative side and Continuous Process Improvement (CPI) on the technical side, both defining the needs for Business Process Reengineering (BPR).

New issues in enterprise systems engineering come from the pressure to pay more attention to environmental and energy issues in product and production system design and operation all over the system life cycle. Recycling of product components as well as production system dismantlement become important issues in systems engineering.

The major phases of enterprise system life cycle have been identified and documented by the Purdue Enterprise Reference Architecture (PERA) and the European pre-norm ENV 40 003 - Framework for Enterprise Modelling produced by Comité Européen de Normalisation (CEN). They involve:

- *Business Objectives and Mission Definition*: This concerns strategic planning defined by top management and must answer such questions as: what will be produced, for which market segments, in what quantities and where.
- *Requirements Definition*: In this phase, business users must express in detail what has to be done to achieve business objectives.
- *Design Specification*: This phase covers preliminary design and detailed design of business processes, resource needs and management, information systems and infrastructure and organisation structures. Technology aspects as well as human factors must be considered in terms of required capabilities. Formal description and simulation techniques are extensively used. Exception handling mechanisms must be planned and analysed.
- *Implementation Description*: Decisions on resource selection and layout, data distribution, computer network configuration, etc. are made and documented in this phase.
- *Installation*: Components and programs of the system are implemented and tested.
- *Operation*: This phase is concerned with actual exploitation of the system in its business environment.
- *Management of change/Continuous Process Improvement*: This concerns the identification of shortcomings or defficiencies in the system and their correction.
- *Dismantlement* or shut down.

5. THE DIMENSION OF INTEGRATION

Integration means putting system components together to create a synergistic whole, the capabilities of which encompass the capabilities of each of its components alone. The benefits of integration rely on improved communications, cooperation and coordination (C3) of the business processes of the system.

In the case of integrated manufacturing systems, integration can be achieved at several levels:

- Computer systems integration: This is concerned with physical systems integration in terms of systems interconnections by means of computer networks. The 7-layer ISO-OSI architecture has been the technical reference but Ethernet is dominating the market of local area networks. In the near future, ATM is supposed to be the new standard both for local and wide area networks because of its high-speed and multimedia capabilities.

- Shop-floor integration: This is concerned with manufacturing systems integration and was originally initiated by the MAP (Manufacturing Automation Protocol) initiative in the US followed by the CNMA ESPRIT project in Europe, both based on the ISO-OSI architecture. The MMS (Manufacturing Message System) language is an important result of this work as well as fieldbus definitions (such as FIP and Profibus), which are simplified and faster computer network architectures dedicated for shop-floor applications.

- Plant integration: This form of integration goes one step beyond shop-floor integration and may involve the exchange of engineering data such as drawings and bills of materials. Thus, standard data exchange formats (such as IGES or STEP/EXPRESS for exchange of product and process data and EDI for administrative electronic data interchange) as well as common services are required to make applications communicate or even interoperate. Different kinds of integration platforms or integrating infrastructures known as middleware platforms are being proposed. Examples of generic integration platforms are OSF/DCE and OMG/CORBA. CCE-CNMA is an integration platform dedicated to CIM environments.

- Enterprise Integration: Enterprise Integration concerns the integration of the various business processes of the enterprise to facilitate information, material and decision flows and therefore increase reactivity and productivity. It deals with integration at the business level of the enterprise. It requires a model of the business processes (as provided by Enterprise

Modelling) and an information infrastructure to monitor execution of this model and coordination of the business processes. These are the ideas promoted by CIMOSA, the European open system architecture for CIM.

- Extended Enterprise or inter-enterprise integration: Finally, the last level of integration concerns the Extended Enterprise. While CIM mostly concerns intra-enterprise integration (i.e. integration of the business processes of a given enterprise), the Extended Enterprise is concerned with inter-enterprise integration (i.e. internetworking of enterprises interacting along a common supply chain).

MMS and CIMOSA are covered in the Enterprise Engineering section of the book while the coordination, CCE-CNMA and integrating infrastructure issues are covered in the Manufacturing System Coordination and Integration section. The Extended Enterprise has a section of its own.

6. THE DIMENSION OF COMPLEXITY

Due to requirements for flexibility/agility, reactivity, integration and automation, manufacturing systems, and especially discrete manufacturing systems, become extremely complex systems to design and to control. Their complexity often largely exceeds the capabilities of one person. For instance, let us think about a manufacturing system producing gear-boxes for cars or trucks or an automated system producing electronic chips. Such systems are made of 10 to 20 workcenters, can produce over a hundred of different product types and may involve dozens of different concurrent business processes.

The complexity may come from the large number of system components and their interactions. It may also come from the sophistication of the system.

In terms of systems engineering, we have outlined the interdisciplinary aspects and the need for different types of models to correctly assess the systems. The mixing of competences (engineering, organisational, socio-economic, etc.) is also a factor of complexity in the design of such systems as well as in the design of relevant Enterprise Engineering methodologies.

Another aspect related to complexity concerns *scalability* of models. Currently, various techniques are proposed to model manufacturing systems, their business processes or their information systems. They are usually presented on small toy-examples in the literature. However, in industry people must deal with large models representing real systems. It is therefore important to consider the scalability aspects when proposing a new modelling approach.

7. CONCLUSION

Enterprise Engineering as defined in this paper is a collection of many different activities put together to engineer enterprise systems, and especially integrated manufacturing systems. The challenge of Enterprise Engineering is to provide industry with methods and guidelines for building efficient enterprise systems (processes and components) faster and better. There is therefore a need for developing advanced tools (models and computer tools) as well as suitable methodologies.

Development of these tools and methodologies must take into account the interdisciplinarity nature of such an engineering process, must cover the entire enterprise system life cycle, pay special attention to integration issues and provide mechanisms to face the complexity dimension of the problem.

The aim of this book is to provide a first step in this direction. Especially, it is shown how recent developments in Enterprise Modelling and Enterprise Integration can be combined with Business Process Reengineering (BPR) and manufacturing system specification and analysis techniques (such as formal description techniques and Petri nets) to achieve some of the goals of Enterprise Engineering.

More developments remain to be done to bring Enterprise Engineering to the level of maturity of Software Engineering for example, from which it borrows many ideas and adopts a similar approach. Especially, it is the authors' opinion that more work on formal techniques should be pursued to design robust and reliable systems so that their qualitative and quantitative properties can be more precisely analysed before they are installed and operated.

8 BIOGRAPHY

Pierre Ladet is a French citizen and lives in Grenoble. He got his Doctorat d'Etat degree in 1982 from the Institut National Polytechnique de Grenoble (INPG) in Physical sciences (Automatic control). He his currently a professor at INPG. He has been a research officer at the Automatic Control Laboratory of Grenoble (1982-1995) where he created a research team on discrete events systems.

His research interests include discrete events systems and decision rules modelling using Petri nets. With his research team, he developed a Petri nets methodology and automatic control architecture including scheduling, decision and control aspects.

Professor Pierre Ladet is currently head of the national programme of CNRS « Conception des systèmes de production », an interdisciplinary research action which puts together disciplines from Ingineering Sciences and Human Sciences.

He also a mission officer for the French Ministry of Research and Education.

François Vernadat is a French and Canadian citizen. He got his Ph.D. degree in 1981 from the University of Clermont, France in Electronics and Automatic Control. From 1981 till 1988, he has been a research officer at the Division of Electrical Engineering of the National Research Council, Ottawa, Canada. In 1988, he joined INRIA, a French research institute in computer science and automatic control.

His research interests include CIM database technology and information systems, enterprise modeling and integration, knowledge representation, formal description techniques, Petri nets and manufacturing plant layout design. He has been involved in several ESPRIT projects and was one of the chief architects of CIMOSA. He has published over 90 scientific papers and is co-author of three books. He is the European editor for the *International Journal of CIM* and serves the scientific committee of several journals.

Dr. Vernadat served several times as a technical expert for the CIM program of the Commission of the European Communities (EU DG III). He acts as a French expert for national, European and international standardisation bodies on Enterprise Modelling and Integration (AFNOR, CEN TC 310, ISO TC 184). He is a member of the IEEE Computer Society, ACM and SME.

PART TWO

The Extended Enterprise

2

Industry Requirements and Associated Research Issues in the Extended Enterprise

Jim Browne,
CIMRU,
University College Galway,
Galway, Ireland.
Tel: +353.91.750414
Fax: +353.91.562894
Jimmie.Browne@ucg.ie.

Peter Sackett,
The CIM Institute,
Cranfield University,
Bedford,
U.K.
Tel: +44.1234.754073
Fax: +44.1234.750852
Sackett@cim.cran.ac.uk

Hans Wortmann,
Technical University of Eindhoven,
Eindhoven,
The Netherlands.
Tel: +31.40.472290
Fax: +31.40.451275
hwo@bdk.tue.nl

Abstract

In this paper, we consider the difficulties faced by manufacturing companies and their response in terms of the emergence of the Extended Enterprise. We argue that the Extended Enterprise represents the context within which manufacturing systems research must be conducted and we identify what we consider to be the key topics for future manufacturing systems research and development.

1. THE MANUFACTURING SYSTEMS ENVIRONMENT

Today's manufacturing enterprise operates in a tremendously competitive environment. This competitive environment arises from a series of underlying realities viz:

- Global markets
- Customers demanding high quality, low cost and fast delivery of increasingly customised products - mass customisation.
- The need to develop environmentally benign products and processes.

1.1 Global Markets

Changes in the international trading relationships, manufacturing and information technology and infrastructure developments are leading to the realisation of the global market. In the past, companies operated in local markets, were subject to local cost and pricing structures and competed with local competitors. Trading conditions, customs controls and tariffs provided further barriers to global trade. Today, a vastly superior transport infrastructure, global communications systems, and increasingly open markets (e.g. the recent GATT agreement) have changed all of that.

1.2 The Rise of Mass Customisation

With the development of hard automation in the early 20th century (the transfer line etc.) the age of mass produced standard products began. The relatively low cost of such products allowed significant proportions of the population of the western world to acquire consumer goods. The development of information technology, and it's application to manufacturing technology in the 1960s and 1970s changed everything. Automation based on information technology is inherently different to the older "hard" automation in that it is flexible. Computer based automation facilitates the production of a wider range of products in lower volumes economically. Today we are approaching the stage where "mass customisation" is becoming a reality. In short, manufacturers are increasingly required to produce customised products at mass production prices.

1.3 Environmentally Benign Products and Processes

Society is putting pressure on manufacturers, to create production systems which are neutral with respect to the environment. This pressure may take the form of legal regulations, economic and marketing requirements.

Today progressive manufacturing companies are developing a total life cycle approach to their products. One multinational supplier of telecommunications equipment and services to the European market has developed a “Product Life Cycle Management” program which currently includes six major activities, namely: design and technology, purchasing of supplies and materials, manufacturing processes, waste reduction and energy management, packaging and post consumer materials management. This company now includes environmental considerations as part of it’s supplier qualification process. The “post consumer materials management” activity represents a long term challenge but is based on the following ideas : in the near future, manufacturing will refurbish and reuse products and recycle as much of their contents as possible; finally companies will source new markets for the recycled material and safely dispose of residual materials when necessary.

Tipnis (1993) makes reference to the "expanded responsibility" of the manufacturer over the entire life cycle and talks about the design of products for sustainability. In his view design for sustainability provides specific targets for design for manufacture, assembly, service, disassembly, and recycling.

In Europe, the Council of the European Union has issued several directives regarding the environmentally sound production, distribution, use and disposal of products. These directives include the principles of the civil liability of the manufacturer for environmental damage caused by his products, financial incentives to achieve effective protection of the environment such as taxes on the use of damaging materials, material disposal charges, independent environmental audits and charges on non-biodegradable materials such as certain plastics. It is likely that in the near future manufacturers will be required to recycle obsolete products themselves or face high public disposal charges.

2. THE EMERGENCE OF THE EXTENDED ENTERPRISE

How can a manufacturing company respond to the pressures placed on it by global competition? In the past the response was to make the factory more efficient and responsive by developing CIM (Computer Integrated Manufacturing) solutions. Now, however, the challenge is greater and requires that a degree of integration takes place across the whole value chain.

De Meyer suggests that manufacturing must "see itself as *a link in an integrated value added chain*, whose goal is to serve the customer" (my emphasis). Coming from a very different perspective, namely that of environmentally benign production, Tipnis (1993) also suggests a similar view when he subtitles his paper "How to design products that are environmentally safe to manufacture/assemble, distribute, use, service/repair, discard/collect, disassemble, recover/recycle and dispose?". It is clear that the manufacturing function must look beyond "the four walls of the manufacturing plant". Global competition and the emerging pressures to develop environmentally benign products and processes, force manufacturing professionals to take a broader view. We term this broader view the "Extended Enterprise".

We believe that the market place to which manufacturing businesses must respond includes:-

- Business processes which cross enterprise boundaries to interface with functional areas in other companies, for example, product design or manufacturing process definition.
- Supplier/customer integration (people and processes) through interchange of commercial/technical data.
- Ability to function effectively as links for information and product in unbuffered supply/distribution chains.

The ability to network the activities of a number of entities to produce and sell manufactured products profitably depends on the relationship of these entities and the communication that passes between them. We are accustomed to thinking about this in the context of a single enterprise with different departments, Sales, Design, Engineering, Manufacturing, Distribution etc. However, within a global market-place, entities from many different enterprises, or entities which in themselves are nominally independent enterprises, relate via a single product to produce a designed result. An example might be a merchandising entity recognising a business opportunity and requesting :-

- a design entity to design it;
- a manufacturing entity to build it;
- a distribution entity to distribute it and;
- a marketing entity to sell it.

The implication of such an example is that all the entities can be considered as "flexible" or "programmable" within their expertise envelope.

This view of the Extended Enterprise is represented in Figures 1 and 2. These figures also serve to illustrate the context for manufacturing research and development. In the past the emphasis was on integration inside the four walls of the manufacturing plant [See Figure 1].

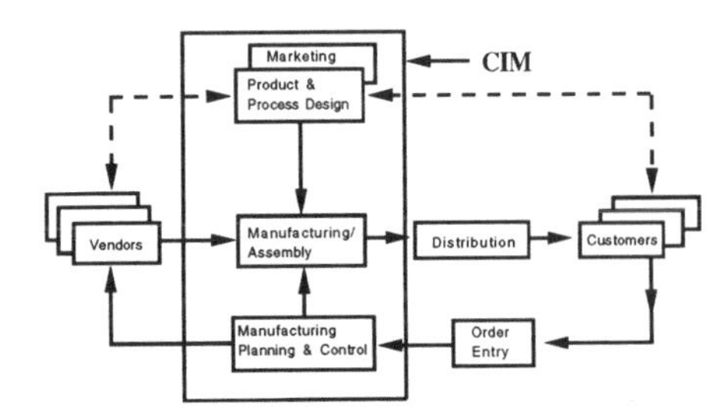

Figure 1 : CIM

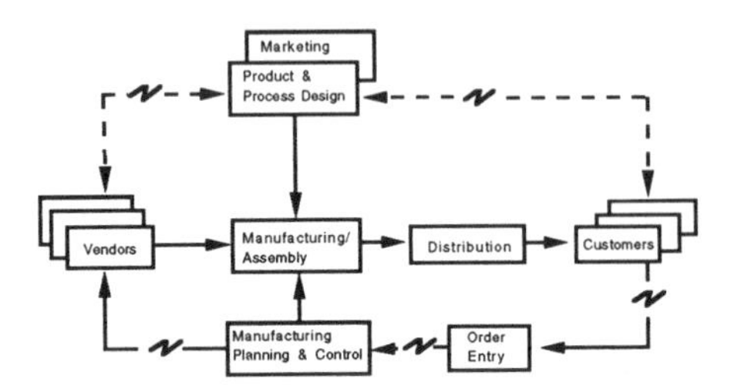

Figure 2 : The Extended Enterprise

CAD/CAM Integration, Integration of Production Planning & Control Systems, the development of sophisticated manufacturing processes and their control through sophisticated shop floor control systems, formed the agenda. Today the emphasis has changed to include the supply chain (integration of the supply chain through EDI & JIT), customer driven manufacturing including for example, the integration of manufacturing and distribution planning and control systems. In fact this issue of the integration of manufacturing and distribution is a good example. In the past, the two systems were seen to be quite separate and decoupled by warehouses and storage points of various descriptions. (See Figure 3).

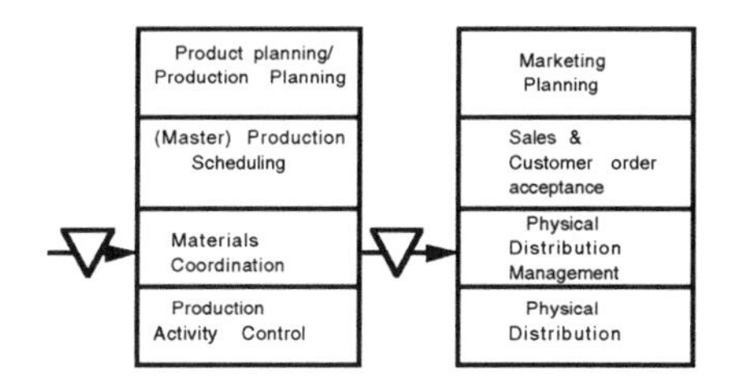

Figure 3 Manufacturing Decoupled from Marketing and Physical

Today with the need to support customization and to realize shorter product delivery lead times, we need to develop integrated systems (See Figure 4).

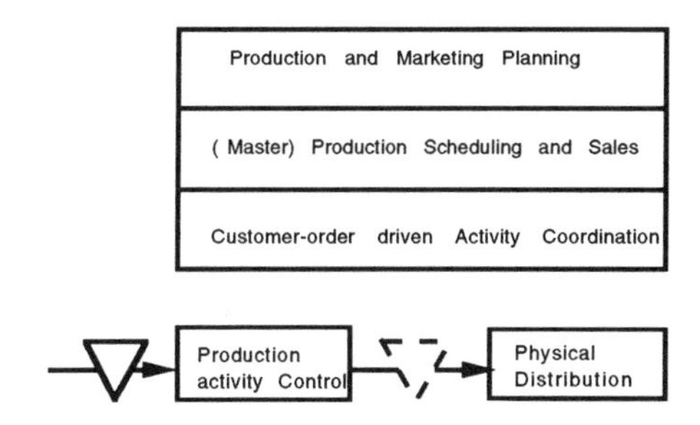

Figure 4 Manufacturing and Physical Distribution driven by Customer Orders.

The emergence of the Extended or Networked Enterprise is of course facilitated by today's emerging computing and telecommunications technologies. EDI is one such technology as indicated in Figure 2. (See Browne, Sackett and Wortmann, 1994).

3. THE EXTENDED ENTERPRISE-EVOLUTION OR REVOLUTION?

We believe that this notion of the Extended (or Networked) Enterprise represents in some ways, a logical development of much of the efforts of manufacturing specialists over the past ten years or so. For the past ten years manufacturing systems specialists have evolved a series of approaches to the design and operation of manufacturing systems. In our view, the drive to inter enterprise networking represents an extension and a synthesis of many of these approaches. Approaches, such as MRP, JIT, EDI, WCM, Concurrent Engineering, Lean Production, Benchmarking and Business Process Redesign, essentially synthesise into the inter enterprise networking model. In this section we briefly outline these approaches and argue that they culminate in Inter Enterprise networking.

The ideas of MRP, originally developed by Orlicky of IBM in the 1960s were important not least because they taught us the importance of hierarchical planning and the involvement of many department and functions within manufacturing to solve materials planning and control problems. Later in the late 1970s and early 1980s, JIT ideas originally developed in the Japanese automotive industry, became known in Europe and the U.S. JIT emphasised customer involvement in the final scheduling of production systems and close co-operation with suppliers to achieve quality and timely delivery. Thus already in the early 1980s JIT began to focus the view of manufacturing systems specialists on issues outside the four walls of the manufacturing plant, viz. customer requirements of timely delivery and supplier involvement.

In many ways the ideas of the WCM (World Class Manufacturing) school (See for example Schonberger 1990) developed from the experience of JIT implementation studies in the U.S.A. Issues of continuous improvement, training of people, integration of product design and process design to facilitate efficient manufacturing were also emphasised. Hayes, Wheelwright & Clark for example identified the key characteristics of world class manufacturing as follows :

1. Becoming the best competitor. 'Being better than almost every other company in your industry in at least one aspect of manufacturing'.
2. Growing more rapidly and being more profitable than competitors. World-class companies can measure their superior performance by 'observing how their products do in the marketplace and by observing their cashbox'.
3. Hiring and retaining the best people. 'Having workers and managers who are so skilled and effective that other companies are continually seeking to attract them away from your organisation'.
4. Developing engineering staff. 'Being so expert in the design and manufacture of production equipment that equipment suppliers are continually seeking one's advice about possible modifications to their equipment, one's suggestions for new equipment, and one's agreement to be a test site for one of their pilot models'.
5. Being able to respond quickly and decisively to changing market conditions. 'Being more nimble than one's competitors in responding to market shifts or pricing changes, and in getting new products out into the market faster than they can'.
6. Adopting a product and process engineering approach which maximises the performance

of both. 'Intertwining the design of a new product so closely with the design of its manufacturing process that when competitors "reverse engineer" the product they find that they cannot produce a comparable one in their own factories without major retooling and redesign expenses'.

7. Continually improving. 'Continually improving facilities, support systems and skills that were considered to be "optimal" or "state of the art" when first introduced, so that they increasingly surpass their initial capabilities'. Hayes, Wheelwright, and Clark also go on to say that 'the emphasis on continual improvement is the ultimate test of a world class organisation'.

Meanwhile the proponents of Concurrent Engineering (CE) begin to emphasise the issue of product design time and to research business, technological and organisational themes in order to reduce time to market It is interesting that one of the important themes in CE is the development of standards to support the exchange of products data between the CAD systems of suppliers and their customers. The work on CE is of course ongoing within the ESPRIT and BRITE-EURAM programmes of the EU and the CALS initiative in the U.S.A.

EDI technology emerged in the late 1980s. Initially used to support business transactions (invoicing, purchase order call off) between suppliers an their customers, it is now beginning to be used to exchange technological product data. Today's most advanced manufacturing companies use EDI to exchange production and purchasing information and to support joint (with suppliers and/or customers) engineering development teams. The STEP initiative is now rapidly moving into industrial practice. Tools are under development to facilitate the use of STEP data including engineering data exchange, data storage and archiving across distributed databases.

Womack et al (1990) defined 'Lean Production' as the successor to mass production. Like mass production, the ideas of lean production were initially developed in the auto industry. Mass production arose in the U.S.A. in the early 20th century in the Ford Motor Company. Lean production developed initially in the Toyota plants in Japan at the end of the 20th century. Clearly our ideas on Lean Production are still under development. According to Jones (1992) the essential characteristics of Lean Production can be summarised as follows :

1. It is customer driven, not driven by the needs of manufacturing.
2. All activities are organised and focused on a product line basis led by a product champion, with functional departments playing a secondary, servicing role.
3. All activities are team based and the organi sation is horizontally and not vertically oriented.
4. The whole system involves fewer actors, all of whom are integrated with each other - 330 engineers in the product-development team versus 1400, 340 suppliers versus 1500, about 300 dealer principals versus 3600 (to sell 2 million vehicles) and 2000 assembly employees versus between 3000 and 5500 (for plant assembling 250 000 units a year).
5. There is a high level of information exchanged between all the actors and a transparent and real cost structure.
6. The activities are co-ordinated and evaluated by the flow of work through the team or

the plant, rather than by each department meeting its plan targets in isolation.

7. The discipline necessary for the system to function and expose problems is provided by JIT and Total Quality in the plant and supplier and dealer performance evaluation.
8. Wherever possible responsibility is devolved to the lowest level possible, in the plant or to suppliers.
9. The system is based on stable production volumes but with a great deal of flexibility.
10. Relations with employees, suppliers and dealers are based on reciprocal obligations that are the result of treating them as fixed costs.

Thus Lean Production addresses Product Strategy, Product Development, the Supply Chain, Manufacturing and Product Distribution. Essentially it addresses many issues in the value chain which results in a networking of customers, assemblers and suppliers.

In more recent years Benchmarking and Business Process Redesign have emerged. The central idea of benchmarking is to decompose a business into it's essential processes, identify examples of best practice for each of these practices and then define an approach to achieve best practice. Thus benchmarking seeks to improve existing processes. Business process redesign (sometimes termed business process re-engineering) offers a more radical approach. It seeks to identify each business process, question it's relevance to the achievement of business objectives and redesign the overall business to incorporate only appropriate processes. In fact benchmarking may well be used following business process redesign to identify and achieve best practice for each individual process. For a detailed discussion on business process redesign and benchmarking see Hammer and Champy (1993). It is worth noting also that most of the work on benchmarking and business process redesign was done in the U.S.A. Also the software tools for business process redesign come primarily from North America.

Clearly many of the ideas and approaches outlined here support the manufacturing enterprise in it's evolution towards the integrated enterprise model we have defined earlier. However the Integrated Enterprise represents a quantum leap for manufacturing systems. The changes which it requires are so far reaching that, in our view, research and development work is necessary if industry is to be successful in the 21st century. Our view of the topics which require research and development effort is presented in Section 4 following.

4. RESEARCH ISSUES IN THE EXTENDED ENTERPRISE

Here we will try to present the topics which we believe need to be addressed by manufacturing systems researchers, in order to facilitate the emergence of the Extended Enterprise. We will identify the topics through five headings, namely :

- Inter-Enterprise Networking
- Concurrent Engineering
- Organisational Learning
- Appraisal Methodology
- Manufacturing Information Systems

Further, we will try to identify whether the work to be done is short term, medium term or long term.

4.1 Inter Enterprise Networking

The extended enterprise is an expression of the market driven requirement to embrace external resources in the enterprise without owning them. Core business focus is the route to excellence but product/service delivery requires the amalgam of multiple world class capabilities. Changing markets require fluctuating mixes of resources. The extended enterprise, which can be likened to the ultimate in customisable, reconfigurable manufacturing resource, is the goal. The process is applicable even within large organisations as they increasingly metamorphasise into umbrellas for smaller business units/focused factories.

The operation of the extended enterprise requires take up of communications and database technologies which are near to the current state of the art. However the main challenge is organisational rather than technological. Concerns experienced in the flatter organisations developed in Computer Integrated Manufacturing type business enterprises such as trust, credibility and project management assume a much higher profile in the extended enterprise.

Research and Development Opportunities

Short Term

1. A methodology to determine and support the information processes in the extended enterprise. This work will build on process modelling but extend beyond the traditional single business boundary. It will provide the basis for effective extended enterprise operation by defining what information should be communicated and when and where decisions must be made.

Medium Term

1. Reverse Logistics. Today very sophisticated systems exist to distribute product from the producer to consumers. Virtually no systems exist to return products at end of life. Procedures and supporting systems and tools are needed to enable product return and reuse.

2. Pre Qualification of Partners. The extended enterprise is a volatile environment; partners will regularly join and rejoin a new 'consortium' of enterprises. Pre qualification tools to determine entry level and acceptability are required. These need to embrace physical process and knowledge work capability.

3. Architecture for engineering partnership, an abstract representation of an extended enterprise engineering partnership and associated systems architecture. This would support concurrent engineering across the extended enterprise and, through specification to participating partners, enable realisation of short Time to Market in this environment.

Long Term

1. Business and legal framework to facilitate the emergence of transient integrated enterprises. Issues to be researched, understood and articulated include product liability responsibility across the value chain, recyclability issues, ownership of individual enterprises and products, intellectual property rights etc.

4.2 Concurrent Engineering (CE)

The UK CE industrial forum has defined Concurrent Engineering as 'delivery of better, cheaper faster products to market by a lean way of working using multidisciplinary teams, right first time methods and parallel processing activities to continuously consider all constraints'. CE must adapt a true life cycle view. This implies consideration of issues such as maintainability, upgradability, recycleability, use of environmentally benign processes, reuse of scarce materials etc. at the earliest stages of design.

Tools in CE can be broadly classified under three categories; management based encompassing such things as team working, project management, and formal methods of geometric manipulation of drawings and models. There has been a focus on tools and techniques and a lack of research into how, when and in what order to implement them. The extended enterprise poses special challenges for the realisation of Concurrent Engineering.

Research and Development Opportunities

Short Term

1. New organisational structures for the organisation of concurrent engineering. The distribution of tasks to various members of a product development team is a very important issue. A related question is how to create cross-functional task teams in such a way that:
- traditional accounting procedures do not hamper teamwork;
- the reward system can be based on an individual's performance in projects rather than in the functional organisation;
- reuse of experience is guaranteed;
- learning is part of the task assignment.

In general, the transition from a traditional hierarchical organisation to a networking organisation is especially important for concurrent engineering.

2. Design Tools to Support the Creation of Environmentally Benign Products and Processes. Today we design for manufacture and assembly. In the near future we need to embrace design for disassembly, design for reduced environmental impact, design for reverse logistics, design for reuse and refurbishment, design for a total life cycle etc.

3. Training Systems for use in extended enterprise rapid configuring teams. Multimedia also offers significant potential benefits here.

Medium Term.

1. Self analysis of design performance. This includes methodologies and systems which prompt the user for information on the design process, map this against established benchmarks and offer specific advice of action, consequences, problems, solutions.

2. Multi function design team support, for use in managing, setting up and participating in extended enterprise design teams.

3. IT tools for the large team environment. Classical CE work has envisaged the compact team. Large multi site, multi sub system design problems require a new infrastructure. Effective operations requires attention to the design process and information management and presentation to minimise information overload. Mutimedia may be effective here.

Long Term

1. Development of a Negotiation Environment in Concurrent Engineering. This environment will facilitate distributed design activity and will manage the interaction between designers by facilitating appropriate local action.

2. The development of design methodologies to incorporate a true product life cycle view at the earliest (i.e. conceptual) stage in design. [Long Term, Industry and Research Institutes].

4.3 Organisational Learning

Competitive advantage is provided by the ability to learn faster than one's rivals. In an environment of uncertainty training does not equip staff with the necessary manufacturing problem solving capacity either in individual or collaborative mode. The central idea of organisational learning is that organisational capability to deal with changes and uncertainty can be fostered, its formal basis defined and best practice replicated and augmented. So in the Learning Organisation, 'organisations', including individuals, can learn from experience to recognise problems and either solve them or initiate successful solution methodologies. In many ways the Learning Organisation adopts the characteristics of a living system. The key difficulties with the concept are what constitutes learning and how to embody it and how to define problems. It is inappropriate to adopt human learning processes and theories directly as a metaphor for the learning which occurs in organisations.

Research and Development Opportunities

Short Term

1. Formation of a learning process designed to maintain an organisation's competitive advantage, enabling it to selectively customise and develop its own emergent manufacturing techniques.

Medium Term

1. Development of a process for managing learning strategically and tactically within an organisation, capitalising on determining the sources of success and failure and hence identifying best practice.

2. Metrics for assessing knowledge generation and use. Knowledge is a key asset of the extended enterprise. Today no widely accepted examples exist with which to measure this important resource.

3. Development of a programme of action-learning research in which the process under development in the previous item is implemented in up to five organisations over a 2-year period. This will give the host organisations a chance to be at the leading edge of organisational development through sharing the necessary teaming, problem-solving and learning skills to implement a strategic learning approach within their organisations.

4.4 Appraisal Methodology within the Extended Enterprise

Decision support aid in option planning offers help to managers in making effective and defensible decisions about organisation and technology. It concentrates on value rather than mechanism: on why to change manufacturing systems or establish new ones rather than how to do it.

Economic models for manufacturing organisation assume that the two basic forms of combination are markets and hierarchies. In the Extended Enterprise this analysis needs to be revised. Extended enterprises retain the disciplines of the market, yet establish a relationship between supplier and customer that outlasts and goes far beyond single commercial transactions.

Investment Appraisal includes a number of related areas of work. The first, concerns the explanation of various kinds of manufacturing organisation: why supply chains should be integrated within the umbrella of a single firm, how the concentration of suppliers and buyers in a particular market can be explained, where economies and diseconomies of scale lie, and so on. A second area of work covers the development of decision making tools such as scoring and finance models. These are sometimes derived from theories about how people make choices, and sometimes rest simply on the basis of their intuitive appeal. The main difficulty is that the parameters are difficult to estimate with any degree of objectivity. A further area of interest is the analysis of the risk associated with new industrial technologies. Probabilistic models are the more traditional approach. In essence they capture the idea that the future will take one of a number of possible states, and that a decision maker will choose one of a number of alternative courses of action based on their anticipated payoffs in each of those states. Semantic models substitute terms in natural language for both the probabilities of future states occurring and for the payoffs. One example of such work is the AMBITE (Advanced Manufacturing Business Integration Tool for Europe) project with BRITE-EURAM program of the EU. This project is using semantic modelling approaches to analyse the impact of technology investment decisions on the manufacturing performance of an enterprise.

Research and Development Opportunities

Medium Term

1. A rigorous methodology to determine the basis of partnership in extended enterprises should be established. This will lead to a prescription of how industrialists need to conduct their business in the resulting organisation.

2. Development of financial value estimating systems. Current financial value estimates are too simplistic and do not properly consider growth options. This is particularly significant in the Extended Enterprise.

Long Term

1. Risk analysis. The organisation structure proposed requires better tools to assess the degree to which qualifications of costs and benefits is subject to error. Simple probabilities are less useful than methodologies which address how people react in the face of uncertainty and how they act to mitigate the effects of uncertainty.

2. Value chain costing and appraisal. The combination of environmental and society concerns relating to manufactured product life cycles and the extended enterprise is a new area. This is likely to have significant impact on the development of competitive and responsive extended enterprise operators.

3. Integrated methodologies. Prescriptive and descriptive models which connect general themes such as economies of scope, environmental impact, with the search for new investment opportunities, with the metrics by which these opportunities are assessed, and the way in which uncertainty about these opportunities can be expressed and managed.

4.5 Manufacturing Information Systems

The extended enterprise manufacturing system requires channels that convey information from one system to another. Without these connections the gains in productivity and flexibility that we look for are compromised by mismatches and rigidities at their boundaries.

One form of this problem involves the integration of existing, proprietary applications that have been built on heterogeneous infrastructures: differing machines, operating systems and database management systems. We have to retrofit to relatively impenetrable human and technical systems the mechanisms that will make their operation cohesive in an extended enterprise environment. In particular we have to provide for product traceability and potential liability arising from environmental legislation and value chain costing.

Distributed database mechanisms offer a potential solution. By presenting the data on which applications work in a uniform, self-consistent way we can make sure they share the same view of the factory. Where the same data is stored in more than one of the local systems it should have the same meaning, if not an identical form, in all of them. The object transcrip-

tion approach, avoids some of the intricacies of distributed databases. The movement of data from one local system to another takes place in complex objects of pre-defined content. Such schemes avoid having to reconcile the different data models and manipulation languages underlying the different applications.
Apart from data infrastructure the extended enterprise, extra enterprise teaming, and prosumer product configuration, require new information communication mechanisms.

Developments in communications technology such as mobile telephones, fax, electronic mail, video conference and even express delivery of conventional mail have been widely adopted. Advances in computer communications have been driven by the need to transfer data from one computer to another. Only electronic mail has attempted to exploit the capability to allow one computer user to communicate directly with another user. Multimedia is the combination of appropriate communications media to best represent human readable information. It can be particularly useful to non specialist users. It allows customer input to the manufacturing system in a novel way. It allows remote site concurrent engineering teams to hold meaningful dialogue. It offers scope for bandwith effective communication of graphical information. It is near language independent.

The potential for multimedia as an extended enterprise user-to-user communications medium is largely unexplored. Most of the physical infrastructure required is being put in place. Telecommunications utilities are proceeding with the implementation of ISDN services, and there is a large base of existing users with demanding communications needs.

Research and Development Opportunities

Medium Term

1. Alternative integration methods. Different business situations place differing demands on information channel cost, quality, volume, graphical and textural content. Methodologies appropriate to these environments need to be readily mapped to the need and the implications for business strategy.

2. Integrating remote applications. Methodologies to resolve inconsistent data meanings and structures across the extended enterprise.

3. Authoring tools for manufacturing multimedia messages. Currently comprehensive tools and author expertise are expected, and development time and cost are justified by having many receivers and an extended lifetime for every message. This is not appropriate for a communications application, where there may be one receiver and a short message lifetime.

Long Term

1. Information Systems to support reverse logistics. In the long term it may be necessary to track each product throughout it's total life cycle and to maintain data on this product to support refurbishment, remanufacturing and ultimately reuse. This presents an enormous challenge to information technology.

2. Product traceability. The determination of a process to cost effectively allow environmentally sensitive product to be tracked in field. [Medium

3. Product description communication. Methodologies for cost effective transmission and control of appropriate graphical, pictorial and textural data to support the business and product development process. This would allow best use of language independent moving image technology.

CONCLUSIONS

In this paper we have argued that the Extended Enterprise forms the context for today's research and development efforts in manufacturing systems. We believe that the developments over the past ten years or so in manufacturing systems engineering, including for example Lean Production and World Class Manufacturing facilitate the emergence of the Extended Enterprise. Finally we presented our ideas on the research agenda for the future in manufacturing systems.

REFERENCES

- Browne, J., Sackett, P.J., Wortmann, J.C. (1993, November), "The System of Manufacturing: A perspective study", Study for DGXII of the EU.

- Browne, J., Sackett, P.J., Wortmann, J.C. (1994, June) "The Extended Enterprise - A Context for Benchmarking", Presented at the IFIP WG5.7 Workshop on Benchmarking - Theory & Practice in Trondheim, Norway.

- De Meyer, A. (1992, December), "Creating the Virtual Factory" - Report of the 1992 European Manufacturing Futures Survey", INSEAD, France.

- R. Hayes, S. Wheelwright & W. Clark (1988), "Dynamic Manufacturing - Creating the Learning Organisation", Collier MacMillian, London.

- M. Hammer & J. Champy (1993), "Re-engineering the Corporation - A Manifesto for Business Revolution", Nicholas Brealey Publishing.

- D.T. Jones (1992): "Beyond the Toyota Production System : the era of Lean Production" in "Manufacturing Strategy - Process and Content", Chapman & Hall, 1993, Editor C. Voss.

- R.J. Schonberger (1990), "Building a Chain of Suppliers", Hutchinson Business Books, USA.

- Tipnis, V.J. (1993), "Evolving Issues in Product Life Cycle Design", Annals of the CIRP, Volume 42, No.1.

BIOGRAPHY

Professor Jim Browne is Director of the CIM Research Unit (CIMRU) at University College Galway in Ireland. He has been engaged as a consultant by companies such as Digital Equipment Corporation, Westinghouse Electric, Apple Computers, De Beers, AT Cross and a number of Irish manufacturing companies, to work in areas such as manufacturing strategy, manufacturing systems simulation and production planning and control. He has also been engaged by the European Union to work on the development of the ESPRIT programme (DGXII) and the BRITE programme (DGXII). He has been engaged in EU research projects with industrial companies such as COMAU in Italy, Renault in France, DEC in Munich, Alcatel in France and a number of European Universities, including the University of Bordeaux, the University of Eindhoven, the University of Berlin and the University of Trondheim.

For the past two years, he has been engaged as a consultant on a project funded jointly by the EU and the State Scientific and Technical Committee of China (SSTCC) - (Peoples Republic of China) to support the National CIMS (Computer Integrated Manufacturing Systems) project of China.

Note :
The authors would like to thank Mr Antonio Colaco of DGXII in the Commission of the European Union for insightful comments on an earlier draft of this paper.

PART THREE

Enterprise Engineering

3

Enterprise Engineering with CIMOSA - Application at FIAT

K. Kosanke
CIMOSA Association (e.V.)
Stockholmer Str. 7, D-71034 Böblingen, Germany,
Tel (49) 7031 27 76 65 Fax (49) 7031 27 66 98 e-mail ko@ipa.fhg.de
M. Mollo
SESAM
Corso Frattelli Cervi 79, I-10095 Grugliasco, Italy,
Tel (39) 11 78 04 667, Fax (39) 11 68 63 708
F. Naccari
FIAT Servizi per Industtria SPA
Via Marochetti 11, I-10126 Torino, Italy,
Tel (39) 11 68 62 505, Fax (39) 11 68 63 708
C. Reyneri
FIAT Auto - MAINS S.r.l.
Via Issiglio 63/A, I-10141 Torino, Italy,
Tel (39) 11 68 52 167, Fax (39) 11 68 53 099

Abstract

CIMOSA will support the increased need for real time and up-to-date information through identification of available information and its use in the operational processes of the manufacturing enterprise. Process based enterprise modelling also allows to describe the operational dynamics as well as resources needed and related organisational aspects. Such models may be used in decision support designing and evaluating operation alternatives through simulation. Experiences from model engineering obtained in applying CIMOSA in FIAT manufacturing are reported in the second part of the paper. The business as well as engineering value of CIMOSA has been identified in terms of significant reduced model engineering time, very much increased modelling flexibility and much more meaningful analytical capability.

Keywords

Open System Architecture, CIMOSA, CIMOSA Application, Enterprise Integration, Enterprise Modelling. Simulation

PART I: CIMOSA ENTERPRISE ENGINEERING

1 INTRODUCTION

The manufacturing industry is deeply in the paradigm shift from manufacturing of scale and manufacturing automation to flexible manufacturing and management of change. This shift brings about an increasing need for up-to-date information, available at the right place, in the right time, with the right amount and the right format. This in turn requires improved orientation in the ocean of information which is available in any sizeable manufacturing enterprise. Information which is created at many different places during the product life cycle and which is needed in many more places to allow efficient continuation and completion of the business processes.

The most effective way to identify the needed information and obtain access to it is via models of the enterprise operation which cover not only its business processes but identify both their internal and external relations as well. However, to achieve sufficient flexibility for maintaining the models and keeping them really up-to-date modelling has to become a tool not only for planning but for operational support as well. For this goal model engineering has to be based on industry standards providing for industry-wide understandability and even more important interchangeability.

The ESPRIT Consortium AMICE[1] has developed, validated, verified and introduced into industry as well as standardisation the CIMOSA (Open System Architecture for CIM) concept of enterprise engineering and model driven enterprise operation control and monitoring (ESPRIT Consortium AMICE, 1993). A concept which has been evaluated and validated by other ESPRIT projects (CIMPRES, CODE, VOICE), professional societies (IFIP/IFAC, others), independent organisations in many countries (China, France, Germany, Hungary, Japan, Switzerland, others) and last but not least by AMICE member organisations. National, European and international standardisation bodies (CEN, ISO) have started normative work to establish standards based on these concepts (CEN, 1990).

CIMOSA is an ESPRIT[2] supported pre-normative development aimed on process based enterprise modelling and application of these models in model driven enterprise operation control and monitoring. Process based enterprise models are an excellent means to structure and identify the information created and used during the production processes and thereby make it available from any place in the enterprise at any time with the amount and in the format needed during a particular process step. CIMOSA consists of an Enterprise Modelling Framework and an Integrating Infrastructure.

2 CIMOSA MODELLING FRAMEWORK

CIMOSA is an ESPRIT supported pre-normative development aimed on process based enterprise modelling and application of these models in model based operation control and monitoring. Process based enterprise models are an excellent means to structure and identify the information created and used during the production processes and thereby make it available from any place at any time with the amount and in the format needed at a particular process

---

1 European CIM Architecture - in reverse.

2 European Strategic Programme for Research and Development in Information Technology.

step. Such models may be used in decision support for engineering operation alternatives and evaluating those through simulation.

The modelling framework shown in Figure I.1 structures the CIMOSA Reference Architecture into a generic and a partial modelling level each one supporting different views on the particular enterprise model. This concept of views allows to work with a subset of the model rather than with the complete model providing especially the business user with a reduced complexity for his particular area of concern. CIMOSA has defined four different modelling views (Function, Information, Resource and Organisation). However this set of views may be extended if needed.

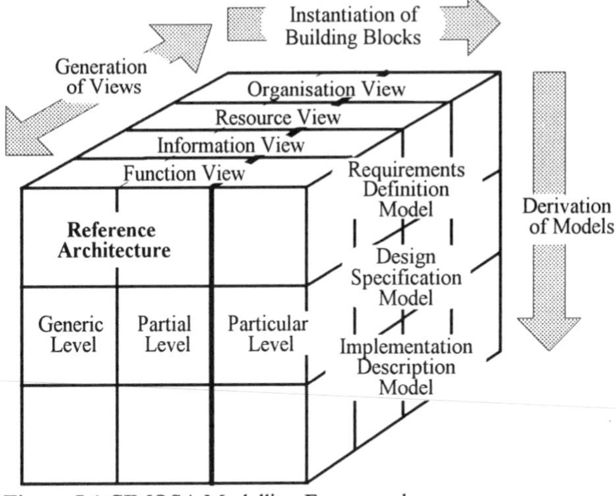

Figure I.1 CIMOSA Modelling Framework.

The CIMOSA Reference Architecture supports modelling of the complete life cycle of enterprise operations (Requirements Definition, System Specification and Implementation Description). Again the sequence of modelling is optional. Modelling may start at any of the life cycle phases and may be iterative as well. Depending on the intention of model engineering only some of the life cycle phases may be covered.

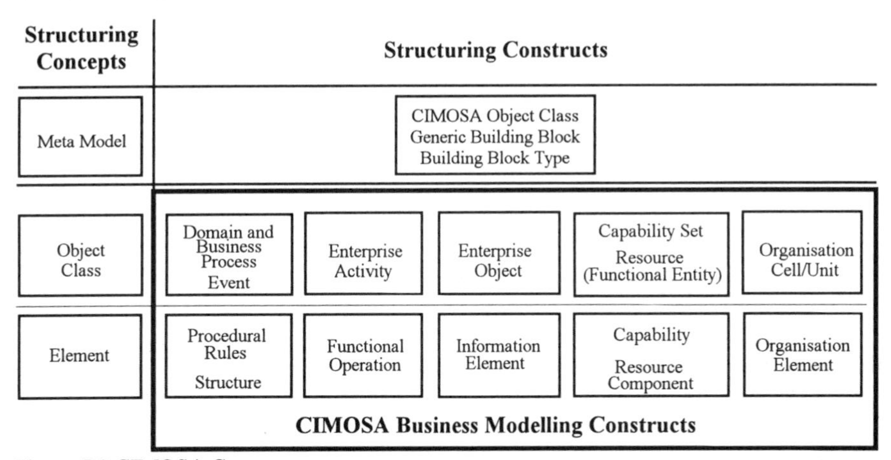

Figure I.2 CIMOSA Constructs.

Enterprise operation should not be modelled as a large monolithic model but rather as a set of co-operating processes. With a set of common modelling building blocks, the CIMOSA Reference Architecture provides the base for evolutionary enterprise modelling. This allows different people to model different areas of the enterprise but provides the integrity of the overall model. Figure I.2 shows the basic set of the common building blocks for business modelling. Processes, Events and Enterprise Activities are the object classes which describe functionality and behaviour (dynamics) of the enterprise operation. Inputs and outputs of Enterprise Activities define the information (Enterprise Object) and resources needed. Organisational aspects are defined in terms of responsibilities and authorisation (Organisation Elements) for functionalities, information, resources and organisation and are structured in Organisational Units or Cells. CIMOSA employs the object oriented concepts of inheritance, structuring its constructs in recursive sets of object classes.

3 CIMOSA - PROCESS-BASED ENTERPRISE MODELLING

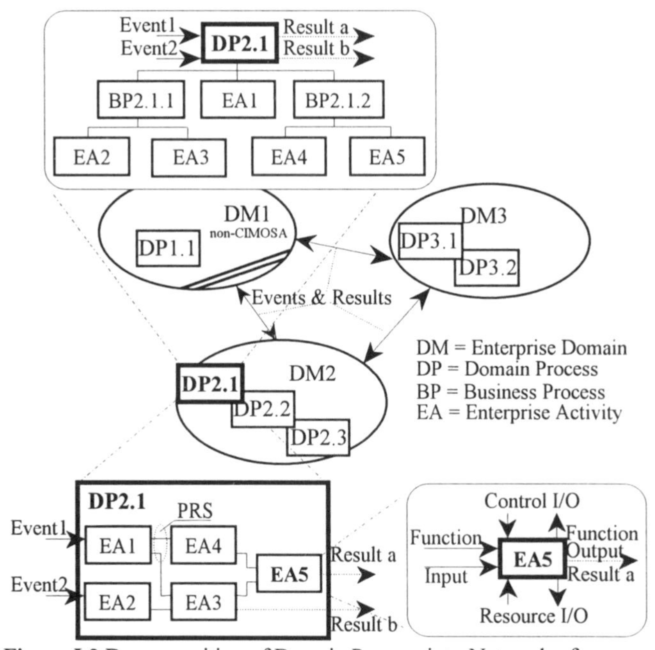

Figure I.3 Decomposition of Domain Process into Network of Enterprise Activities.

CIMOSA model engineering is demonstrated in Figure I.3 which shows three enterprise Domains (DM1-3) each one represented by its functionality - a set of Domain Processes (Domain Processes). Domain Processes communicate with each other via Events and Results. Decomposition of Domains Processes (DP2.1) via Business Processes (Business Processes) leads to identification of Enterprise Activities (EA1-5) and their connecting control flow represented by a set of Procedural Rules (PRS). The network of these Enterprise Activities is the functional and dynamic representation of the Domain Process DP2.1. Events (1-2) and Results (a-b) which relate to Domain Process DP2.1 actual trigger EA1 and EA2 and are produced by EA3 and EA5 respectively. The different Inputs and Outputs identified for each Enterprise Activity are shown in Figure I.3 as well without identification of Resource and Control I/O's and any identification of organisational aspects (for those details please refer to CIMOSA references).

At system design level Enterprise Activities are further decomposed into Functional Operation (Figure I.4). Such CIMOSA Functional Operations are defined in relations to their executing resource types; the Functional Entities. Each Functional Operation will be completely executed by one Functional Entity, but a Functional Entity may be capable to execute more than one type of Functional Operation. Functional Entities are resources which are capable to receive, send, process and (optional) store information.

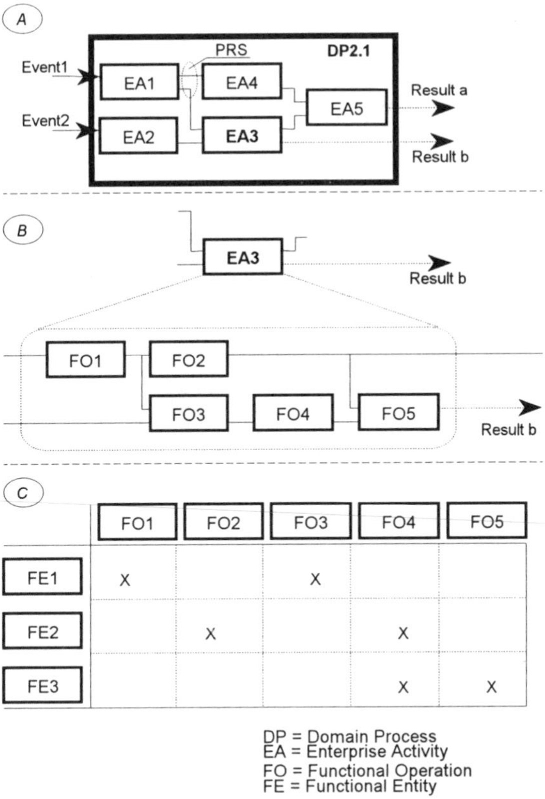

DP = Domain Process
EA = Enterprise Activity
FO = Functional Operation
FE = Functional Entity

Figure I.4 Decomposition of Enterprise Activity and Relation between Functional Operations and Functional Entities.

4 CIMOSA INTEGRATING INFRASTRUCTURE

For model engineering and model driven enterprise operation control and monitoring especially in heterogeneous environments the Integrating Infrastructure provides a set of generic IT service entities for model engineering and execution (see Figure I.5 for the latter).

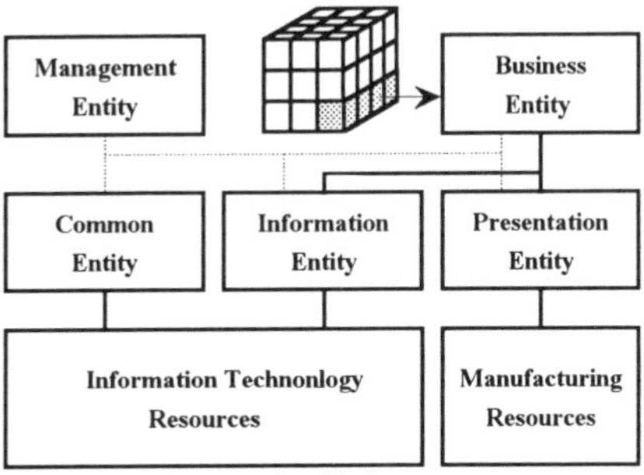

Figure I.5 Integrating Infrastructure.

Control on execution of the Implementation Description Model is provided by the Business Entity which receives the Events and creates occurrences of the related Domain process and all its contents. Process Control, Resource Management and Activity Control (all part of the Business Entity) analyse the model contents, assign the resources, identify the required information and connect to the necessary information technology and manufacturing resources via the Common, Information and Presentation Entities. Finally the Business Entity controls the execution of the Domain Process.

5 CIMOSA AND THE REAL WORLD

CIMOSA distinguishes explicitly between enterprise engineering and enterprise operation placing emphasis on the need for enterprise engineering in a similar mode of operation as for product engineering. Therefore the life cycle phases for enterprise engineering should be followed by an explicit release for operation processes including both model and implementation validation.

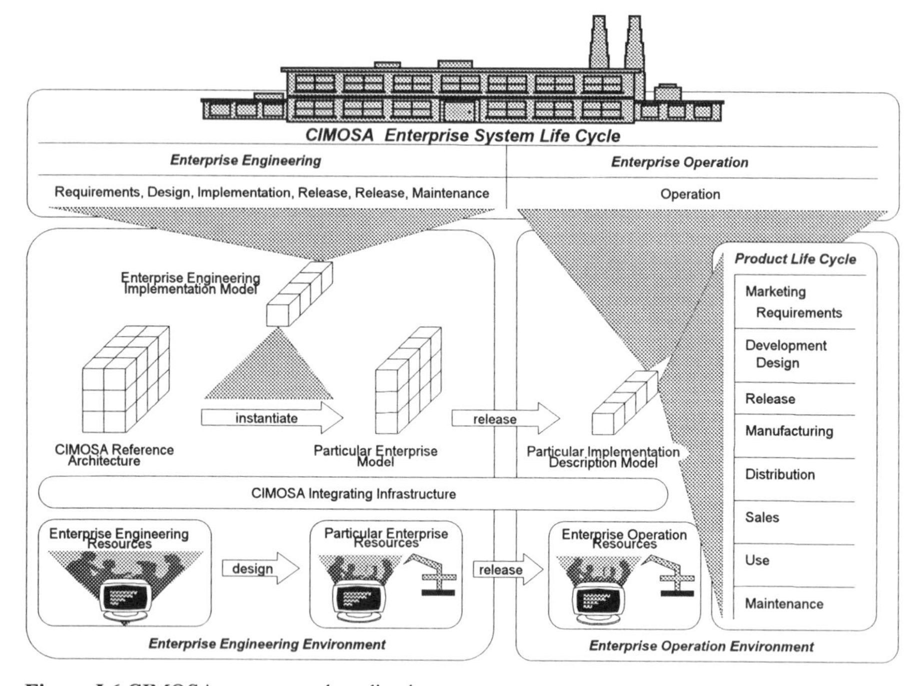

Figure I.6 CIMOSA concept and application.

Figure I.6 demonstrates the use of CIMOSA in model engineering as well as in operation control and monitoring. Using the CIMOSA Reference Architecture Particular Enterprise Models are engineered under control of the Enterprise Engineering Implementation Model. The latter will be implemented in a CAE tool guiding through the engineering phases of the CIMOSA System Life Cycle. The released Particular Implementation Model is then used to directly drive the operation through monitoring and controlling the relevant product life cycles or parts thereof and their business process implementations. The Integrating Infrastructure links to the enterprise resources. This link is required for model creation as well for real time model maintenance (model up-dates and extensions).

6 CIMOSA VALIDATIONS, APPLICATIONS AND EXPLOITATIONS

CIMOSA models provide for very high flexibility in enterprise engineering through fast modelling and evaluation via simulation of operation alternatives and direct implementation of the final solution. This has been explicitly verified by the AMICE project in the FIAT model engineering pilot implementation where CIMOSA has been used to model and evaluate operational alternatives in gearbox production and assembly. Compared with State of the Art methods currently applied by the FIAT Auto Division CIMOSA benefits are in considerable enhanced analysis capability (better structuring and more details on information, resources and organisation), much lower modelling time (reduced by factor of 3) and significantly improved re-usability of model parts (modelling time for similar application reduced by factor of 8).

The CIMOSA concept has been validated and verified in a total of 8 pilot implementation done by the ESPRIT projects AMICE, CIMPRES, CODE and VOICE covering both model engineering and model based operation control and monitoring. Results have been presented to the public in numerous publications and presentations (e.g. Petrie, C.J. Jr 1992)as well as demonstrations at several CIM Europe workshops. General descriptions of CIMOSA, user guides and technical specification have been made publicly available (CIMOSA Association 1994)and up-dates representing latest results will be presented to the public in the near future.

Exploitation of this work is done in both standardisation and product developments. Standardisation efforts on enterprise integration and based on CIMOSA results are currently in progress on ISO level (Framework for Enterprise Modelling), CEN level (Modelling Constructs/Building Blocks and Framework for Integrating Infrastructure) and various national organisations supporting international and European standardisation. Product developments especially on CIMOSA modelling tools have been reported by several IT vendor companies.

7 REFERENCES

CEN (1990) CEN/CENELEC/AMT, *Computer Integrated Manufacturing (CIM): CIM systems architecture framework for modelling*, DD194: 1990, ENV 40 003: 1990, Draft for Development.

Petrie, C.J. Jr (1992), *Enterprise Integration Modelling.* ICEIMT Conference Proceedings, MIT Press.

ESPRIT Consortium AMICE (1993), *CIMOSA - Open System Architecture for CIM.* Springer-Verlag.

CIMOSA Association (1994), *CIMOSA - Open System Architecture for CIM, Technical Baseline.* private publication.

PART II:
CIMOSA ENTERPRISE ENGINEERING - APPLICATION AT FIAT

1 INTRODUCTION

FIAT has applied CIMOSA model engineering in a gearbox assembly plant analysis (see Figure II.1). This application includes the global modelling of the FIAT logistics processes for ordering parts and products (gearboxes) as well as the material flow control in the parts manufacturing-assembly-warehouse environment. Modelling the different processes in the Gearbox Manufacturing Domain has allowed to simulate different logistics scenarios and to analyse and evaluate the results in terms of throughput and turn-around time. Starting with Kanban and Push-Type production control, the optimum solution is a combination of both Push-Type in the purchasing and parts manufacturing areas and Pull Type in the gearbox assembly shop.

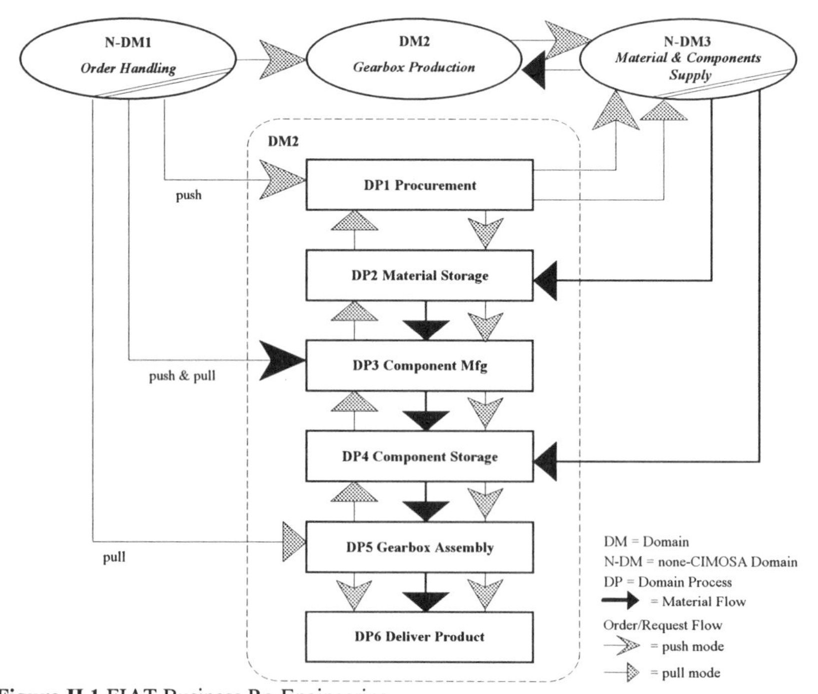

Figure II.1 FIAT Business Re-Engineering.

The CIMOSA Modelling Framework and Modelling Paradigms have been used to perform the modelling of this FIAT production system with the purpose of giving answers to the following three questions:

1) does CIMOSA improve the replies we get from modelling and simulation?
2) does CIMOSA improve the efficiency of the modelling process?
3) is CIMOSA implementable using existing technologies?

The following will demonstrate that three times yes are the answer to the above questions.

To understand the benefits which can be obtained by using CIMOSA it is necessary first to examine what is expected in the manufacturing environment from the processes modelling activities. As depicted in Figure II.2, today's strategy in design and manufacturing operations is to perform validation operations on the designed processes as early and as efficiently as possible.

Why to Model and Simulate Manufacturing and Business Processes

Today: new systems build from scratch and validated in real life

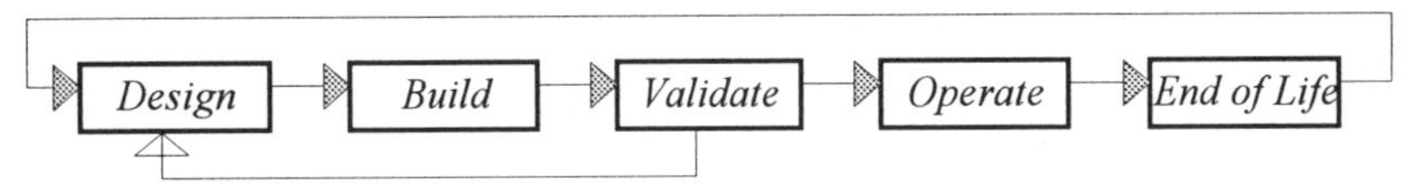

Tomorrow: new systems build by modifying old ones and validated via computer simulated models

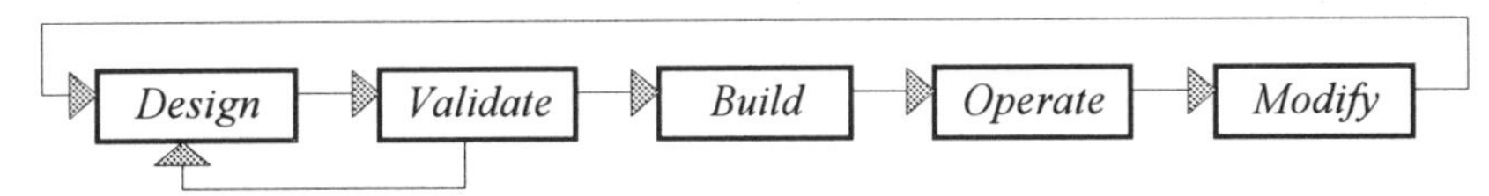

Figure II.2 Changes in the system design process.

2 VALIDATING CIMOSA AT FIAT

In the last years, FIAT, as well as other companies, has experienced the benefits of modelling and validating its manufacturing systems before building and operating them. The experience has shown the need of a structured and integrated approach to better organise modelling activities. Such need has been expressed in a set of requirements the models must fulfil. Depending on the purpose the models are used for, such requirements are the following:

Process Verification: design of enterprise processes requires frequent on-going verification steps. Models must describe the dynamic behaviour of the enterprise and must be capable to be verified by simulation.

Enterprise Process Integration: design of the enterprise involves many interacting processes:

- Production and Maintenance,
- Logistics,
- IT Systems,
- Suppliers,
- Personnel Management and Organisation,
- Marketing and
- Administration.

A good design is optimised over the interaction and the integration across these different processes. Models integrating the descriptions of the different processes of the enterprise are required to allow co-design and design optimisation.

Process Views Integration: Enterprise needs to be described in terms of:

- Performed Functions and their behaviour,
- Used Information,
- Employed Resources and
- Organisation.

Models must contain views on Functions (Processes), Information, Resource and Organisation. Such views must be structured, conceptually separated, comprehensive but physically integrated. Other requirements for industrial models are:

Modelling Efficiency: Models must maximise the ratio between the obtained benefits (accurate description of the enterprise, simulation capabilities, ...) and the efforts required to produce the models. Models must be built as much as possible by re-usable building blocks having different levels of complexity.

Coverage: The building blocks used for modelling must be able to accommodate all kind of information required to model specific processes of the enterprise. Processes emulation for verification and acceptance of equipments and controllers, operators training, re-engineering processes require the description of totally different enterprise data and behaviours that have to be re presentable in the models.

3 BENEFITS OF USING CIMOSA

Looking at the state of the art in the process modelling methodologies to find a methodology matching most of the above requirements, the actual scenario doesn't give satisfactory answers. There are many methodologies but no one is a world-wide standard and is really comprehensive. Often methodologies are just tools linked to methodologies assumed as reference. Often too, the methodologies cover specific aspects of the reality and not the whole complexity: they develop primarily a process or a view of the enterprise and a marginal relevance is left to other processes and views. As a consequence, when an enterprise process is to be analysed from different points of view, many models (information model, functional model, organisation/chart model) have to be created using different tools.

CIMOSA has been demonstrated to be capable to offer the solutions of all the above problems through the concepts of:

- Modularity,
- Integrated Approach by Views,
- Stepwise Derivation and
- Re-Usability.

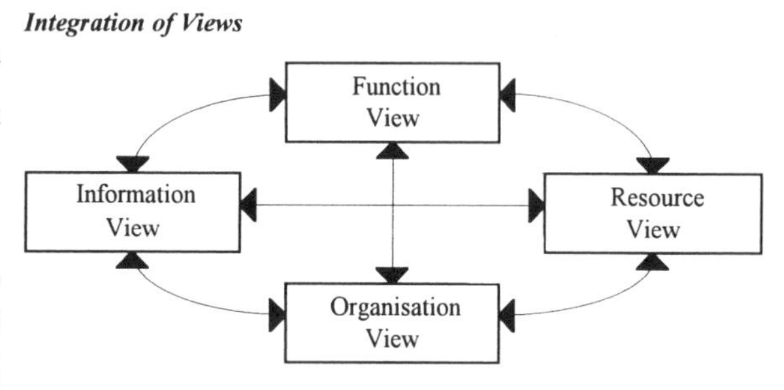

Figure II.3a The Integration of CIMOSA Views.

3.1 Modularity

Modularity is supported by the CIMOSA Building Blocks (especially Domain, DP, BP and EA) and by the definition of standard contents and standard interfaces among the Building Block's. The benefits due to the modularity that had been experienced in the FIAT pilot implementation were modelling easiness, reduction of modelling efforts and costs, maintainability and re-usability of models and models parts.

3.2 Integrated Approach by Views

The integrated approach is related to the capabilities offered by CIMOSA to describe a system in terms of Function, Information, Resource and Organisation Views, all related to each others. Such integrated approach is based on the concept of a structured integration and a conceptual separation of the Views. This allows co-design activities among experts of the different disciplines giving to each of them, in the mean time, a separated description of the information pertaining its particular field of expertise.

Examles of an integrated reply

Necessary Information	*Dynamic Behaviour*	*Process Re-Engineering*	*Resource Utilisation*	*Department Organisation*
Lead Time Throughput	*Cost*	*Productivity*	......	......

Figure II.3b The Integration of CIMOSA Views.

The obtained benefits are a better understandability of the different aspects related to the enterprise processes and the views related to the enterprise functionalities and an enterprise model optimisation across views and processes. Figures II.3a and II.3b explains the application of the above concepts in the solution of a simple problem: getting integrated answers to question concerning different processes and different views. Another advantage of the integration of the views is shown in Figure II.4.

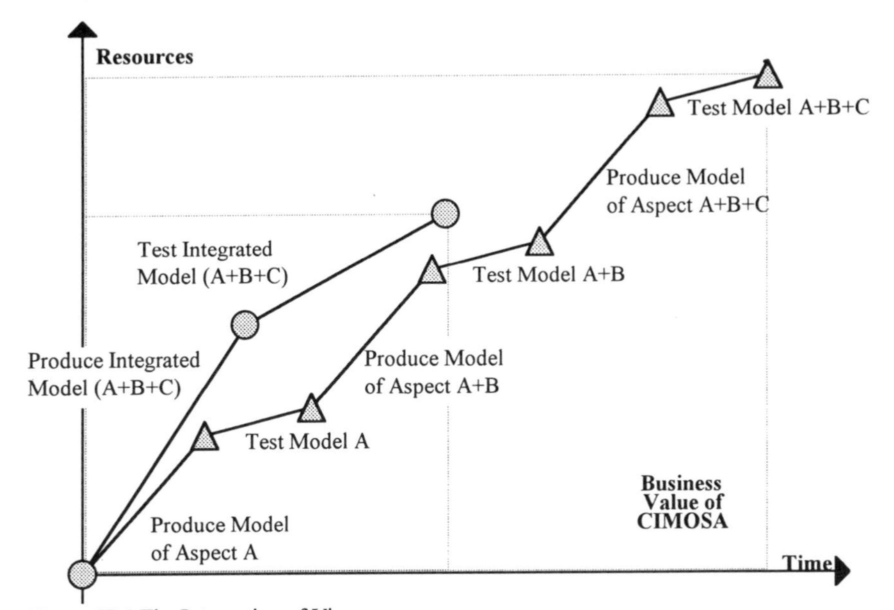

Figure II.4 The Integration of Views.

Often, in actual modelling practice, the analysis of a reality starts with the analysis of just one view (Function or Information or Resources or Organisation). Then the same reality needs to be analysed from another point of view: a new model is to be built and validated. It may be easily understood that making many models that focus on just one view and develop the other views as marginal "attributes" of the main one, duplicate efforts so, the time-by-resources area that in Figure II.4 is saved by making just one integrated model, represents a Business Value of the CIMOSA approach.

3.3 Stepwise Instantiation

The stepwise instantiation of building blocks allows to obtain models of simple and complex functionalities derived by customising the corresponding building blocks of the model at higher level of generality. In the way shown by Figure II.5 and II.6, it was possible to derive, in the implementation of the FIAT Gearbox Assembly System Model, all the 340 Particular Building Blocks used to model the pilot at any level of complexity (Domain Processes, Business Processes, Enterprise Activities) just from 6 Generic Basic Function Types and 17 Partial Models of Standard Function Types. Figure II.6 also shows the relations between the different parts of the model and the different contents of the particular model of the gearbox assembly and its runtime version (Model Occurences).

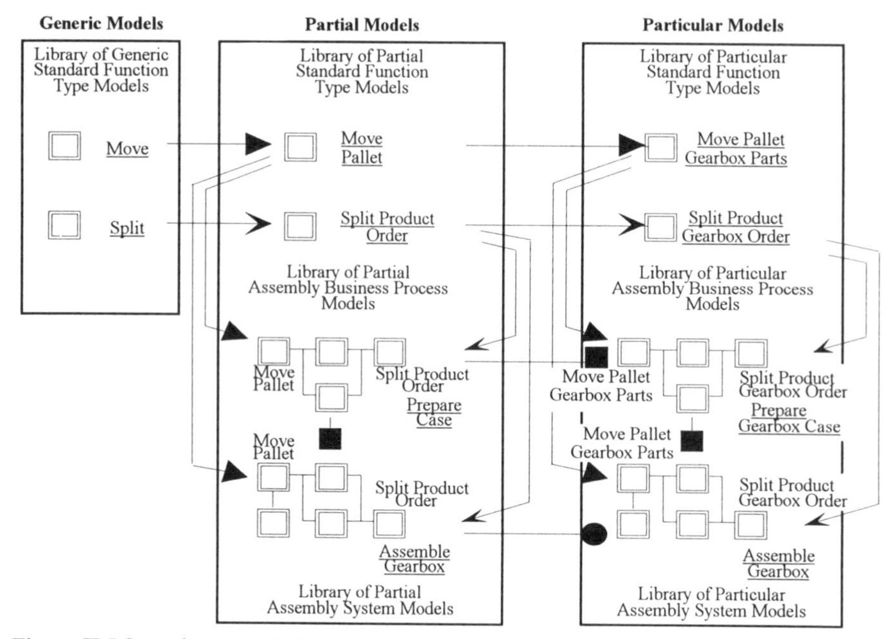

Figure II.5 Stepwise Instantiation and creation of Partial Models.

3.4 Re-Usability

This performance was also due to the application of the concept of RE-USABILITY that allowed to re-use the same Building Block's (even complex Building Block's like Domain Processes and Business Processes) in different parts of the model, instantiated with different names. The benefits of the re-usability are shown in Figure II.7.

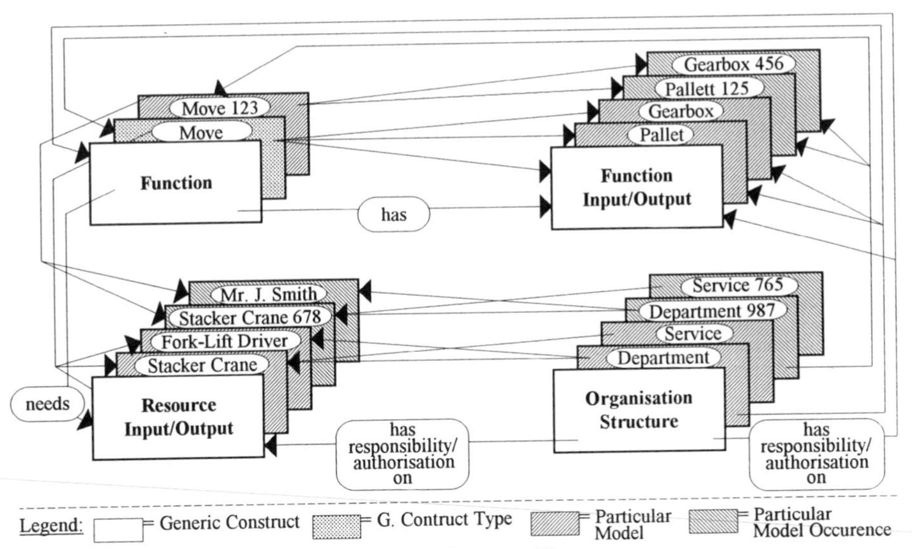

Figure II.6 Stepwise Instantiation and Relation between Views.

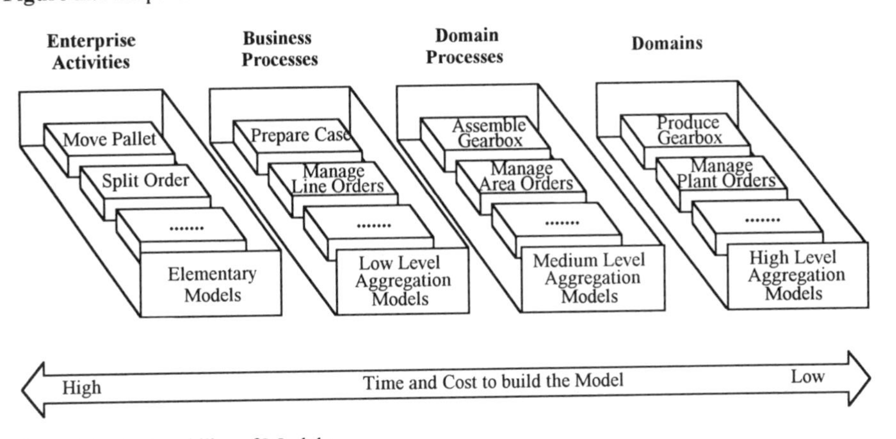

Figure II.7 Re-Usability of Models.

3.5 Business Benefits

Another benefit of the utilisation of the CIMOSA Methodology is the feasibility of the model optimisation by Objectives and Constraints. CIMOSA requires the definition of the modelled functionalities in terms of planned Objectives, imposed Constraints and the if-then-else combination among them. Such set of information has been recognised to be useful for a computerised evaluation of the simulation results by a decision making tool in order to ensure that the modelled production system meet the predefined goals (in terms of productivity, quality, costs, resource availability).

The experience of utilisation of CIMOSA Concepts applied to the ARTIFEX[3] modelling of a real environment has provided evidence of the following Business Values of CIMOSA:

- Enhancement of the quality of the analysis,
- Reduction of the analysis and validation lead times and costs and
- Additional targets for application of CIMOSA.

Table 1 General Experience of using CIMOSA.

Enhancement of the Quality of Model Analysis	
Integration of the Views provides better Simulation Capabilities	
State of the Art	**CIMOSA Models**
Information embedded in Functionality	Information structured by specific Building Blocks described at Data Base level
Organisational Structure treated as Technological Structure	Organisational Structure modelled by specific Building Blocks
Resource description embedded in Functionality	Resources described and structured by specific Building Blocks

As shown in Table 1, the integration of the Views allowed by CIMOSA has brought enhancement of the quality of the analysis with respect to previous modelling practices regarding the Information, Resources and Organisational descriptions. A comparison among the lead-times and efforts required in the past to model a productivity system and the lead times and costs to model the same system, in a "coteris paribus" situation, using re-usability of step-wise derived models has given the figures shown in Table 2.

As important as the general improvements of enterprise modelling obtained with CIMOSA are the significant reduction of modelling lead-time which have has been obtained as well. A reduction factor of 3 times in the time to build an ex-novo model and a reduction factor of 8 time to adapt a model to a similar plant have been experienced.

Table 2 Specific Experience of using CIMOSA.

Reduction of Model Creation Time and Lowering of Analysis Cost		
Benefits are Proportional to the size of Libraries of Re-Usable Models		
	State of the Art	**CIMOSA Models**
Time to build ex-novo the FIAT Model	about 6 Month	about 2 Month
Reduction in Model Engineering time		3:1
Time to re-use the Model in a similar Plant	about 4 Month	about 15 Days
Reduction in Model Engineering time		8:1
Obtaining these Factors depends on	Kind of Enterprise Reality to be modelled (and the know-how about it)	
	Availability of Elementary and re-usable Models	

3 ARTIFEX is a software product of ARTIS s.r.l. Corso Cairoli 8, 10123 Torino, Italy

4 CONCLUSIONS - EXPLOITATION IN BUSINESS ENVIRONMENT

Using CIMOSA in FIAT in a pilot implementation has shown significant business benefits both in model engineering time and in model analysis capabilities.

The structuring concepts of CIMOSA enable a clear separation of the different operational aspects (function, information, resources and organisation) during the model engineering process, but still keeping everything as part of the same model. This concept of views reduces model complexity and thereby enhances significantly model engineering efficiency. In addition, re-usability of modelling constructs has contributed very heavily to reduction in model engineering time for modelling similar or related operations. Establishing a library of standard building blocks (or building block types) as well as partial models has proven to be key in enhancing model engineering efficiency.

This experience leads to the recommendation for development and subsequent standardisation of industry-wide applicable building blocks and partial models. Only if sufficient standards have been established can models be used efficiently and effectively in enterprise operation across enterprise boundaries. This will enable optimisation of the own enterprise operation recognising the impact of supplier and customer operation as well. Standardisation in the area of enterprise modelling is also expected to decrease computer illiteracy and increase awareness on the enterprise operation and recognition of the impact of specific contributions by the individual employee.

CIMOSA based enterprise models provide a much wider scope for analysis of business alternatives. The specific model of the FIAT gearbox manufacturing operation could be used to evaluate different manufacturing scenarios. Pure push and pull-type (Kanban) strategies as well as combinations of both strategies with different change-over points could be analysed with the same model without model re-engineering. The FIAT operation can be optimised for a given set of constraints on Turn-Around-Time, Work in Process and Work in Stock by identifying the change over point from push to pull at a particular process step.

The experience made has outlined that all the above benefits may also be obtained in other fields rather than CIM only. The Experiments under execution are showing that the generated Elementary Standards Function Types are suitable to be easily adapted to other non-Manufacturing Environments (Accounting, Logistics, ... by changing the associated structure of the data.

4

Making CIMOSA operational

Giorgio Bruno, Rakesh Agarwal
Politecnico di Torino, Dip. di Automatica e Informatica
Corso Duca degli Abruzzi 24, 10129 Torino, Italy, Email:bruno@polito.it

Carla Reyneri
Fiat Auto Mains
via Issiglio 63/a
10141 Torino

Bernardino Chiavola
Alenia
corso Marche 41
10146 Torino

Mauro Varani
Artis
corso Cairoli 8
10123 Torino

Abstract

This paper shows how the CIMOSA modeling approach can be made operational using existing object-oriented techniques and tools, such as Artifex and Quid. The main benefits include: a graphical representation of functional and informational aspects, the simulation and graphical animation of the model, the automatic generation of prototypes as well as of distributed applications. Three examples concerning the analysis of a production system, the testing of information systems using emulators and the monitoring of an engineering process are discussed.

1 INTRODUCTION

Increasing efforts in the domain of CIM are being devoted to the integration of the enterprise's processes in order to improve quality and competitiveness. Such integration is based on the definition of suitable models and of a standard system life cycle.

The CIMOSA approach ESPRIT (1993), which has been carried out by the Amice Consortium within the Esprit Programme, provides a sound framework consisting of:

- three modeling phases. These phases, Requirements Definition, Design Specification and Implementation Description, which identify the major tasks in the system life cycle, are associated with appropriate methods and building blocks.
- four modeling views. CIMOSA organizes the modeling task into four sets of constructs, called views, each focusing on a different aspect of the problem. Such views address the functions to be modeled, the information structure operated on by the functions, the resources associated with the functions and the organization responsible for the functions.
- three genericity levels. The first level, called the Generic Level, provides the basic language specialized by view and modeling phase. The aim of the second level, the Partial Level, is to provide the user with standard building blocks suitable for specific

kinds of manufacturing enterprises. Finally, the third level, the Particular Level, refers to the representation of a specific reality.

Using models to study the properties of complex systems is common to all disciplines. In particular, models have long been investigated in the domain of software engineering, thus it is important to consider the issues and the results coming from this discipline.

Models are largely used to specify software requirements: in fact, if the analyst builds a model that formalizes the requirements, he or she can get an insight into the behavior of the system being developed in order to work out whether there are any possible inconsistencies or whether any information is missing, before the actual development takes place.

However, the study of a model yields limited results if it is only based on inspection. On the contrary, if the model can be executed so that traces of the system's behavior are obtained, then a thorough analysis can be performed and the risk of delivering an unsatisfactory product is minimized.

Growing interest is being shown in operational models Zave (1984), i.e. models that can be executed using a suitable support environment. Most operational models are graphical and can be considered as high-level programs which are developed using high-level modeling languages.

An operational model can be modified and tuned until the behavioral traces it generates match those expected. In this way, the model is a reference point for the development of the system and, what is more, the purchaser feels confident that the system, being developed according to the model, will behave properly.

Operational models often allow timing constraints to be expressed and, consequently, a discrete-event simulation of the model can be performed. In this way, statistical estimates of the system's parameters can be collected in order to support decision making. When a formal proof would be too expensive, such statistics can confidently be used to determine some properties of the system.

Operational and evolutionary principles can be brought together to form a powerful software development paradigm. For this reason, the final system can be seen as the final step of an evolutionary transformation which enriches the initial abstract model with details and progressively turns it into the deliverable system.

This approach provides important benefits, such as

- minimizing the risk of finding out that information is missing or inconsistent at the time the system is brought into operation;
- maximizing the reuse of software modules (this is because their corresponding models are reused and reusing models is much easier than reusing programs);
- improving productivity, because the final code can be generated from the model automatically.

This paper illustrates how the CIMOSA approach can take advantage of operational techniques and tools. The goals of the work presented in this paper are the following:

- to provide CIMOSA constructs with a graphical (but rigorous) representation which facilitates the description of complex systems and promotes the reuse of models.
- to enrich CIMOSA constructs with features that allow models to be executed and

simulated so that the analyst can validate the model (check its logical correctness) and also assess the performance of the intended system (e.g. the throughput of a plant or the size of a buffer).
- to allow the user to obtain from the model a prototype of the information system that will support the system/process being investigated.

To reach the above goals, some existing operational techniques and tools, namely the Protob language Bruno (1995) and the Artifex and Quid Bruno (1992) toolsets have been used and extended to cope with CIMOSA requirements.

The next sections contain an overview of the above-mentioned techniques and tools, the mapping of CIMOSA constructs onto them, and the description of some actual applications developed in this way.

2 OVERVIEW OF PROTOB, ARTIFEX AND QUID

This section presents an overview of the techniques and toolsets which have been used to make CIMOSA operational.

2.1 Protob and Artifex

Protob is both a modeling and a development language for event-driven systems. It combines the most important features of high-level timed Petri nets with those of extended dataflows and organizes them within an object-oriented framework.

A detailed description can be found in references Bruno (1995), while the paper by Murata (1989) is an excellent survey on Petri nets and the book edited by Jensen and Rozenberg (1991) is a collection of recent papers on high-level Petri nets.

The application domain of Protob mainly concerns discrete-event concurrent systems, such as real-time embedded systems, telecommunications systems and manufacturing systems.

An application written in Protob is a collection of communicating objects, each object being an instance of a class.

Protob objects are also called *actors* to emphasize that they represent components which play an active role, as they can take decisions and react to external events autonomously.

A Protob class has a graphical part, the net: it is made up of *places*, depicted as circles, *transitions*, depicted as rectangles, and oriented *arcs*, which connect places to transitions and transitions to places.

Places contain units of information called *tokens*, which are mobile information packets. Tokens contain structured data (records) or references to objects.

A place can contain several tokens at a time; all the tokens contained in a given place are of the same type. Each place has three attributes: the place name, the place type and the number of tokens in the initial marking (such tokens are called initial tokens). The first two attributes are strings, while the third is an integer number which can be omitted if the place has no initial tokens. For the sake of expressivity, initial tokens are usually depicted in the illustrations as dots inside places.

Places are queues (not sets) of tokens, thus when a token is put into a place, it is added to the end of the queue. Tokens are ordered in places on the basis of their arrival times.

Transitions are the processing units of the model. They carry out token-driven computations.

If predicates, priorities and delays are ignored for the moment, a transition fires as soon as it is enabled (i.e. none of its input places is empty) and firing consists in removing one token from each of its input places and adding one token to each of its output places. The tokens taken from the input places are called input tokens, while those delivered to the output places are called output tokens.

Since tokens usually contain information, applying the standard firing rule strictly, i.e. removing and destroying input tokens and generating and delivering output tokens, would be awkward. In fact, if an input token carries information which must not be lost, the transition, before destroying it, has to perform an action to copy its contents into an output token; this action has to be written by the programmer. In such a case, it would be more convenient to turn that input token into an output token. When an input token becomes an output token, we say that it is propagated by the transition. In most cases, the propagation of tokens is done automatically in Protob, so no action at all has to be written. An input token can be either moved into an output place (in this case it is a propagated token) or destroyed and an output token that is not a propagated token is generated by the transition itself.

When a transition fires, it can execute an action. The action is a piece of C code which has visibility on the tokens (propagated, destroyed and generated) acted on by the transition. The action can modify the contents of propagated tokens and initialize the contents of generated tokens, as well as invoke external subprograms.

Tokens are usually taken from places in FIFO order (i.e. the oldest token first) unless the transition has a predicate.

A simple example showing the interaction between a sender and a receiver is presented in figure 1.

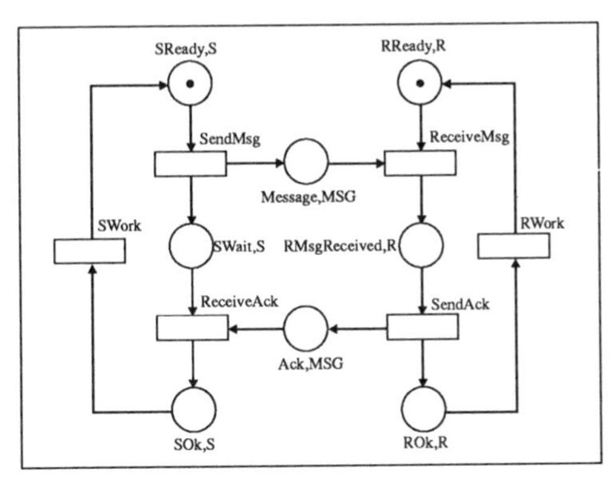

Figure 1 The interaction between a sender and a receiver

1. Initially, there are two tokens in the net, one in place SReady and the other in place RReady, indicating that both the sender and the receiver are ready to start their activities.
2. In this situation, SendMsg is the only transition that can fire: it generates a new token

(representing the message produced) and puts it into place Message, and moves the token from place SReady to place SWait. Now, the sender is waiting for the acknowledgement from the receiver.
3. Transition ReceiveMsg fires, thus consuming the token in place Message and moving the token from place RReady to place RMsgReceived.
4. After that, transition SendAck fires: it moves the token from place RMsgReceived to place ROk and puts a new token (representing the acknowledgement) into place Ack.
5. At this point both transitions ReceiveAck and RWork can fire. When ReceiveAck fires it moves the token from place SWait to place SOk and consumes the token in place Ack.
6. Transition SWork fires.

The introduction of timing constraints into the model enhances its descriptive power and facilitates a careful analysis of the performance of the system being considered Bruno (1995).

Models in Protob can be structured according to the principles of object orientation.

Since objects are based on nets, it is natural that they interact by sending and receiving tokens. Special places, called input ports or input places (of the object), and output ports or output places (of the object), are introduced so that an object is enabled to communicate with other objects.

Input places receive tokens from other objects. An input place is drawn as a double circle; it has a name and a type and contains a queue of tokens.

When an object has to send a token to other objects, it puts the token into an output place. An output place is drawn as a circle with a triangle inscribed. Output places do not hold tokens, because when a token is put into an output place, it is immediately delivered to the destination object(s).

The collection of all the input and output places of an object forms its interface.

Objects are graphically represented by a double square. Composition is graphically represented, too, because the model associated with a class can contain icons which represent objects belonging to other classes, as shown in figure 2.

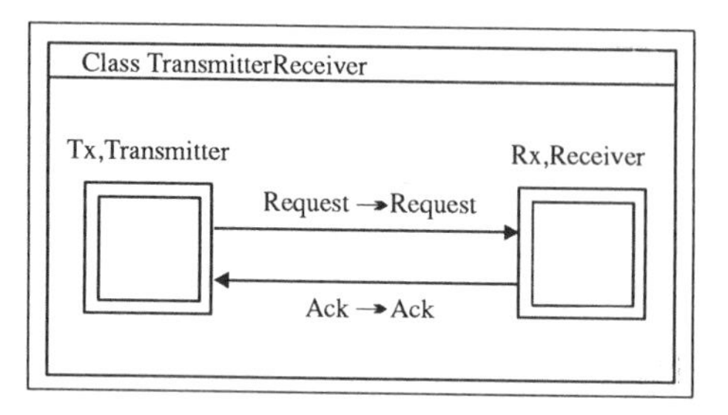

Figure 2 An example of composition

Only fixed composition is considered in this paper, therefore a compound class/object has a number of components which cannot be changed.

An object has two identifiers: the first is the object name, the second is the name of the class to which the object belongs.

The classes, Transmitter and Receiver, of the objects, Tx and Rx, which appear in figure 2 are shown in figure 3.

A compound class has visibility on the interfaces of its component objects, so it can link an output port of an object to an input port of another object, provided that the ports to be connected are of the same type. Communication is graphically defined using links: a link is an oriented arc which connects an output port of an object to an input port of another object.

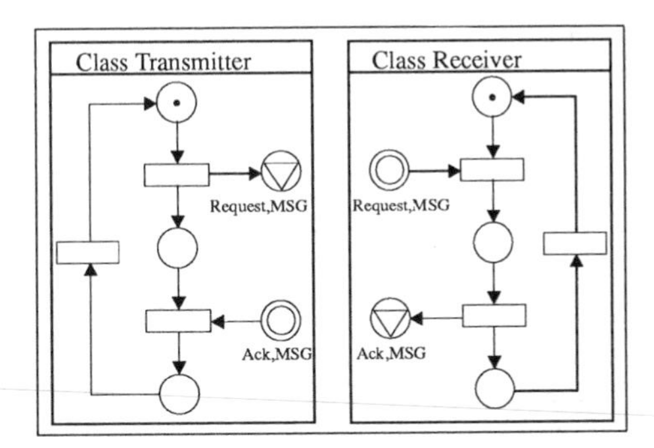

Figure 3 The sender and the receiver modeled as Protob objects

To avoid crowding a model with too many links, if two objects need to be connected by several links, we can group these links into a single connection line, called a compound link.

A compound link stands for a set of links, which can have different directions, and connects a compound port of the first object to a compound port of the second. A compound port is a named sequence of ports. The label of a compound link consists of the name of the source compound port, the arrow symbol and the name of the destination compound port. As an example, if two compound ports are defined in classes Transmitter and Receiver (each compound port consisting of ports Request and Ack), a compound link can be drawn from object Tx to object Rx in figure 2 instead of two links.

Protob objects are easy to put together, in accordance with the metaphor of software chips; in fact, an object does not know the other objects it will interact with and the interaction is only based on the tokens that it sends and receives through its interface, so it is the task of the compound class to set suitable links between its component objects.

Other topics, such as using local variables and parameters, building client server models, extending inheritance to Protob nets, are not covered here for lack of space; the interested reader is referred to the textbook Bruno (1995).

Artifex is a toolset that supports the editing, simulation and animation of Protob models. Moreover, it has an automatic code generation facility which is able to translate a Protob model into a multiprocess application, either located on a single processing unit or distributed over a network of several processing units. Interprocess communication and synchronization is automatically managed. The resulting distributed application can then be monitored and animated so that the state of each object can be observed while it is running on the target architecture.

2.2 Quid

Quid Bruno (1992) is both a language and a toolset and has been developed to make the Entity-Relationship approach operational.

Basically, models developed with Quid are made up of entities and relationships.

An entity represents several individuals, also called instances or objects, which have the same properties; the term property denotes either an attribute or a kind of association with other objects. Every attribute has a name and a type and the notion of type is the same as in conventional programming languages.

The model presented in figure 4 represents the information structure of a cell supervisor.

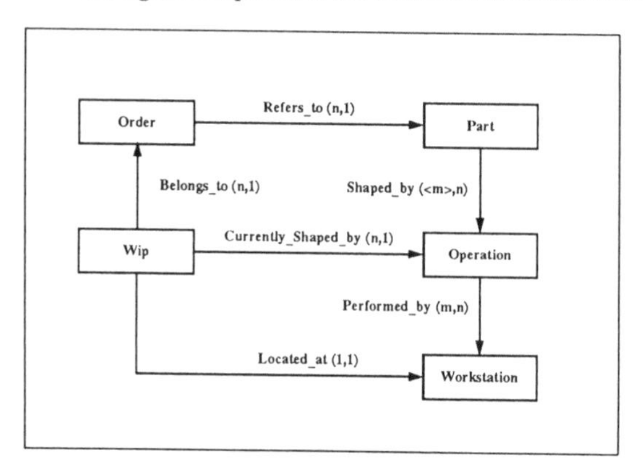

Figure 4 The information model of a cell supervisor

It shows that each order refers to a particular part type; each part type is shaped by a sequence of operations; each operation is performed by a set of workstations; each instance of entity Wip, which models a part present in the cell (i.e. a component of the work in process), is related to a particular order, is located at a particular workstation, and is currently being shaped by (or it has just been shaped by) a particular operation.

Relationships are drawn as oriented arcs for the arrow indicates the direction in which the name of the relationship must be read. For example, relationship Performed_by indicates that operations are carried out by workstations; entity Operation is the source of relationship Performed_by, while entity Workstation is its destination.

A relationship between entities represents the associations that can exist between the objects of the source entity and the objects of the destination entity. Relationships can have cardinality constraints: one-to-one (1,1), one-to-many (1,n), many-to-one (n,1) or many-to-many (m,n).

Associations can be ordered and this constraint can be expressed as follows: if the associations leaving (entering) the source (destination) instances are ordered, then, the first (second) constraint of the cardinality of the corresponding relationship is enclosed within angular brackets.

An example of an ordered relationship is given in figure 4; in fact, relationship Shaped_by associates, with each part, the sequence of operations to be performed on that part.

Recursive and inheritance relationships can be defined in Quid as well.

Objects and associations are unique and can be referred to by pointers, called handles.

Quid provides statements to generate or delete objects or associations as well as to navigate the actual information structure (also called the object graph).

Navigation is the act of traversing the information structure following the paths that are specified with a construct, called a path expression. The objects that are reached during navigation can be acted on.

2.3 Integrating Protob and Quid

When both a complex behavior and a complex information structure have to be represented, the integration of Protob and Quid is needed. Such integration can be achieved in two ways.

1. The transitions of the Protob model can include Quid statements in order to generate or cancel objects or associations as well as to navigate the underlying object graph.
2. Some tokens can be given the meaning of handles to objects in the object graph. This is done in the following way: if there are places in the net whose type name is identical to the name of an entity defined in the object model, then the tokens contained in such places are assumed to be handles to objects belonging to that entity. In many cases, the presence, in a given place, of a token which is a handle to an object indicates that the object is in a particular state. Therefore, the states of objects can effectively be shown using places without it being necessary to add state attributes to the corresponding classes.

As an example, we consider a fragment of a cell supervisor, shown in figure 5. It consists of transition Issue_Mission_2 which has to decide whether a mission of type 2 (i.e. moving a part from the workstation that has completed the current operation on the part to a workstation that is able to perform the next operation on that part) can be started or not.

At a certain instant, the object graph, whose model appears in figure 4, is assumed to be the one depicted in figure 5.

At that time, places Finished and Idle contain one token each: the token in place Finished refers to workstation Wst4 and shows that Wst4 has finished working on a part, whilst the token in place Idle refers to workstation Wst1 and thus indicates that Wst1 is idle. Now, if a token is put into place Mission_Enabled, transition Issue_Mission_2 is allowed to fire only if Wst1 is able to perform the next operation on the part that is located at Wst4. This condition can be checked by associating a suitable Quid navigation with the transition. Details on this example can be found in Bruno (1995).

3 REPRESENTING AND EXECUTING CIMOSA CONSTRUCTS

This section describes how the constructs belonging to the four CIMOSA modeling views can be mapped onto the operational constructs provided by Artifex and Quid. Functional constructs are dealt with first because they are responsible for the architecture of the model, then informational constructs will be examined and, finally, resource and organizational constructs will be jointly represented from an operational point of view.

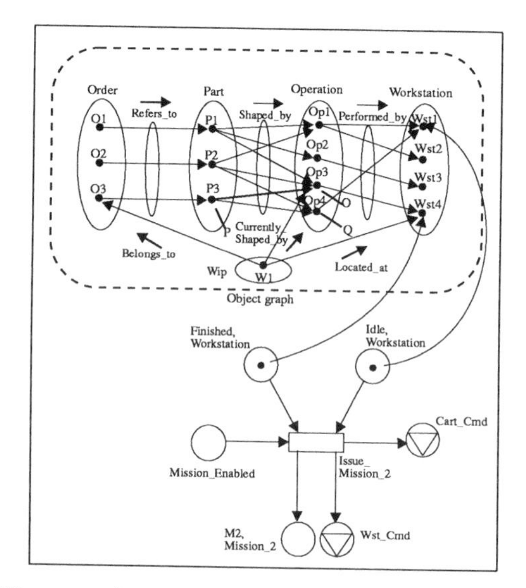

Figure 5 A transition that acts on an object graph

3.1 Modeling the functional view

The CIMOSA functional view is organized hierarchically, starting from a set of communicating domains, each domain representing a major component of the system being considered. Domains send and receive events which can be accompanied by pieces of information called enterprise objects. Some domains may represent the system's environment, thus they are not further decomposed (non-CIMOSA domains).

Each (CIMOSA) domain (DM) is made up of communicating domain processes (DPs) and each DP consists of business processes (BPs) and enterprise activities (EAs). BPs contain other BPs and EAs, EAs being the lowest level of functional decomposition. DPs are event-driven as they react to the events received by the domain of which they are part. BPs contain networks of EAs in the sense that a BP, when all its internal BPs are expanded, turns out to consist of basic functions (i.e. the EAs it contains) and rules; such rules, called procedural rules, control the activation of EAs. Each BP or EA will be started by its parent DP or BP and when it ends it returns a state to its parent so as to inform it of the result of the execution of the activity. Procedural rules, as shown in figure 6, define all the forms in which activities can be organized, such as sequence, parallelism and synchronization, choice and confluence, and event-driven activation.

Using Artifex, DMs, DPs, BPs and EAs are represented by objects, procedural rules by transitions, events are modeled by tokens sent from output ports and received in input ports. The activation of BPs and EAs is managed by associating a pair of places (i.e. a start place and an end place) with each BP and EA, so that when a token (called a start token) is put into the start place of an activity, this will start and when a token (called an end token) is put into the end place of an activity, this is assumed to be completed.

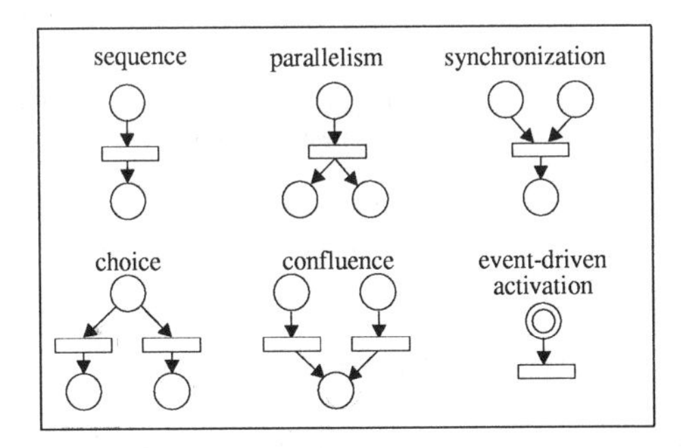

Figure 6 Procedural rules represented by transitions

Transitions modeling procedural rules basically have end places as input places and start places as output places.

EAs must be precisely defined for the model to be executable, so EAs are expanded into Protob nets which provide the necessary details of their behavior. The model of an EA is a Protob net, thus it is based on the simple notions of transitions and places; however, it is possible to structure such a model so that some transitions can be interpreted as elementary (no more decomposable) activities while others have the meaning of procedural rules.

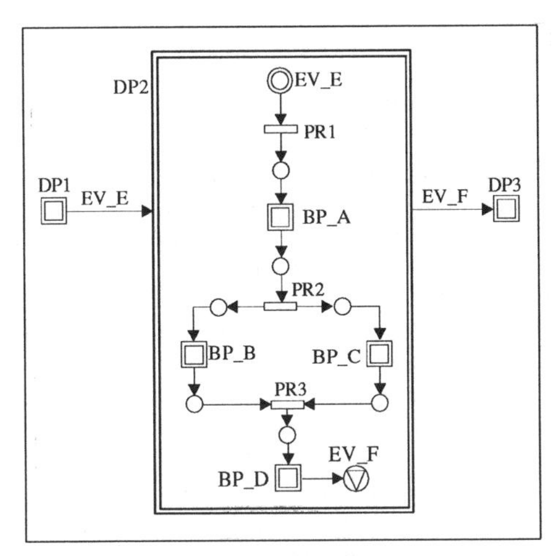

Figure 7 An example of functional model

Modeling DMs, DPs, BPs and EAs with Artifex objects, which belong to specific classes, improves reusability, because the same model can contain several objects of the same class (and this is likely to occur when models of production systems are built). Although different icons might be associated with DMs, DPs, BPs and EAs, we prefer to distinguish among them from a semantic point of view, while, from a graphical (or syntactical) point of view, we emphasize that they are actors, i.e. instances of Protob classes. However, to facilitate the reading of the model, we prefix domain names by DM, DP names by DP, BP names by BP, EA names by EA.

The example in figure 7 shows three domain processes, one of which, i.e. DP2, has been expanded. DP2 is made up of four BPs; the first of them, BP_A, is started by procedural rule PR1 when event EV_E is received from DP DP1. When BP_A is finished, BP_B and BP_C are executed in parallel, then, when both of them are completed, BP_D is activated. BP_D produces an event, EV_F, which is sent to domain process DP3.

3.2 Modeling the information view

Processes and activities exchange and manipulate data which, in general, may have a complex structure. For this reason, CIMOSA defines an information model which is basically an extended Entity-Relationship model consisting of classes of objects and of classes of associations between objects. Using such an information model, CIMOSA establishes which kind of objects are needed and produced by each activity.

A CIMOSA information model is mapped onto a Quid model (since both draw upon the ER formalism) and the link between the functional view and the information view is set by associating handles to Quid objects with start tokens and end tokens. In this way, when an activity is started, it receives all the information it needs through its start token. Owing to the Quid navigational language, an EA, through its internal transitions, is able to traverse and manipulate the actual underlying information structure as shown in the example illustrated in figure 5.

3.3 Modeling the resources and the organization

When an operational model is being built, the resources and the organization are taken into account to the extent they set constraints to the processes being considered. Constraints, as such, are usually defined through a dynamic model that specifies how to check their violation, what to do in such circumstances, how to solve conflicts and so on. For example, if two processes compete for a scarce resource, e.g. a single machine, a policy to assign the resource must be defined. Such a policy can be embedded into a dynamic model which is similar to a functional process. For this reason, we do not introduce special constructs for modeling resource and organization constraints, but we use instead the features of the operational languages, Artifex and Quid. An example is shown in the next section.

We introduce the notion of agent as the entity which is responsible for managing a pool of resources. From an operational point of view, an agent is modeled by a Protob object which provides services to requesting activities through a client-server mechanism similar to the one shown in figure 1. When an activity needs a resource, the object representing it sends a request to the agent that manages the resource and waits for the reply. The agent contains the logic necessary to appropriately manage the resources under its control. Agents may be simple, such as the receiver in figure 1, or complex. Further, they may interact with each other when authority is hierarchically structured, thus complex organizations can effectively be handled.

4 EXAMPLES

This section presents three examples referring to different areas and featuring different goals.

4.1 Analysis of a production system

The model being considered refers to the Fiat gearbox production plant located at Termoli and has been developed as WP1 within the Esprit project 7110. A previous model of the same plant has been described in Bruno (1991).

As shown in figure 8 there are four domains, one of which, DM_GBX, is a CIMOSA domain.

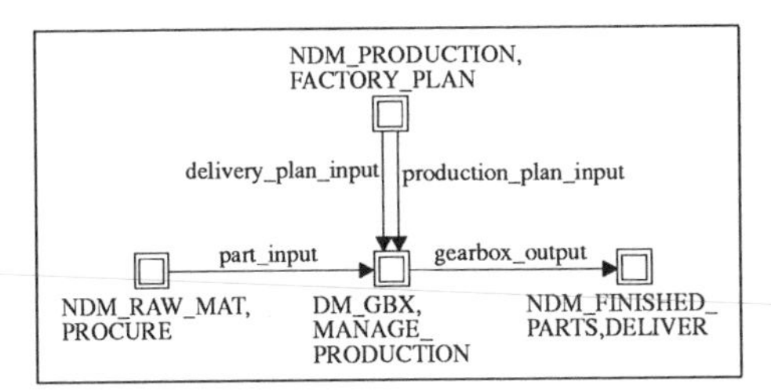

Figure 8 Domains related to a gearbox production plant

Domain NDM_PRODUCTION models the factory production planning system which is responsible for providing production and delivery plans. The prefix NDM denotes a non-CIMOSA domain. Domains NDM_RAW_MAT and NDM_FINISHED_PARTS supply raw materials and receive finished gearboxes, respectively.

Domain DM_GBX, shown in figure 9, encompasses both the physical system and its software supervisor. It consists of five DPs: in particular, DP_PRODUCE_COMPONENTS represents the area in which gearbox components are machined, while DP_ASSEMBLE_-COMPONENTS represents the area in which gearboxes are assembled.

The goals of the above model were the validation of the management logic and the assessment of the material flow. They were attained thanks to the ability of simulating the model and observing its behavior in critical scenarios. In addition, the use of a methodology, such as CIMOSA, enabled the analysts to standardize some building blocks, which correspond to CIMOSA partial models, suitable for modeling a large family of Fiat production systems. Those building blocks are highly reusable, so the time to build a model within that family has been estimated to be 12% of the time to build the model without them.

The object-oriented features of Artifex are essential, since the above model is made up of 339 instances and 23 classes.

4.2 Testing information systems using emulators

Before the information system that governs a plant is installed, it must be thoroughly tested because errors found after its installation are difficult to capture and dramatically

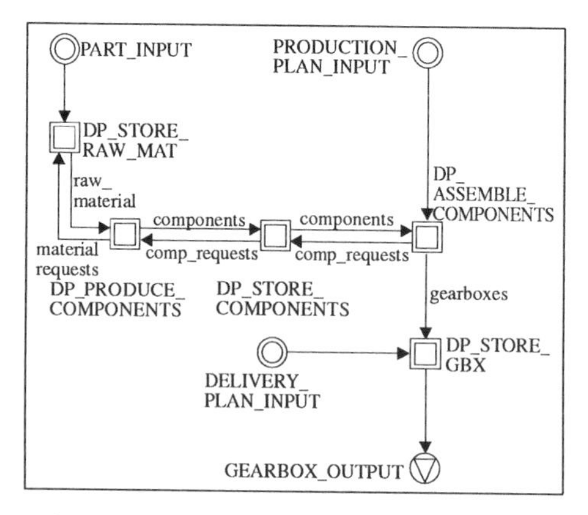

Figure 9 The contents of domain DM_GBX

costly to fix. For this reason, production information systems in Fiat are subjected to a testing phase in which both logical aspects and performance ones are assessed. To do that, a software system (called an emulator) which emulates the plant is built and connected to the information system to be tested so that the former can provide the necessary stimuli (with the appropriate timing constraints) to the latter and check the correctness of the replies. Since the emulator may be very complex, because the plant and the interactions with the information system are complex, it is best represented by a dynamic model which is structured according to the operational approach described in the previous section.

We can think of the overall model as composed of two domains: one, the emulator, is a CIMOSA domain, while the other, the information system, is not because it is developed with traditional techniques.

In addition to simulating the emulator (this is done to validate it), in this case, using the Artifex toolset, the model is translated into a distributed application which interacts with the information system through the actual communication mechanisms. Therefore, the information system is not modified and it is sufficient to replace the emulator with the actual interface to the physical devices in order to have the information system manage the plant.

The use of models permitted to substantially decrease the cost of the testing activity carried out in the laboratory compared with the traditional testing practices and, what is more important, the application (i.e. the information system) was installed with practically zero defects.

4.3 Monitoring engineering processes

The case study being considered has been proposed by Alenia as a test case for the AIT Consortium. The goal was to investigate the requirements of an information system (simply called a monitor) suitable for monitoring an actual process which implements changes to the assembly of aircrafts. The monitor has to detect and notify delays as well as suggest possible recovery actions or revisions of the date of the point of embodiment and of the aircraft delivery.

In order to study such a monitor, a model of the process to be monitored, although it is not automatic but mainly involves manual operations, must be built. Therefore, the overall model, shown in figure 10, consists of two domain processes, one, DP_CHP, is the process, the other, DP_MON, is the monitor.

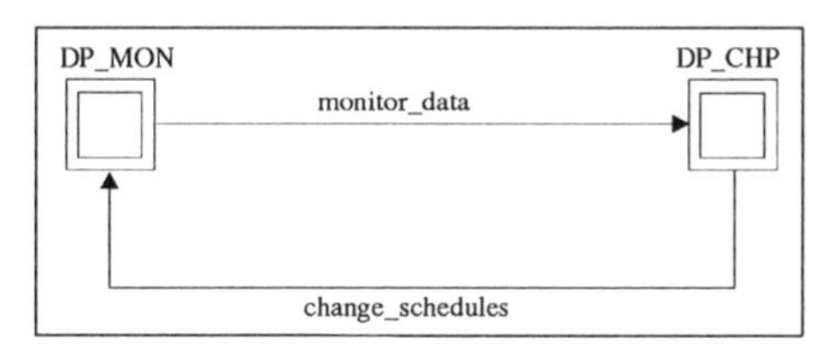

Figure 10 Two domains representing an engineering process and its monitor

A fragment of the engineering process is shown in figure 11.

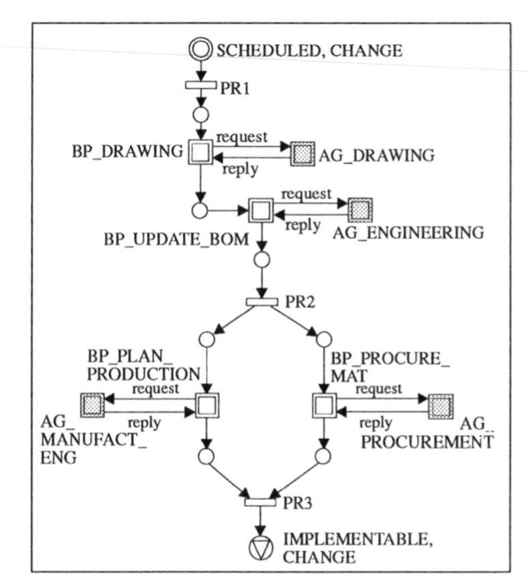

Figure 11 A fragment of the engineering process

It consists of five activities, which, instead of being performed automatically, are carried out by people belonging to specific departments. Since the goal of the model is to study a monitor system, it is important to trace when activities are committed to the departments and when they are declared as completed. For this reason, each activity is an object of the same class as the major concern here is to interface activities with agents, each agent managing the resources of a different department.

Agents are event-driven, so they are similar to domain processes; in order to distinguish them from the other objects, we prefix their names by AG.

The other domain process of the model, DP_MON, is similar to the actual process, but its activities have the purposes of monitoring the corresponding actual activities. When a delay is detected, the process manager is notified so he/she can take corrective actions. DP_MON shows the current situation and the forecast about the next activities through a graphical user interface which helps the analyst understand and specify the requirements of the information system that will actually support such monitoring. The model in this case is used as a rapid prototype of the intended system.

5 CONCLUSION

This paper focused on how the CIMOSA modeling approach to CIM systems can be made operational using (and extending) existing techniques and tools.

The aim of this work is twofold. On the one hand, CIMOSA can really be put into practice because there are tools that support it and, what is more, they allow CIMOSA models to be validated through simulation and graphical animation. In addition, they provide object-oriented structuring mechanisms that allow analysts to build reusable models. On the other hand, the tools themselves can take advantage of a methodology that guides developers in building models which are easier to understand and that is likely to have a broader scope than the initial one.

6 ACKNOWLEDGMENTS

The work presented in this paper took advantage of the support, cooperation and suggestions of many people and organizations. We wish to thank R. Canavese, M. Corsaro and Franco Naccari of Fiat for their support, K. Kosanke and the members of the CEN/TC 310/WG1 for their encouragement.

REFERENCES

Bruno, G. and al. (1991) The impact of software engineering in the development of CIM systems. In *7th CIM-Europe Conference, Turin, Italy*, pages 101–113. Springer-Verlag, London.

Bruno, G., Grammatica, A., and Macario, G. (1992) Operational Entity-Relationship with Quid. In *5th Int. Conf. on Software Engineering and its Applications, Toulouse, EC2, Nanterre, France*, pages 433–442.

Bruno, G. (1995) *Model-based Software Engineering.* Chapman & Hall, London.

ESPRIT Consortium AMICE (1993) *CIMOSA: Open System Architecture for CIM.* Springer-Verlag, Berlin.

Jensen, K. and Rozenberg, G. (eds) (1991) *High-level Petri nets: theory and applications.* Springer-Verlag, London.

Murata, T. (1989) Petri nets: Properties, analysis and applications. *Proceedings of IEEE*, vol. 77: 541–580.

Zave, P. (1984) The operational versus the conventional approach to software development. *Communications of the ACM*, 27:104–118.

BIOGRAPHY

Giorgio Bruno received the doctorate degree in electronic engineering from Politecnico di Torino, Torino, Italy, in 1977. Since 1986 he has been associate professor at Politecnico di Torino (Faculty of Engineering, Dipartimento di Automatica e Informatica), where he teaches undergraduate and PhD courses on software engineering and CIM (Computer Integrated Manufacturing). His interests concern software engineering and software quality, discrete event simulation, distributed systems and languages, Petri nets and graphical interfaces. Bruno participated in several international (ESPRIT and ESA) and national research projects. He has authored over ninety technical papers and performed extensive consulting activities on the above-mentioned subjects. His latest book "Model-based Software Engineering" published by Chapman & Hall (London, 1995) illustrates the results of his research on operational specification and design techniques and tools.

Rakesh Agarwal has done his Masters in Computer Application from BIT, India and since 1987 has been working as assistant professor in the College of Engineering and Technology, Bhubaneswar, India. He is currently doing his PhD at the Politecnico di Torino. His interests concern software architecture and object orientation techniques. Agarwal has (co)authored around 15 papers on the above-mentioned topics.

Carla Reyneri is an electronic engineer graduated at the Politecnico di Torino. In 1986 she was appointed by FIAT AUTO as the head of the design of the integrated factory information system and the Termoli Plant production system. She is currently responsible for the introduction of advanced techniques for modelling, designing and testing company processes and control systems for production processes. She successfully led the implementation of the FIAT CIMOSA Pilot Project which demonstrated the business value of adopting the CIMOSA modelling framework. She has been appointed by UNINFO (a member of UNI) as the modelling expert to represent Italy at CEN TC 310.

Bernardino Chiavola graduated in Mathematics at the University of Torino in 1972. Since 1980, he has been responsible of the Configuration Management function within the Engineering Department of Alenia. Currently, he is also the technical coordinator of a new European Research Project called AIT (Advanced Information Technology in Design and Manufacturing) within Alenia Aeronautics. In the Engineering Department he is focal point of the analysis and definition of the user's requirements for information technology application development. He is the AECMA national delegate in the technical committee dealing with the drawing preparation and operating procedures, as well as Alenia Representative in several EUROFIGHTER technical working groups.

Mauro Varani graduated in Computer Science at the University of Milano, in 1987. He is software engineering consultant and project leader at ARTIS S.r.l. He is expert in manufacturing process modeling, simulation and software testing techniques. He published papers on process modeling and software system testing. He made simulation studies of production plants for several major national and international companies. He contributed to the Esprit project CIMOSA and developed the model of a production control system in the CIMOSA pilot phase.

5

Rapid prototyping of integrated manufacturing systems by accomplishing model-enactment

M. W. C. Aguiar, I. A. Coutts and R. H. Weston
Manufacturing Systems Integration Research Institute
Loughborough, Leics. LE11 3TU, England, I.A.Coutts@lut.ac.uk

Abstract

This paper describes a model-driven approach to the flexible integration of software system components. This formalised approach is based on a combination and extension of state-of-the-art reference architectures centred on CIM-OSA (i.e. Open Systems Architecture for CIM). Also described within the paper are the techniques which have been devised to generate code of new software system components.

1 INTRODUCTION

Modelling provides a formalised way of coping with high levels of complexity commonly found when using computer systems to integrate manufacturing activities on an enterprise-wide scale. Many research initiatives have focused on defining models, reference models and reference architectures to capture design guidelines which help structure and support one or more aspects[1] and phases[2] of the life cycle of an Integrated Manufacturing Enterprise (IME)[3]. An analysis of the findings of such initiatives conducted by the authors in 1992 (Aguiar 1995b)

---

1. A modelling aspect is "an abstraction viewpoint of one total view, which emphasises some particular [view] of the model and disregards others for ease of analysis" (ESPRIT/AMICE 1993a).
2. In this paper, the life cycle of an IME is considered to encompass the following phases (or stages in the life of an enterprise): strategy definition, conceptual analysis, design, implementation, operation and maintenance. A more thorough discussion about the IME life cycle is presented in (Aguiar 1995b).
3. The terms integrated manufacturing enterprise, integrated manufacturing systems, CIM system are used interchangeably in this paper, as are CIM, system integration and enterprise integration. The difference between the use of these terms often lies in the scope or focus of their application.

concluded that: (1) considerable gaps exist in the formalism and completeness of any single reference architectures available at that time, hence no single architecture could lend structured support to the life cycle of an IME; (2) the most promising approaches were not yet directly usable by industry, due mainly to a lack of usable and comprehensive software tools to support the architectures; (3) potentially, a combination of architectures centred on CIM-OSA offered a way forward, if appropriate software tools could be conceived to facilitate their use. Here, the authors considered that CIM-OSA could be used as an overarching framework or skeleton which would support all key aspects and phases of enterprise-wide integration, but that this framework would need to be populated by other reference architectures and software tools created to deliver the technology to industry.

Although the CIM-OSA specification, conceived by the AMICE consortium (ESPRIT/AMICE 1993a) can be implemented to provide solutions to major issues of Enterprise Integration, there are important issues which it does not address. This is particularly evident with respect to its 'function', 'resource' and 'organisation' modelling views at the 'design specification' and 'implementation description' modelling levels[4] (Aguiar 1993). Additionally, despite effort in on-going initiatives (Aguiar 1994a) (ESPRIT/VOICE 1995), CIM-OSA has yet to be implemented as an organised method supported by wide scope CASE tools which can progressively support modelling through to realisation of a software system running upon an integrating infrastructure.

Based on these findings, the authors proposed and have implemented SEW-OSA, which is a formal systems engineering workbench that combines CIM-OSA, generalised stochastic time Petri-nets, predicate-action Petri-nets, object-oriented design, and the services of the CIM-BIOSYS[5] integrating infrastructure to support systems integration projects.

This selection of architectures provides means of overcoming primary limitations encountered in the CIM-OSA architecture, namely:

- Petri-nets and object-oriented design were adopted to populate design specification and implementation description modelling levels of CIM-OSA, thus enabling complete support for model-building;
- Petri-nets were also adopted as a means of enabling analysis and simulation;
- CIM-BIOSYS was the only integrating infrastructure available (at the time that this research started) which could be enhanced in order to support model-enactment, thus enabling rapid-prototyping of an integrated system.

The acronym SEW-OSA (i.e. "system engineering workbench centred on CIM-OSA") was given to the workbench due to its orientation, namely: (1) it aims to address the **engineering** of an enterprise from a **systems perspective**; (2) its underlying framework is structured chiefly on the formalism of **CIM-OSA**; (3) it aims to provide the level of life-cycle support of a **workbench**, consisting of integrated tools for automating the entire design process, which

4. The CIM-OSA (Open Systems Architecture for CIM) architecture defines a modelling framework for model-building which embraces the definition of the modelling constructs required for modelling four views (i.e. function, information, resource and organisation), along three modelling levels or stages (i.e. requirements definition, design specification and implementation description) based on three levels of generality or detail (i.e. generic, partial and particular). CIM-OSA also provides the specification of an integrating infrastructure for model execution. A detailed description of the CIM-OSA constructs is presented in (Aguiar 1995a) (ESPRIT/AMICE 1993b)].
5. CIM Building Integrated Open SYStems is an integrating infrastructure produced by the Manufacturing Systems Integration (MSI) Research Institute at Loughborough University (Coutts 1992) (Weston 1990).

encapsulates: the steps to be followed during the process (i.e. a method); the constructs to be created and manipulated in order to formalise the design (i.e. a language); the considerations, analyses and decisions to be made at each step (i.e. a framework); and associated documentation of design activities.

A workbench (such as SEW-OSA) is necessary in order to enable the investigation of a plethora of integration 'problems' and possible 'solutions'[6], by providing means of: (1) describing each problem and defining possible solutions through the application of a formal language; and (2) testing the solutions. These requirements lead, respectively, to the proposition of the two main capabilities for such a workbench, namely: a **model-building capability** and a **model-enactment capability**[7] (as shown in Figure 1). Basically, each class of capability is required to support more than one modelling level of CIM-OSA, in a way which bridges gaps that exists within and between these modelling levels. Such capabilities are required to enable graceful migration from a modelling description of 'what' the system should do, to a description of 'how' the system should do it, by means of the actions executed by its components. Indeed, an important contribution envisaged for these two capabilities is that of defining a method for the organised application of the architectures, where their amalgamation would be under the framework defined by CIM-OSA.

Thus, SEW-OSA provides the first instance of an organised method for the application of CIM-OSA, by providing two classes of capability associated with its design methodology. At each stage of the design methodology, constructs[8] are manipulated in the form of diagrams and templates, thereby providing a means of organising design information as it is created.

Essentially, the method proposed is based on the application of a model-building capability to generate models (i.e. the business model shown in Figure 1) which formalise a possible relationship between problem and solution, which could then be made available to a model-enactment capability in order to test this relationship. The integrated use of these two capabilities is the fundamental support provided by SEW-OSA which this research has realised for encapsulating the formalism of selected architectures. Indeed the method entails the application of the model building capability outlined in Figure 2 which generates models which leads to the generation of models which can be input to and processed by the model enactment capability illustrated by Figure 1.

Between '92 and '95, the realisation of SEW-OSA provided a major focus of research effort within the "Model-Driven CIM programme"[9]. Indeed the aim of such an effort was to offer in SEW-OSA a framework of methods encapsulated into toolset within a scope which embraces the complete IME life cycle. SEW-OSA does not attempt to provide automated support for all design decisions, but it does provide a structured framework for decision making. As our detailed understanding of 'IME design best practice' grows, SEW-OSA can be advanced and

6. The term 'problem' is used here to describe a set of business requirements to be met or objectives to be achieved which motivated a system integration project. The term 'solution' is used to describe the final organisation of a system and specification of its components, constructed to address related requirements and objectives.
7. Model-enactment is viewed in this paper as the ability to let a model evolve over time (i.e. the evolution of its dynamic behaviour).
8. A construct (or a modelling construct) is a generic building block which characterises an element of the formalism of a modelling method or language.
9. Model-Driven CIM is a major programme of UK funded research (SERC/ACME research grant) at the MSI Research Institute.

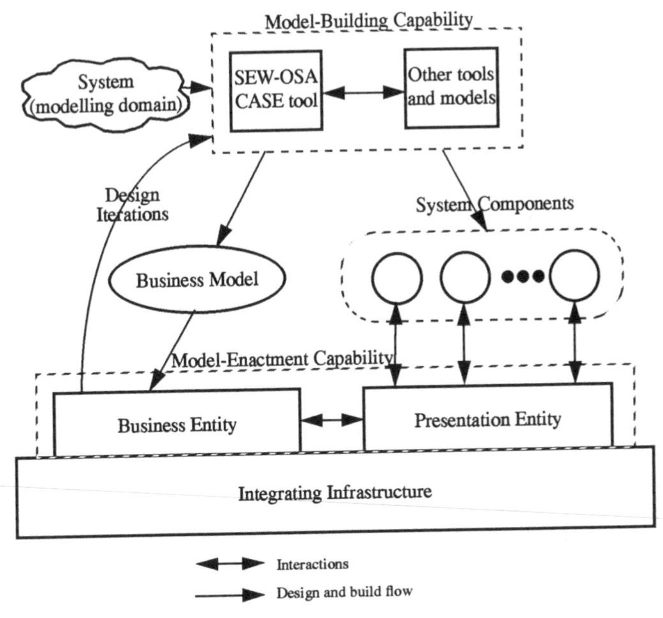

Figure 1 - SEW-OSA - a conceptual view

design knowledge[10] gradually fleshed into the framework. Here, the inclusion of design knowledge can be realised through use of models of integration 'problems' and 'solutions' to address them or by enhancing the modelling constructs of the SEW-OSA workbench.

This paper addresses two main issues. Sections 2 to 4 describe the way in which system prototyping is facilitated by the model-building and model-enactment capabilities of SEW-OSA. Here an overall description of the structure of the workbench is included. Section 5 describes the methodology adopted and implemented to generate the code of software components. Sections 6 to 7 outline major aspects of an evaluation of SEW-OSA through a case study.

2 Model Building Capability

The main contribution of the SEW-OSA model-building capability is the realisation of a modelling method which associates business process and object-oriented descriptions (as represented by the two clouds in Figure 2) in order to produce a complete business model. Such a method captures the essential definitions of the CIM-OSA modelling constructs,

10. Design knowledge is related to knowledge about 'problems', 'solutions' and possible relationships between them.

organises them into a usable form and enhances them with additional constructs from object-oriented formalisms and Petri-nets.

Following is an overview of the main diagrams which comprise the model-building capability of SEW-OSA. Correspondence between the acronyms used in the text below and the constructs manipulated in each diagram of the SEW-OSA CASE[11] tool is indicated in the model example represented in Figure 2.

Requirements definition modelling level

This modelling level embraces a business process description of the enterprise domains under consideration. This description formalises 'what' the system to integrate the domains is expected to achieve.

- **Context diagram**. This diagram defines the domains (i.e. main areas of an enterprise) under consideration, and the relationships between the domains. One context diagram is created for each individual enterprise model.
- **Domain diagram**. This diagram defines the major domain processes of a domain. One domain diagram is created for each CIM-OSA-compliant domain[12].
- **Structure diagram**. This diagram defines the functional decomposition of a domain process[13] in terms of enterprise activities and business processes[14] in a structured manner. One structure diagram is created for each domain process.
- **Behaviour diagrams**. These diagrams define the flow of control used to execute the functionality of a domain process and its business processes. Behaviour diagrams can also be referred to as process diagrams. One behaviour diagram is created for each domain process and subsequently for business processes within the domain process structure.
- **Functional diagram**. This diagram defines flows of material, information and control through the atomic building blocks of the domain process (i.e. enterprise activities[15]). One functional diagram is created for each domain process.

---

11. Computer-Aided Software Engineering.
12. According to the ESPRIT/AMICE consortium (ESPRIT/AMICE 1993b), domains "identify well-defined, totally integrated, functional areas of the enterprise". Domains are modelled either as CIM-OSA-compliant domains or non-CIM-OSA-compliant domains. A CIM-OSA-compliant domain identifies the area to be engineered [or integrated] within the enterprise. A non-CIM-OSA-compliant domain identifies other areas with which the area to be engineered interacts.
13. According to the ESPRIT/AMICE consortium (ESPRIT/AMICE 1993b), domain processes (or DP's) "are high-level processes [...] triggered by some events and producing a defined end result (function output). Domain processes are at the highest level of functional decomposition of domains. They must be triggered by nothing else than events" [12]. Domain processes can also be viewed as objects which communicate via exchange of information, material and events.
14. According to the ESPRIT/AMICE consortium (ESPRIT/AMICE 1993b), a business process (or BP) "is a subprocess of a domain process. It cannot be directly triggered by events and is always called by a parent process." A business process works as an intermediate construct between domain processes and enterprise activities. It should be noted that the term 'business process' is used in this paper in a particular sense (as adopted by CIM-OSA), and should be distinguished from 'business process' as used in a more general sense by Davenport, Shourt and Hickman (Childe 1993).
15. According to the ESPRIT/AMICE consortium (ESPRIT/AMICE 1993b), enterprise activities (or EA's) "describe [the] basic enterprise functionality (i.e. things to be done). They are defined by their function input, function output, control input, control output, resource input, resource output and ending status and have no behaviour defined at the requirements definition modelling level. They are always called by a parent process."

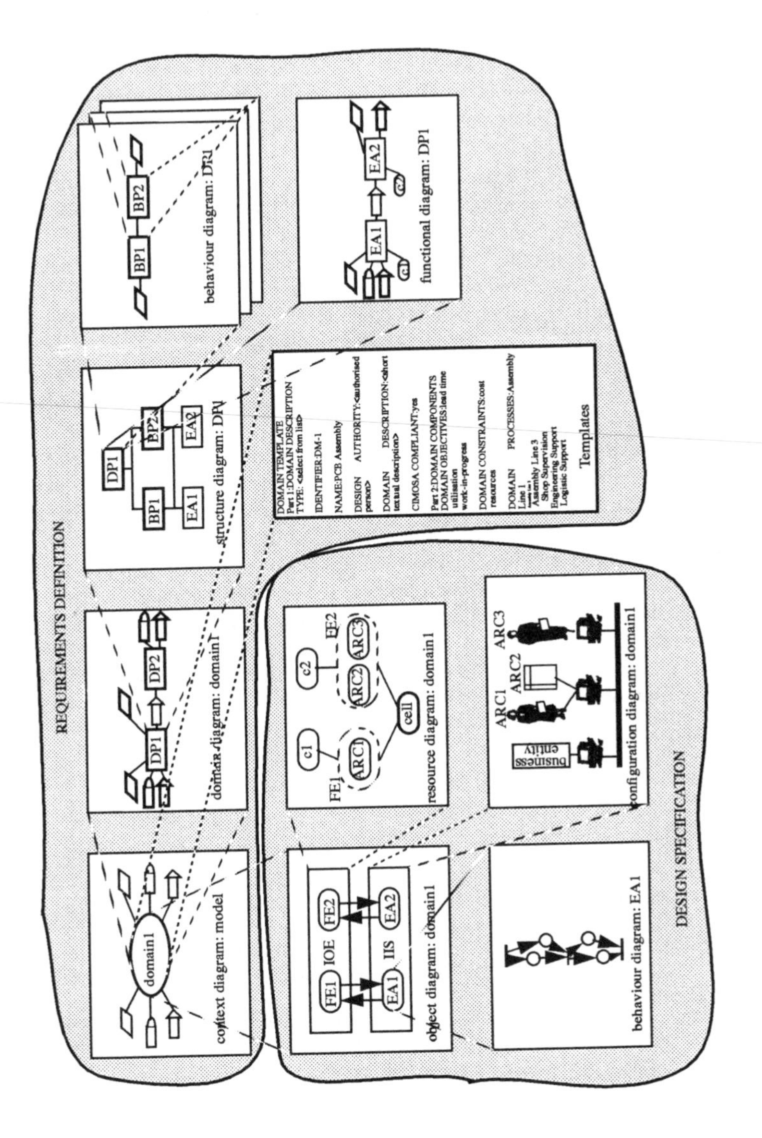

Figure 2 - SEW-OSA model-building capability

Design specification modelling level

This modelling level embraces a object-oriented description of the enterprise domains under consideration. This description formalises 'how' the integrated system achieves its purpose through interactions among its components and actions internal to each component (i.e. functions that each component performs on its own).

- **Object diagram**. This diagram defines the flow of messages between functional entities[16], the business entity[17] (i.e. enterprise activities) and the information entity of CIM-OSA. One object diagram is created for each CIM-OSA-compliant domain.
- **Activity behaviour diagram**. This diagram defines how the enterprise activity uses (via message exchanges) the integrated operation environment (i.e. a grouping of functional entities, or FE's), in order to perform its basic functionality (i.e. functional operations[18]). One activity behaviour diagram is created for each enterprise activity.
- **Entity behaviour diagram**. This diagram defines the expected external behaviour of a functional entity as perceived by the enterprise activities. Such a description is used to emulate the behaviour of an active resource component during the rapid-prototyping stage of a system. One entity behaviour diagram is created for each functional entity.
- **Resource diagram**. This diagram specifies instances of active and passive resource components (i.e. ARC and PRC)[19] associated with the classes specified by their functional entities which are able to execute the functional operations required by an enterprise activity. These functional operations are related to the capability required by the enterprise activity; the resource capability being defined in the functional diagram (see Figure 2). One resource diagram is created for each integrated operation environment associated with a particular domain (stemming from the IOE construct, as illustrated in Figure 2).
- **Configuration diagram**. This diagram defines the computer configuration of the system (i.e. where each active resource component will be executed or interfaced with). One configuration diagram is created for each segment of the IIS which serves a particular enterprise domain.

A text template is associated with each symbol represented in the diagrams of Figure 2, this prompts the user to define all necessary attributes. The text templates implemented in SEW-OSA comply with the format and attributes of the templates specified by CIM-OSA. The use of diagrams is not explicitly defined in the CIM-OSA specifications but was found by the authors to be very important to improve the usability of the method.

---

16. According to the ESPRIT/AMICE consortium (ESPRIT/AMICE 1993b), a functional entity "is a resource able to perform, completely on its own, a (class of) functional operation(s)". functional entities are viewed in this thesis as functional representations of active resources components required to fulfil the capability associated with enterprise activities identified in the functional diagram.
17. The business entity includes functions required to control the enterprise operation as described in its business model (i.e. function, resource and organisation models). This entity plays a key role in enabling business integration (i.e. coordination of interaction among system components).
18. According to the ESPRIT/AMICE consortium (ESPRIT/AMICE 1993b), a functional operation "is a basic unit of work defined at the design specification modelling level (i.e. lowest level of granularity in the function view). At run-time, [a functional operation is] fully executed or not at all."
19. An active resource component identifies a component of a system which is able to execute functional operation(s) on its own. It can also be a modelling description which characterises either a human being, an application program or a machine that possess a computerised controller (i.e. human functional entity, application functional entity and machine functional entity (ESPRIT/AMICE 1993b)). A passive resource component is an object used by the active resource component when performing functional operations.

3 Model Enactment Capability

Design information formalised in the form of models produced by the SEW-OSA model-building capability are passed in the form of a number of pieces of interpreted code (i.e. the business model in Figure 1) to the SEW-OSA model enactment capability. The model-enactment capability manipulates the design model to enable simulation and rapid prototyping of solutions. Model execution is achieved through populating the "design specification" and "implementation description" modelling levels of CIM-OSA with the elements defined in the following.

The design specification modelling level:
A capability has been implemented to facilitate analysis of Generalised-Stochastic Time Petri-nets. These nets are generated automatically by a software tool created to process CIM-OSA functional models. Also, tools have been produced to simulate the dynamic behaviour and to carry out performance evaluation of the system based on the execution of the Petri-Net model in simulated time (here various metrics can be used, such as: lead-time and cycle-time values, level of utilisation of resources, profile of work-flow, work-in-progress and cost). The methodology devised and supported to facilitate such evaluation exercises is discussed more thoroughly in (Aguiar 1993).

A capability was also conceived and implemented to facilitate rapid prototyping of system solutions and their (emulated) components. This capability was included to facilitate testing of the system structure, this by providing means of executing the business model generated by the CASE tool. In conformance with CIM-OSA ideas, the rapid-prototyping capability achieved model execution via a business entity.

The implementation description modelling level
This capability facilitates configuration of the physical system by gradually replacing its (emulated) components by physical ones which will be used in the final system (i.e. machines, application programs, data storing devices and human beings). As illustrated in Figure 1, physical components gain access to the integrating infrastructure via appropriate interfaces (i.e. the presentation entity[20]).

Before a physical system is finally commissioned, typically there will be a series of iterations between the use of the model-building and the model-enactment capabilities. Indeed, one of the most important contributions of SEW-OSA is that it facilitates these iterative processes in a structured and consistent manner, thereby traversing different phases of the IME life cycle. This can lead to much improved solutions and much shorter design-to-build lead-times via what essentially becomes an integrated systems engineering process. Thus, SEW-OSA provides the support of an organised method based on model-enactment, which brings together conventionally separate life cycle phases, from 'requirements definition' to the

---

20. The presentation entity provides means of integrating enterprise components (including legacy components). This entity maps the CIM-OSA internal protocol into protocols that are understood by enterprise components (e.g. proprietary machine commands). This provides the remaining entities of the CIM-OSA IIS (i.e. the business entity, the information entity, the common services entity and the system management entity (ESPRIT/AMICE 1993a)) with a uniform means of interacting with heterogeneous system components. Such a mapping is defined based on the models manipulated by the function and resource views.

'configuration and execution' of computer integrated manufacturing systems upon an industrially tested integrating infrastructure (i.e. CIM-BIOSYS).

4 Realisation of SEW-OSA

Realising, applying and validating the model-building and the model-enactment capabilities involved the development activities identified by a tag number in Figure 3, namely: (1) developing a CASE tool that encapsulates the CIM-OSA model-building method; (2) establishing a link to a simulator to enable dynamic analysis based on use of a business model; (3) creating a business entity for CIM-BIOSYS which enables rapid prototyping of systems; (4) establishing links to other tools and services (created by other researchers working within the 'Model-Driven CIM' research programme) this to address design aspects not covered by SEW-OSA (e.g. detailed information and resource modelling) (Clements 1993), (5) developing a presentation entity for CIM-BIOSYS; (6) evaluating use of SEW-OSA through a case study application to solve shop-floor integration and coordination problems at an industrial site; and (7) validating the results obtained. Implementing the elements tagged (1) and (3) (which are of focus of interest in this paper) involved the following research activities:

CASE tool
In order to achieve such a method, two major tasks have been accomplished, namely:

- (1) to **review CIM-OSA** (by analysing, correcting and complementing those parts that were not well defined) particularly in regard to design specification and implementation description modelling levels and
- (2) to **enhance CIM-OSA** with modelling constructs from the other architectures, this leading to the implementation of part of the CIM-OSA modelling methodology in combination with predicate-action Petri-nets and an object-oriented representation in a CASE tool.

As a result of these tasks, a number of incremental contributions (in their own right) have been made (refer to (Aguiar 1995a) for details).

Business Entity
Realising the business entity involved to:

- develop a model interpreter and debugger which incorporates a capability to drive interactions between system components. This is realised by interpreting business models generated by the CASE tool. Here, the interpreter and debugger comprises the following four components, namely: an Event Handler, a Process Controller, an Activity Controller and a Resource Manager (see Figure 3). Each of these components addresses a particular aspect of the business model. These components interact with one another via the CIM-BIOSYS infrastructure. Further information about the Business Entity of SEW-OSA is presented in (Aguiar 1994b). The business entity consists of a layer of services (written in "C" which execute in an X-Windows/Unix environment) which draw upon the general integration services of the CIM-BIOSYS infrastructure (which functions as a distributed operating system, flexibly mapping application software onto manufacturing and computing resources). This layer of services achieves a link to the CASE tool and, in so doing, enacts models which it helps create, thereby achieving structured and flexible integration of system components;

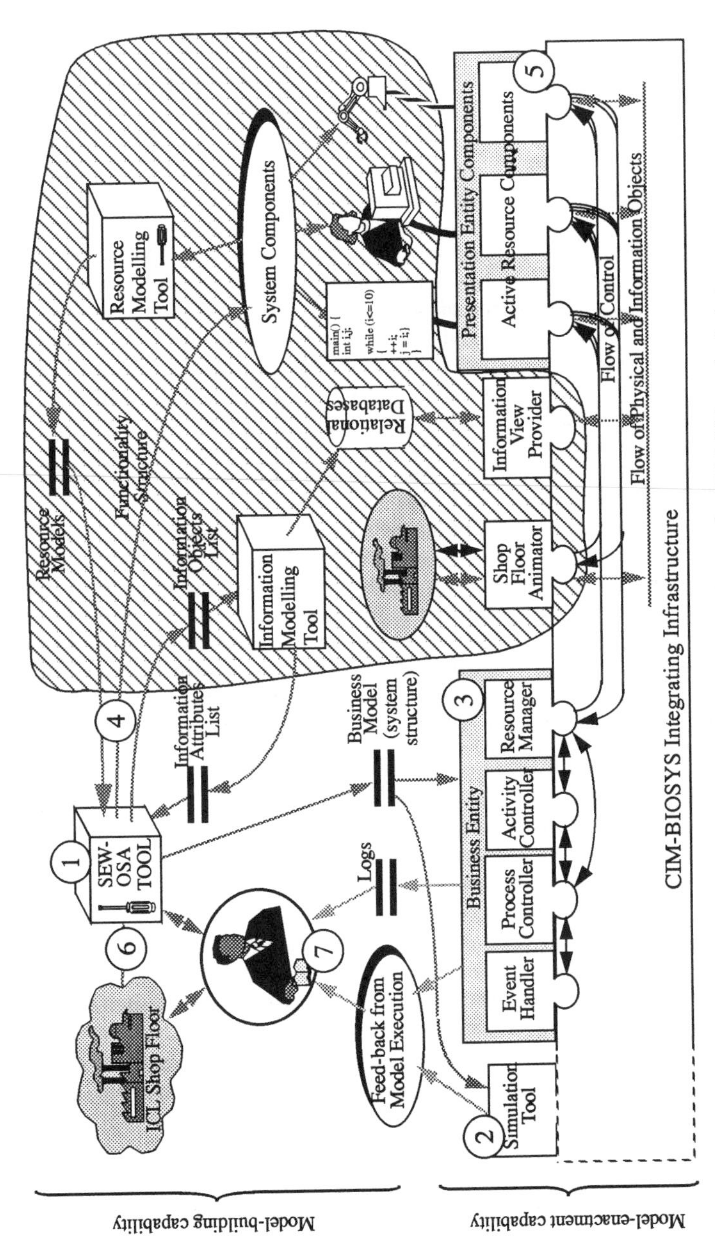

Figure 3 - System Engineering Workbench centred on CIM-OSA - a physical view

- enable rapid generation of prototypes of the system structure and its components, where the latter are generated in an emulated form. In the early stages of rapid prototyping, Active Resource Components (represented in Figure 3) consist of interpreters of predicate-action Petri-net models, which emulate their internal behaviour. The structure of the Active Resource Components is used as a basis for generating software application components, as discussed in the following sections.

5 Application Software[21] Generation

This section describes the infrastructural basis upon which SEW-OSA can be applied to the production of flexibly integrated distributed systems i.e. systems which comprises distributed interacting software components which are hardware and operating system independent.

Previous research at the MSI Research Institute (Coutts 1992) highlighted the need to facilitate interactions between distinct functional elements via the use of an integrating infrastructure. Here the infrastructure should provide means of achieving interaction between applications but should not define the manner in which they occur i.e. contain and prescribe the behaviour of a system. Recent advances in system modelling have provided methodologies and tools which can describe the behaviour of separate interacting elements, but to date have not provided a clear path to the automatic creation of code which realises such behaviour.

Current initiatives within MSI (Clements 1993) have lead to the definition of a structure for software components which separates internal functionality and behaviour whilst enabling interaction between applications. Such a separation is of vital importance in establishing formal and generalised methods of using a combination of high level modelling techniques and "bottom-up" implementation expertise within computer aided environments for rapid prototyping of systems, this including the generation of code for final system implementation.

Figure 4 depicts the application decomposition proposed by the authors. It should be stressed that the focus here is on the generation of code which achieves both the behavioural and interaction aspects of the application, not on its internal functionality.

The previous section discussed the generation of "Enterprise Activities" and "Active Resource Components" using SEW-OSA. The behaviour of these activities and components whilst interacting with the external environment is described (within the CASE tool) by predicate action Petri-net models. Using MSI's application structure the requirement for rapid system prototyping can be met by a means of enacting these models. This approach provides a means of mapping modelled systems (represented in some abstract form) onto implemented systems.

5.1 Current Implementation of the Model Enactment Facility

A Prolog execution environment (Clocksin 1984) was chosen to execute the internal behaviour of enterprise activities and active resource components described by predicate-action Petri-nets for the following reasons:

- it provided a natural environment for predicate logic;

---

21. The terms 'application', 'application software' and 'software component' are used interchangeably in this paper.

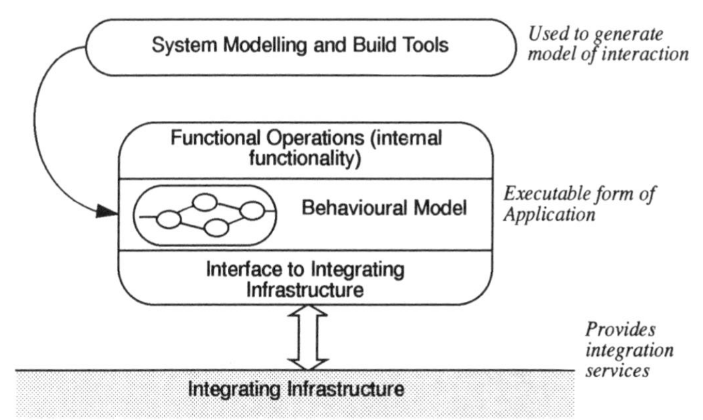

Figure 4 - Proposed Structure of an Application

- it offered the capability to extend its parser to allow user defined syntax (i.e. the transition syntax of Petri-net models);
- it facilitated incorporation of additional functionality (i.e. support for Petri-net token counts);
- it possessed an inherent ability to deal with the asynchronous nature of interacting elements (such as incoming message arrival instantiated in a Prolog database).

A link between the execution environment and the CIM-BIOSYS integrating infrastructure was established, so that the interactions described by the Petri-net models could be realised. A functional decomposition of the environment created is illustrated in Figure 5. A brief description of each function block follows.

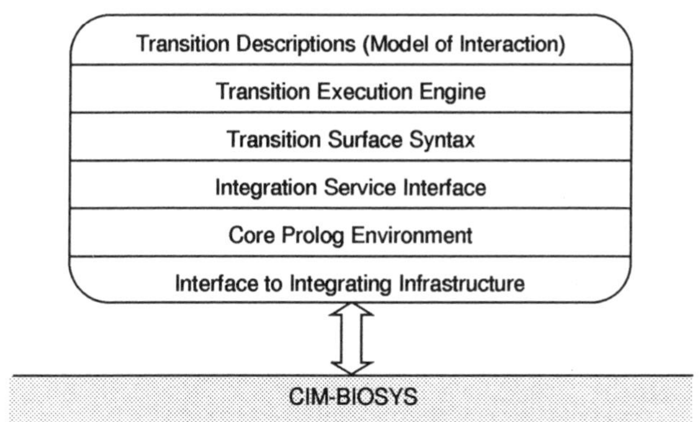

Figure 5 - Predicate-Action Petri-net Execution Environment

Interface to the Integrating Infrastructure

This comprises an inter-process communications mechanism, packet encode and decode and low level 'handshaking' functions required to interact with the CIM-BIOSYS integrating infrastructure. These have been coded in 'C' and incorporated within the runtime Prolog system. Incoming packets from other CIM-BIOSYS applications are instantiated as facts in the

Prolog database.

Core Prolog Environment
This is the "C Prolog" system as defined by Fernando Pereira, July 1982, EdCAAD, Dept. of Architecture, University of Edinburgh.

Integration Service Interface
Within this layer, the CIM-BIOSYS integration services are represented as Prolog facts. By instantiating a given fact, integration service requests are sent to CIM-BIOSYS. Responses from CIM-BIOSYS are instantiated in the Prolog database. A couple of examples follow.

Send message "start function 3" (sequence number 21) to an application called "tom".

```
send_app(tom,21, "start function 3").
```

Retrieve data regarding object "pad_positions" where attribute "no_of_legs" is equal to 2.

```
sel (pad_positions, "where no_of_legs = 2").
```

Transition Surface Syntax
This layer includes a parser which reads and evaluates both conditions and actions. The conditions determine if a particular Petri-net transition will fire. If the condition is true the accompanying action is executed. Global variables are also supported by the parser. An example of the condition syntax follows.

If variable i is equal to 2 and variable j is less than ten, send message "hello world" to an application called "fred" then increment variable i by 1.

```
on i = 2 & j < 10 do send_app (fred, "hello world") @ i is i + 1.
```

Transition Execution Engine
Once an application is invoked this engine tests and fires the transitions represented by the Petri-net model.

Transition Descriptions (Model of Interaction)
This comprises of the list of transitions and global variables which form the Petri-net model. An example of the syntax used in this model is presented as follows.

Transition name "link_to_fred", triggered when i is equal to 2 and j is less than 10. This transition establishes a link with an application called "fred" and then increments i by 1.

```
transition(link_to_fred,((i=2)&(j<10)),(est_link(fred) @ (i is i+1))).
```

6 Model Enactment Example

This section briefly outlines an industrial case-study application using the SEW-OSA workbench. The modelling domain chosen for this evaluation work comprised the Surface Mount Technology (SMT) assembly lines of a major UK supplier of electronics products. In this paper focus will be on a 'TO-BE' description of just one of these assembly lines. The model is a 'TO-BE' description in the sense that it anticipates use of an integrating infrastructure, whilst accurately modelling the current mode of operation of the assembly line. Figure 6 shows the basic arrangement and associated material and control flows for the SMT assembly line. Thus, batches of PCB's flow along the line and are inspected at two inspection points, i.e. after the printing process and whilst located on the inspection conveyor. Inspection operations are triggered either by problems detected by the operators or as part of a regular inspection process which is programmed to occur (e. g. every "n" PCB's manufactured at the point in question). Buffers are used to smooth the flow of material and overcome imbalances in the line, thus avoiding inappropriate PCB queue sizes.

This process was modelled using the SEW-OSA model building capability in order to:

- formalise the requirements associated with the process (i.e. populate the requirements definition cloud illustrated by in Figure 2).
- define a system configuration which can meet these requirements (i.e. populate the design specification cloud illustrated by Figure 2).

A fragment of the requirements definition model is shown in Figure 7. This figure presents a simplified version of the domain diagram and is effectively the source of the behaviour diagrams of the business model. The behaviour diagrams so produced encode necessary coordination activities related to system components (such as operators and printers involved in the preparation stage of the assembly line).

A fragment of the design specification model produced during the evaluation study is shown in Figure 8. Here, enterprise activities identified in Figure 7 implement a functional transformation (i.e. act on their information and material inputs, in order to produce desired outputs). They achieve this by requesting the execution of functional operations (i.e. FO's) supplied by the functional entities of the system. Functional operations can also relate to data access operations (as shown in Figure 8).

The internal behaviour of the enterprise activities so defined is described by predicate-action Petri-nets. An example of such a description for the enterprise activity "print" is shown in Figure 9. Here the top and bottom transition represent the initialisation and termination of a run-time occurrence of the enterprise activity. On initialisation, a token is included in pEA-3_1 which enables tEA-3_2. When tEA-3_2 is fired the functional operation FO-5 is sent to FE-2 (see Figure 8). Then, EA-3 waits for the receipt of FO-6 in order to carry out its processing. FO-21 and FO-22 represent a data reading operation from a data-base. This piece of data is essentially a count of the PCBs produced between the two inspection operations (i.e. the number of executions of the enterprise activity "check" illustrated in Figure 7). Conditions tEA-3_3 and tEA-3_4 in Figure 9 indicate alternative ending statuses for this enterprise activity upon which a decision is made as to whether to check the PCB or to finish executing the procedural rules of the business process "prepare" (as shown in Figure 7).

As part of the evaluation study, similar behavioural descriptions were defined for all remaining processes (i.e. Figure 7) and activities (i.e. Figure 9) in the business model. Having completed the business model, interpreted code can be generated for the system (i.e. model-enactment for rapid-prototyping of the system, as illustrated in Figure 1) in order to test the design solution.

Figure 10 depicts example code produced by SEW-OSA, in this case to enact (and hence execute) EA-3 using the predicate-action Petri-net execution engine described in this paper. This code along with other code fragments for the remaining constructs of the business model are executed by the business entity (see Figure 3). The business entity functions as an engine which enacts the business model in order to coordinate all interactions amongst system components.

A number of iterations of model-building and model-enactment were found to be necessary in this evaluation study before a satisfactory design solution was obtained for the system represented in Figure 6. It is expected that such an iteration process would normally be

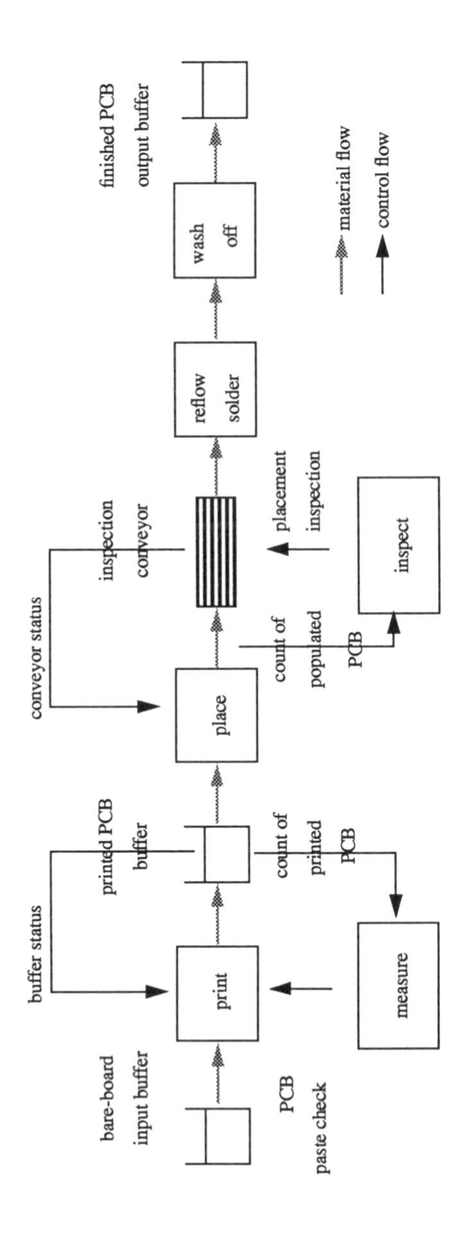

Figure 6 - Surface Mount Technology Assembly Line

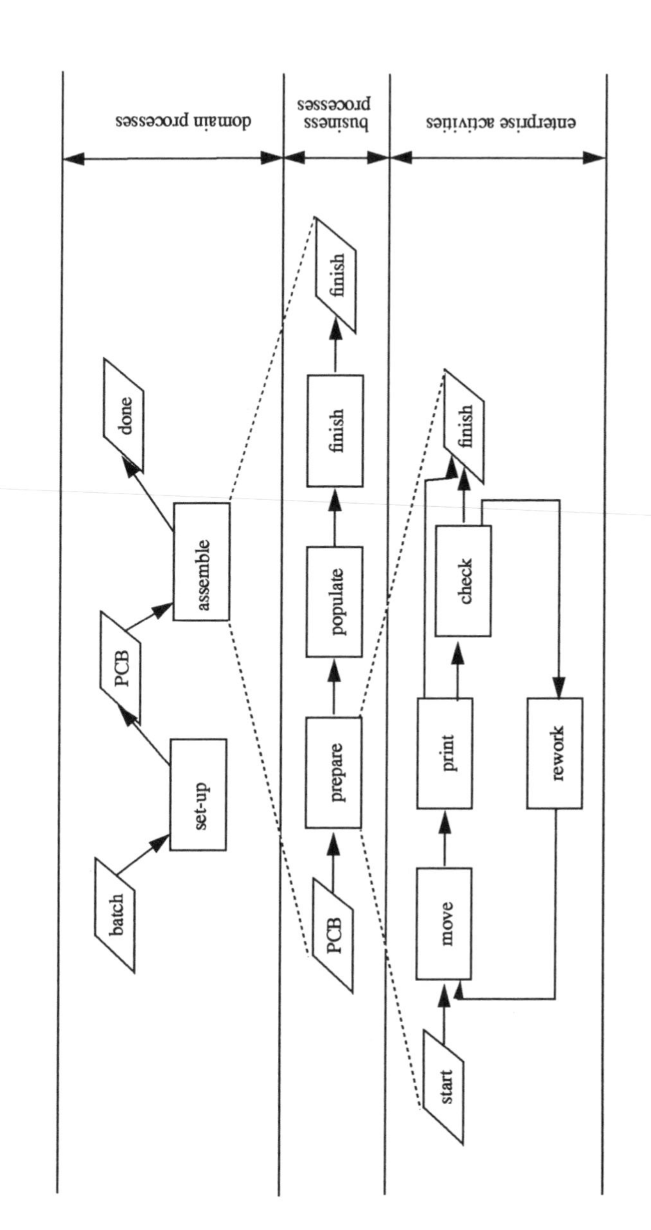

Figure 7 - Fragment of the behavioural model of the SMT assembly line

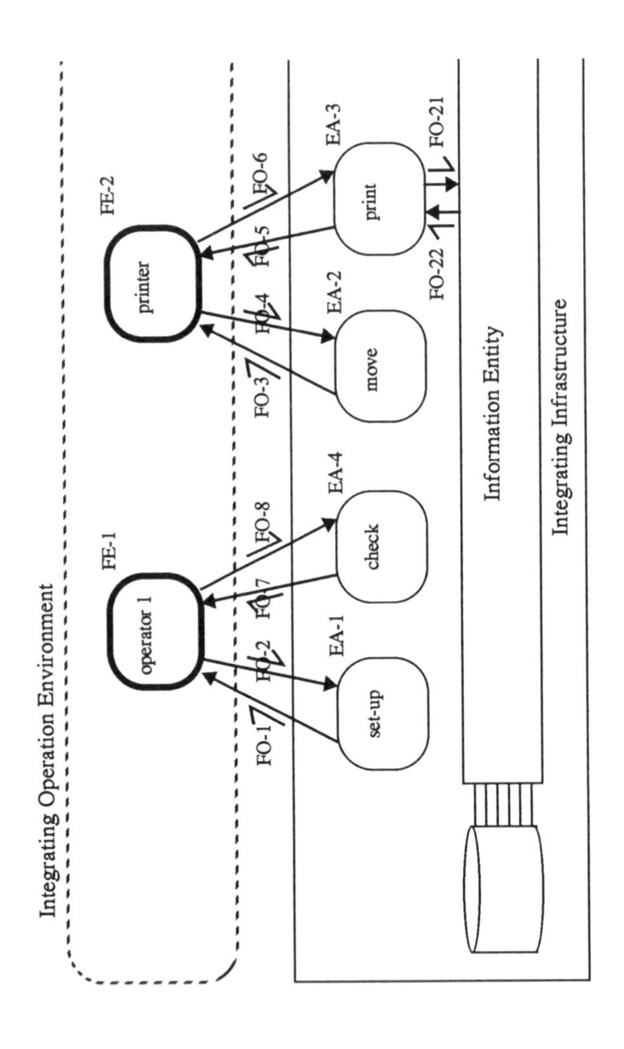

Figure 8 - Object Diagram: smt line

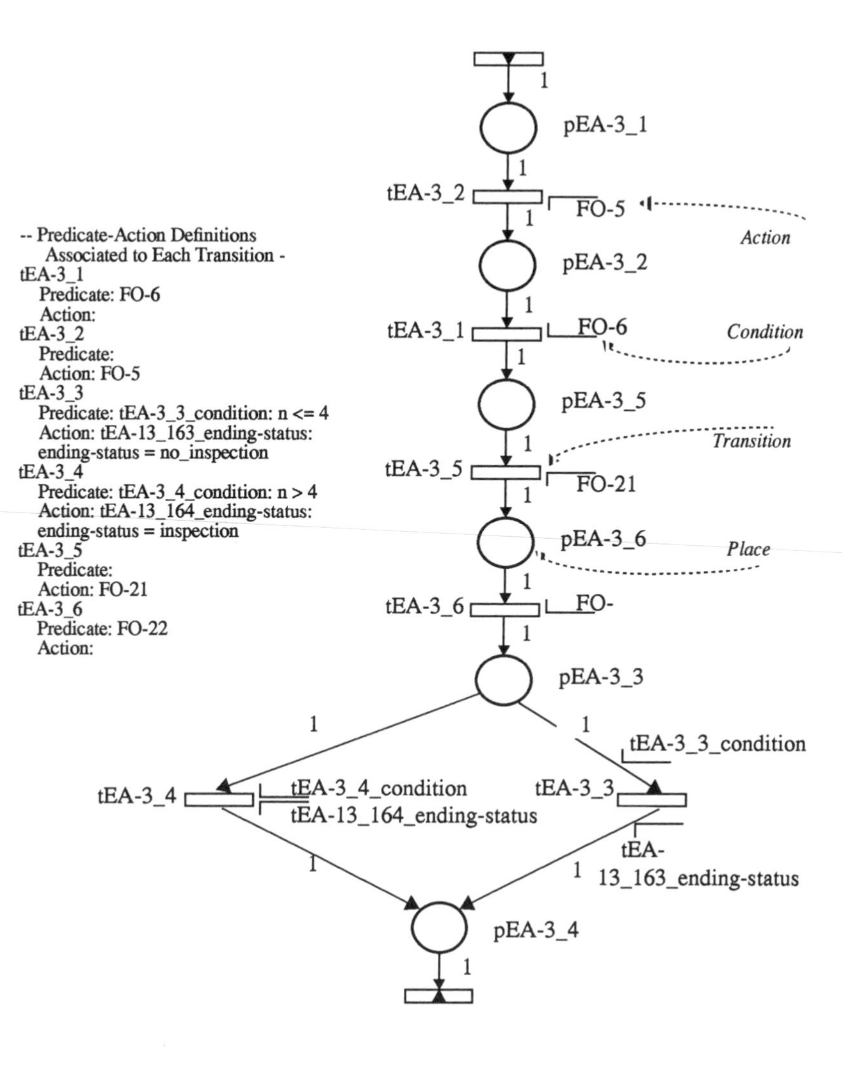

Figure 9 - EA-3: Behaviour Diagram: print

required, particularly where a largely untested system model is being created and used.

Predicate-action Petri-nets have provided a satisfactory way of describing internal functional flow and external infrastructure interaction, but the Prolog environment used for their execution does not possess the runtime performance required to implement complex systems. Hence ongoing work at MSI is focused on re-implementing the model enactment facility in the form of a "compiler/post processor" which will convert the transition descriptions into 'C' code. This code will then be "linked" with user supplied functionality to compose a complete software component.

7 Analysis of Results

In various evaluation exercises similar to that summarised in Section 6, use of the SEW-OSA workbench has led to findings which can be classified under four headings: implementation results, case study results, architectural results and research results.

The **Implementation Results** obtained relate to what has actually been implemented and how well it has worked. Analysis of these results has been based on the definition and use of performance metrics. This has led to an analysis of ability to handle complexity and associated performance issues when addressing typical industrial situations.

Case Study Results include qualitative and quantitative results obtained so far based on the gathering of industrial data. Activities which have led to these results consisted of simulation runs as well as analysis of performance associated with the proposition of a re-engineered system to coordinate the SMT shop-floor. The main result obtained consists of a specification of how shop-floor re-engineering can be achieved in order to operate on a model-driven manner.

The **Architectural Results** obtained relate to architectural definitions which were produced as part of the development of SEW-OSA in the context of the "Model-Driven CIM" project. This includes a technical evaluation of the specifications of SEW-OSA, as proposed in this research. A key issue here is how much synergy can be obtained from the proposed combination of architectures.

Research Results include an examination of the outcome of the research in the light of the issues that were set out to be investigated. This includes an evaluation of the basic axioms that motivated the underlying proposition of SEW-OSA (i.e. the desirable features of an architecture). At this stage, benchmarking against systems and tools stemming from current practice is also being developed. Finally, the methodology proposed in this paper is itself being analysed in regard to its efficacy and efficiency.

Findings under these categories have demonstrated that real industrial problems can be tackled with SEW-OSA and that its inherent methods enable the very rapid-prototyping of solutions. This should significantly reduce engineering effort and timescales involved when realising and changing integrated manufacturing systems. However, current limitations on the widespread industrial use of SEW-OSA exist in respect to:

- performance limitations related to the way in which certain components of the workbench have been implemented, as well as inherent limitations of the CIM-BIOSYS integrating infrastructure. This currently places limits on the complexity of the models that can be enacted, although means of achieving re-implementation have been identified which will extend the complexity of systems that can be handled;

```
variable(ptrans_in, 1).
variable(pEA3_1, 0).
variable(pEA3_2, 0).
variable(pEA3_3, 0).
variable(pEA3_4, 0).
variable(pEA3_5, 0).
variable(pEA3_6, 0).
variable(ending_status).

transition(trans_in,(ptrans_in = 1),(pEA3_1 is pEA3_1 + 1 @ ptrans_in is ptrans_in - 1)).
transition(tEA3_2,(pEA3_1 >= 1),(pEA3_2 is pEA3_2 + 1 @ pEA3_1 is pEA3_1 - 1 @ send_app(ac, “make()
    FO-5”))).
transition(tEA3_1,(pEA3_2 >= 1 & recv_app(ac, “made()”)),(pEA3_5 is pEA3_5 + 1 @ pEA3_2 is pEA3_2 - 1)).
transition(tEA3_5,(pEA3_5 >= 1),(pEA3_6 is pEA3_6 + 1 @ pEA3_5 is pEA3_5 - 1 @ send_app(ac, “obtain(n)
    FO-21”))).
transition(tEA3_6,(pEA3_6 >= 1 & recv_app(ac, “value(n,<value of n>)”)),(pEA3_3 is pEA3_3 + 1 @ pEA3_6 is
    pEA3_6 - 1)).
transition(tEA3_3,(pEA3_3 >= 1 & n <= 4),(pEA3_4 is pEA3_4 + 1 @ pEA3_3 is pEA3_3 - 1 @ ending_status
    is “not_inspect”)).
transition(tEA3_3,(pEA3_3 >= 1 & n > 4),(pEA3_4 is pEA3_4 + 1 @ pEA3_3 is pEA3_3 - 1 @ ending_status is
    “inspect”)).
transition(trans_out,(pEA3_4 >= 1),(pEA3_4 is pEA3_4 - 1 @ send_app(ac,”finish(ending_status)”) @ halt)).
```

Figure 10 - Code fragment for the enterprise activity “print”

- the unavailability of reference models to provide guides related to good practice for system designers and builders. Without these models, users of the workbench must build models and generate solutions from scratch, this requiring greater levels of experience, longer learning and development times and most likely will result in significantly less than optimal outcomes.

8 CONCLUSIONS

This paper has reported progress on a research initiative which is combining and extending reference architectures to support the life cycle of Integrated Manufacturing Enterprises. The research has led to the definition and realisation of SEW-OSA (a Systems Engineering Workbench centred on CIM-OSA), which represents a major advance on currently available formal methods and tools, in terms of the support it provides for the life cycle of integrated manufacturing systems. Primary features of SEW-OSA's model building and model enactment capabilities are described along with an outline of initial results from an industrial case study application.

It was found that SEW-OSA can support life cycle phases from 'conceptual analysis' through to 'system operation (at run-time)', this by bringing together 'process-oriented' and 'object-oriented' techniques in a consistent manner. By so doing, SEW-OSA consolidates various models of an enterprise at different levels of abstraction. Its ability to enact models, by enabling simulation, emulation and execution stages in an iterative and consistent way leads to model-driven solutions which can be advanced and extended to meet the challenges of modern customer-driven markets. Indeed, for the SMT assembly line discussed here, the approach can potentially facilitate important enhancements in system co-ordination and control, whilst positively supporting system 're-configuration and change'.

Through being coupled with other "Model-Driven CIM" methods and tools produced at the MSI Research Institute (Edwards 1995) which enact other modelling perspectives of a manufacturing enterprise, the use of SEW-OSA promises to provide a highly effective way to re-engineering systems, so that they more closely meet high-level business requirements. Indeed, in the not too distant future such requirements may themselves be expressed formally as models which can be enacted via SEW-OSA and other "Model-Driven CIM" tools.

9 REFERENCES

Aguiar, M. W. C. and Weston, R. H. (1993) "CIM-OSA and stochastic time petri nets for behavioural modelling and model handling in CIM systems design and building". Proceedings of the Institution of Mechanical Engineers Part b - Journal of Engineering Manufacture, 1993, vol. 207, no, 3, pp. 147-158, England.

Aguiar, M. W. C. (1994a) "Benchmark of SEW-OSA against other tools and methods" Working Paper. UK.

Aguiar, M. W. C.; Weston, R. H. (1994b) "The business entity of SEW-OSA - Systems Engineering Workbench centred on CIM-OSA". Proceedings of the 10th National Conference on Manufacturing Research. England, pp 245-249.

Aguiar, M. W. C. (1995a) "An approach to enacting business process models in support of the

life cycle of integrated manufacturing systems". Doctor of Philosophy Thesis, Loughborough University, England.
Aguiar, M. W. C. and Weston, R. H. (1995b) "A model-driven approach to enterprise integration". International Journal of Computer Integrated Manufacturing, England.
Childe, S.; Bennett, J.; Maull, R. (1993) "Manufacturing re-engineering around business processes". The International Conference on Managing Integrated Manufacturing - Organization, Strategy & Technology. England: KAMG - Keele University.
Clements, P., Coutts, I., Weston, R. H. (1993) "A life-cycle support environment comprising open systems manufacturing modelling methods and the CIM-BIOSYS infrastructure tools". MAPLE'93 - Symposium on Manufacturing Automation Programming Language Environments. Ottawa, Canada.
Clocksin W.F. and Mellish C.S. (1984) "Programming in Prolog", 2ed., Springer-Verlag.
Coutts, I. A. et al. (1992.) "Open Applications within Soft Integrated Manufacturing Systems", Proc. of Int. Conf. on Manufacturing Automation, Hong Kong, ICMA 92.
Edwards, J. M.; Murgatroyd, I. S.; Gilders, P. Aguiar, M. W. C.; Weston, R. H. (1995) "Methods and tools for manufacturing enterprise modelling and model enactment". Submitted to the Proceedings of the Institution of Electrical Engineers - Special issue on Science, Measurement and Technology. England.
ESPRIT/AMICE. (1993a) "CIM-OSA Architecture Description", AD 1.0. 2. ed 1993.
ESPRIT/AMICE. (1993b) "CIM-OSA formal reference base".
ESPRIT/VOICE. (1995) "Validation of CIM-OSA (Open Systems Architecture) - A joint ESPRIT projects report".
Weston, R. H., Hodgson, A., Coutts, I. A., Murgatroyd, I. S. and Gascoigne, J. D. (1990) "Highly extendable CIM systems based on an integration platform". Proceedings of CIMCON'90, NIST, USA.

10 BIOGRAPHY

M W Aguiar: Seven years as managing director in charge of the Integrated Automation Division of a Research and Development Institute of the Federal University of Santa Caterina/ Brazil. Three years as a member of the MSI Research Institute, involved in the Model-Driven CIM project, with particular responsibility for the conception, realisation, application and evaluation of SEW-OSA.

I A Coutts: Two years at Marconi Research as a research scientist, working on industrial assembly automation and robotics projects. Seven years as a member of the Loughborough SI group. Particular responsibilities have included work on the group's software integration platform.

Prof. RH Weston: Over fifteen years experience of research in areas of machine control and systems integration. Supervisor for over fifty RAs and PhD students during that time. Author (or joint author) of over 200 refereed journal and conference publications in the field.
Chairman of member of various national (PSRC, DTI, LINK and BS) committees (ISO, IFAC, IFIP) serving the area and member of the board of five journals.

6

MMS Virtual Manufacturing Devices generation : The Paris subway example

E. Rondeau, T. Divoux, F. Lepage, M. Veron
Centre de Recherche en Automatique de Nancy (CNRS URA 821)
Université de Nancy I, BP 239, 54506 Vandoeuvre-lès-Nancy Cedex, France.
Tel : (33) 83-91-24-33, Fax : (33) 83-91-23-90,
Email : CRAN@cran.u-nancy.fr

Abstract

European research projects about CIM like CIM-OSA advise for the information system analysis the use of the entity-relationship formalism, and for the communication implementation, the MMS standard. At the logical level of this analysis which describes the data organisation we take into account both database and communication aspects. The logical analysis must generate the database structures and the MMS configuration of all devices, which corresponds to physical step. According to the CIM-OSA concept, we developed a process which allows to generate the VMD of each equipment by the information system study. We finally present its application for the communication system design of an underground station.

Keywords

MMS, CIM-OSA, Information System

1 INTRODUCTION

The CIM (Computerised Integrated Manufacturing) concept implies to reduce the functional independence of the different devices which compose a manufacturing system. So each entity has to take into account the global environment in which it is working for its own processing. This integration leads to many information exchanges between machines and their control systems. Industrial communications specificity and equipment heterogeneousness have required the definition of a standard at the application layer of OSI model (Open System Interconnection). Called MMS (Manufacturing Message Specification), it specifies the format of data supported by equipment and the services which allow their exchanges as well.

The aim of our work is to generate the communication system implementation in manufacturing environments without creating new modelling tools. Thus we follow the recommendations of CIM-OSA project (Open System Architecture) to configure each device modelled by MMS from the information system study.

We first describe MMS principles. Then, we remember the CIM-OSA concept bases, especially its information view and its integrating infrastructure. We then present how to translate an entity-relationship model into MMS syntax and the dedicated tool we developed. In the framework of an important contract between the CRAN and the Paris underground company (RATP : Régie Autonome des Transports Parisiens), we used this approach to define and integrate experimentally virtual devices in an MMS context.

2 THE MMS STANDARD (ISO 9506/1)

2.1 Introduction

ISO (International Standard Organization) has defined the OSI model which allows the heterogeneous equipments interconnection. In this model, the user interacts with the communication system through the application layer where he can find a set of specialized service groups. One of them, MMS, has been defined in order to meet the communication requirements in the manufacturing industry (Brill, 1991). MMS offers services for data communication between manufacturing devices and provides a common set of commands. This guarantees the communication of individual components in open systems.

This standard uses an abstract object modelling technique in order to fully describe the MMS device model and the MMS service procedures. The objects and their attributes as well as their respective operations are described.

This object modelling technique is a formal tool which helps understand the intent and effects of MMS services better than any kind of verbal description. The relationship between services and objects, and vice versa, is formally described in the standard. By implementing MMS, a real system maps the concepts described in the model to the real device.

An object is then represented by a data structure. MMS defines a number of object classes. Each object is a class instance and constitutes an abstract entity which exhibits some specific features and may be affected by some MMS services and operations. For each class a name is given by which it may be referenced.

2.2 The Virtual Manufacturing Device

The Virtual Manufacturing Device (VMD) serves as a model to represent the behavior of a real device. A VMD is an abstract representation of the common characteristics of all real manufacturing devices and represents the externally (from the network) visible behavior relevant to MMS. All standardized services refer to this virtual device which has to be mapped to real functions. MMS only describes the effects of the services on the VMD and does not prescribe a specific implementation or transformation to real functions.

2.3 The ten MMS service groups

A service group comprises those services described in the standard which all refer to the same object or object group. The name of each service is fixed by the standard. A reference to a service therefore consists of the name and the corresponding parameters. A reference to a service makes up the functionality of MMS. For example, variable management services allow to read, write, and send alarms (*InformationReport*); program management services allow the remote control of applications (*start, stop, kill, reset, ...*).

2.4 MMS companion standards

The specificity of industrial communications has led to define a general model allowing devices to dialogue : it is the MMS standard. The continuous evolution of the standardisation has enabled a definition of MMS model subclasses according to the needs of each machine communications : these subclasses are called companion standards. So the behavior of robots (ISO 9506/3), CNC (ISO 9506/4), programmable controllers (ISO 9506/5), ..., is defined. Thus, constructors can unambiguously propose standardized models of their products. Note that companion standards are also defined to describe functionalities such as production management, supervising, ...

2.5 Main interests

MMS is very interesting for RATP. For such a large company, the important number of devices of the same type leads it to be, either dependent on one vendor in order to improve applications homogeneousness, or confronted with interworking problems between several constructor solutions. Then it is difficult to evolve in the first case because one completely depends on the constructor improvements, and in the second case because any modification throws the whole architecture back into question. MMS solves these problems, irrespective of constructor specificities, improving this way applications portability, modularity and re-usability. Development costs are minimized by using the VMDs which are specific to the company, which really become reference models allowing the application multiplication. For example, all programmable controllers, whichever their trademark, will present the same user interface for communication.

3 CIM-OSA METHODOLOGICAL INTEGRATION CONCEPT

3.1 Introduction (Kosanke, 1990) (Vernadat, 1990)

The AMICE consortium (European Computer Integrated Manufacturing Architecture) proposes an open architecture called CIM-OSA and the associated methodology. It is based on a modelling framework which specifies three principles :

• the instantiation which proposes to build a particular model of the firm from partial models, the latter expressed in terms of basic generic constructions.

• the derivation which induces first to model the firm needs, then the design specifications, and finally the implementation description.
• the generation which suggests to model according to four views : functional (Jorysz, 1990a) (Kosanke, 1992), informational (Jorysz, 1990b), resources (Vliestra, 1991) and organisation (Jorysk, 1991) (Russel, 1991).

An Integrating Infra-Structure (IIS) (Klittich, 1990) (Querenet, 1991) allows, according to the instantiation principle, the execution of the particular implementation models, using a set of services (information, business, front-end and common) generally used in manufacturing systems. Specific Functional Entities (SFE) represent all the functions dedicated to one application (database access, application programs management, human interactive dialogue, machine control). Figure 1 shows how a SFE group is connected to an IIS, through front-end services.

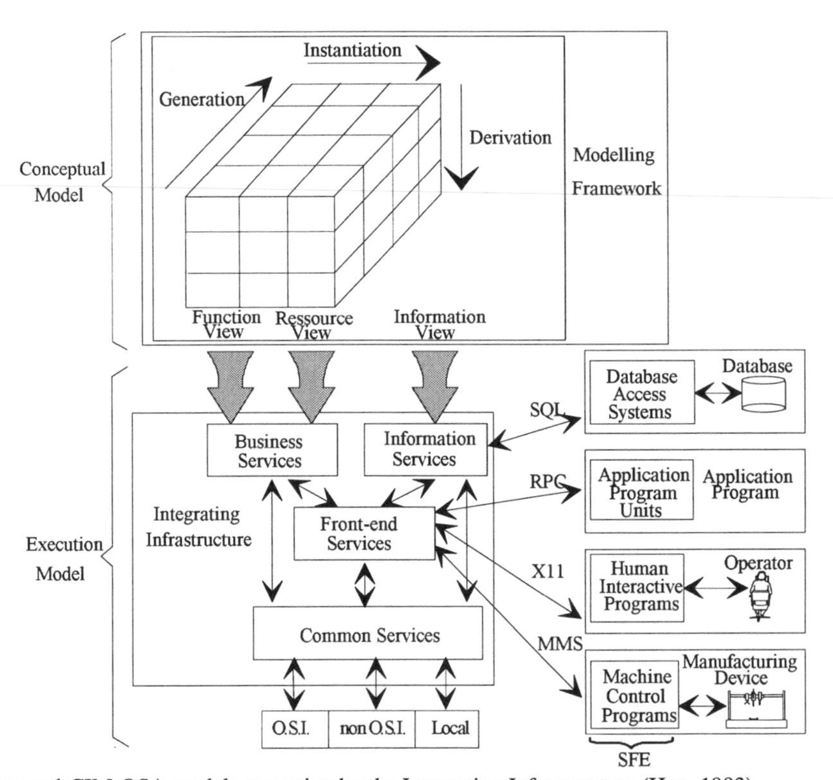

Figure 1 CIM-OSA models execution by the Integrating Infrastructure (Hou, 1993)

3.2 Study location in the CIM-OSA context

Our goal is to define, from the production system informational analysis, the configuration of devices towards the communication system. Coherence between both systems must be guaranteed. So, we focus on implementation description modelling level (grey shading, figure 2).

At this level, the information view specifies the database structure, and the resources view the manufacturing equipment configuration (data addresses, ...). But we can note the lack of relationships between these two views, yet logically linked.

CIM-OSA recommends to describe the information view the use of entity-relationship formalism, especially the M* methodology (Vernadat, 1989). At the level of implementation description, the conceptual model and its external schemes are only expressed with a kind of SQL (Standard Query Language) in order to establish the internal information system scheme. Concerning the resource view, MMS has been selected to represent and access to the communicating production devices. So we propose to establish links allowing to obtain MMS objects from the data logic model (figure 2).

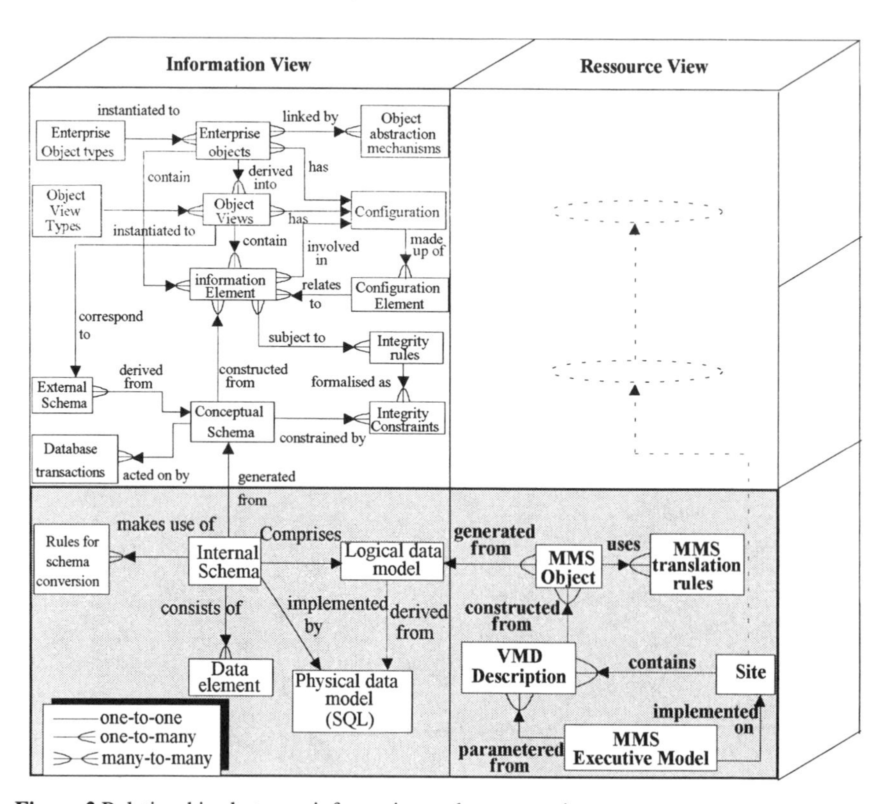

Figure 2 Relationships between information and resource views

A piece of information can also be located on both physical supports, because the database is able to refresh directly its table questioning the VMD and *vice versa*. But how can we interface the MMS and SQL services which respectively access either to a VMD or to a database ? Figure 3 shows how to design an information system which uses both SQL and MMS. It also shows the different accesses. For instance :

• Applications such as statistics or expert systems for diagnosis, only ask the database,

• Industrial devices refresh their VMD in real-time through a simple mapping operation,

• Production system control management uses information located in database and or in VMD.

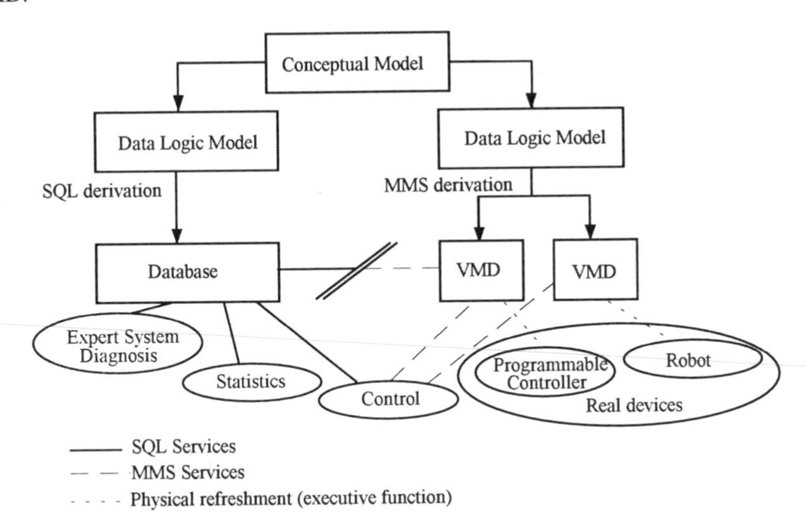

Figure 3 SQL and MMS derivation of the logical data models

4 HARMONIZATION BETWEEN THE COMMUNICATION AND THE INFORMATION SYSTEM

4.1 Introduction

According to CIM-OSA, the information system study is modelled by using the entity-relationship formalism while MMS uses an object modelling. Hence, the survey consists of the interface description between these two kinds of models. A set of translation rules has been proposed to obtain MMS objects through data models and reciprocally : regarding the Reverse Engineering concept, this reverse translation corresponds to the restructuring study which is defined as the transformation of a representation kind into another, at the same abstraction level (Rekoff, 1985). MMS objects transformed into entities can be directly integrated in the information system study at the logical level.

Obtaining MMS objects through entity-relationship models must be achieved without knowing the MMS standard because the information system designer is not advised *a priori* on the MMS standard. So we have developed a software which allows this translation. The MERISE (Tabourier, 1983) (Tardieu, 1983) method has been chosen because it uses the extended entity-relationship formalism and provides communications and data processing models.

4.2 The translator (Rondeau, 1993)

The translator software is integrated in the MEGA tool which respects the MERISE specifications. Two applications are added in the MEGA environment : ERtoMMS and ERtoMMS* :

• ERtoMMS translates the entity-relationship model created with the MEGA tool, in the MMS syntax. This software has allowed to valid the set of the rules of translation .

• ERtoMMS* achieves the same function as ERtoMMS but includes the organization of communications since it is able to analyze MERISE communication models. This software provides the application topology, the configuration needed to MMS equipment and the information flows.

The configuration of each equipment towards the communication is taken dynamically on another software called ConfMMS. ERtoMMS* issues files which contain all information for setting each MMS equipment. These files are downloaded on each equipment. ConfMMS, implemented on each equipment installs VMD, opens channels, configures channels, initiates the communications and sends different messages periodically (as *InformationReport* service) according to the information contained in each file. Figure 4 gives a summary of the global study which allows to build the communication system through the logical study. It illustrates the process between the conception level and the implementation level and its interaction since an evolution of the conception level affects the implementation level directly.

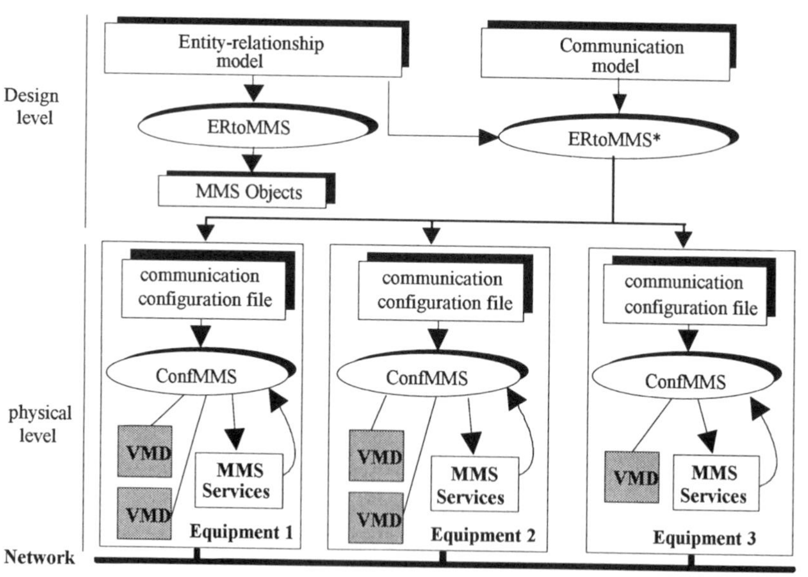

Figure 4 Interface between information and communication systems

5 APPLICATION TO THE PARIS UNDERGROUND

5.1 Introduction

RATP (Régie Autonome des Transports Parisiens) daily accomodates over five million people using several hundred railway kilometres to serve Paris an its immediate suburbs. So, the efficiency of this company greatly conditions social and economic aspects of the French capital. It implies a continuous pursuit of structural, organizational, and technological improvements. Two sets separated by the boarding quay can be briefly identified : the "rail" part which manages the flow of subway trains, and the "station" part which manages the travellers flow. That is our application topic.

The main constituent elements of a subway station are escalators, automatic gates, video security systems, telephonic terminals, travellers information systems, ventilation devices, automatic toll systems, ... Today, RATP has to renew an important part of its station control devices. In fact, they are old design equipments : relay boards, ... dating back to the seventies. The continuous evolution of the stations led to selective and/or unduplicatable upgradings which have induced a huge heterogeneity of these devices, which puts a considerable brake on new developments and especially the maintenance of these devices.

At such an enterprise scale, redesigning station control must be achieved in a particular context. The Paris underground has over two hundred stations which are managed by several thousand control devices. The solutions which will be selected will have to take into account the past experience, have a sufficient life duration and consider the RATP gigantism. A first study made by the company presents the list of requirements which will allow the development of a new network architecture offering New Station Services (N2S project).

In this framework, the CRAN (Research Center for Automatic control of Nancy) and RATP have concluded a contract in order to make optimal choices regarding previous criteria. Main objectives consist in designing a modular, portable and re-usable control, as independent of materials as possible. Whatever and wherever they are, accepted equipments have of course to communicate in a standardized way. So that their possible heterogeneousness do not constraint this communication. Such massive changes naturally require to have a good functional and informational knowledge of the system. Then the first solutions, after a prototyping phase, have to be technologically validated and possibly adjusted in a retro-engineering context.

5.2 The N2S architecture

Remember that we are focused on the "station" part, setting the "rail" part aside. Stations are gathered in autonomous sectors which are integrated in the underground line. One sector is managed from an operational center, called "Link Center". The latter generally integrates one station of its sector. The N2S architecture presents three levels :

• Level 1 : Station. Station controllers oversee equipments from a supervisor, using video screens, telephones, etc, ... They can momentarily delegate their functions to the upper level;

• Level 2 : Link Center. The controller has the same function as the station, but also remote security and control functions at the sector level;

• Level 3 : Operational headquarters. The functions of the control are centralized here : supervision, maintenance and security centers which assume a global management.

5.3 Objectives

This brief presentation shows how large this hierarchical system is, and it also presents the multiplication of identical functionalities. So it is natural to think that control cannot be designed for each station for example, but have to be considered in a global way for them all. It is also this large scale which explains that equipments choices lead to heterogeneous solutions. They guarantee the evolution of the system and the independence of enterprise towards its suppliers.

Therefore, a sufficient abstraction level is necessary for the control to be generic, hence re-usable. Taking into account material considerations must be put off as late as possible. So, equipments have to be virtualized, and working on standardized images is required. Thus, control genericity can be guaranteed, harmonizing the information presentation and access. This virtualization has also to allow to work in an homogeneous communication environment, using a standardized common language.

Then, in this context, the system analysis can be started, in order to understand the existing system and to consider its evolution. Firstly, the functions to be implemented must be identified, before bringing up the informations they need. This analysis, weighty regarding the size of the system, have to follow a rigorous methodology which we present now.

5.4 Informational study (Rondeau, 1993)

A Link Center controls several underground stations. On the other hand, a station works with a single Link Center. Station and Link Center can control several programmable controllers. Each of the latter supports functionalities such as escalators, video, ... management. Figure 5 shows a macroscopic view of the station informational model, but details the escalator object. For the escalator function, the conceptual data model brings up the following information :

Escalator_Name : Exclusively identifies the escalator.

Escalator_Control : Contains a number which identifies the escalator supervisor. This number also allows to know where the programmable controller has to send the escalator image when an alarm is triggered off.

Escalator_Order : Contains the supervisor order to carry out on the escalator (Up order, down order, stop order).

Escalator_Status : Contains the escalator physical state : Stop, Up, Down, and its logical state : available (to receive an order).

Escalator_Failure : Mentions escalator defects (Emergency stop, fire state, fire detection, breakdown, 48V lack, local control).

Escalator_Camera : Number of the camera which affords to visualize the escalator.

5.5 Organisational communications modelling

Arrows which join up boxes, materialize communications between physical entities, and thus the ones which exchange messages two by two can be identified (figure 6). Then informational study exploitation make it possible to specify the real data supported by these messages and the induced constraints (figure 7).

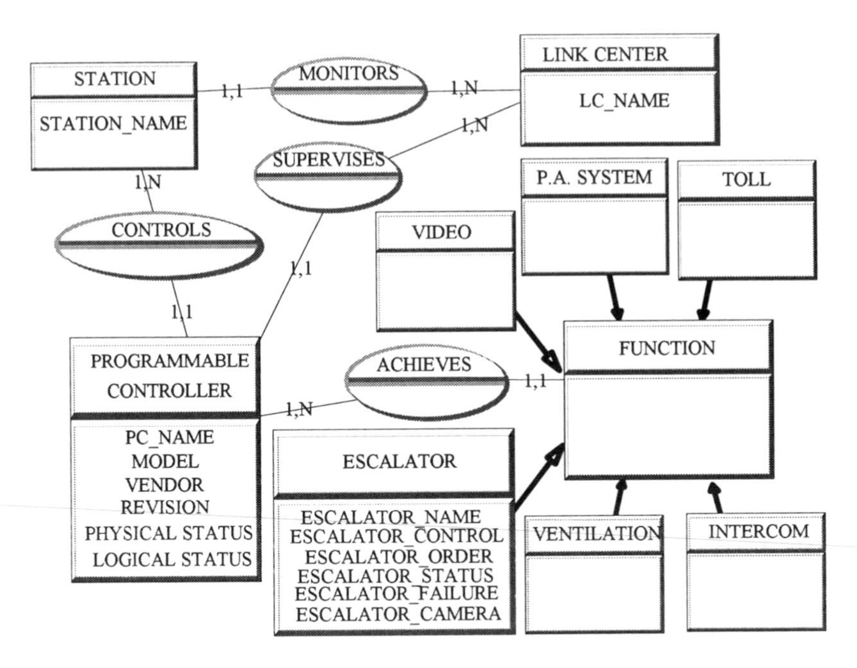

Figure 5 Station informational model

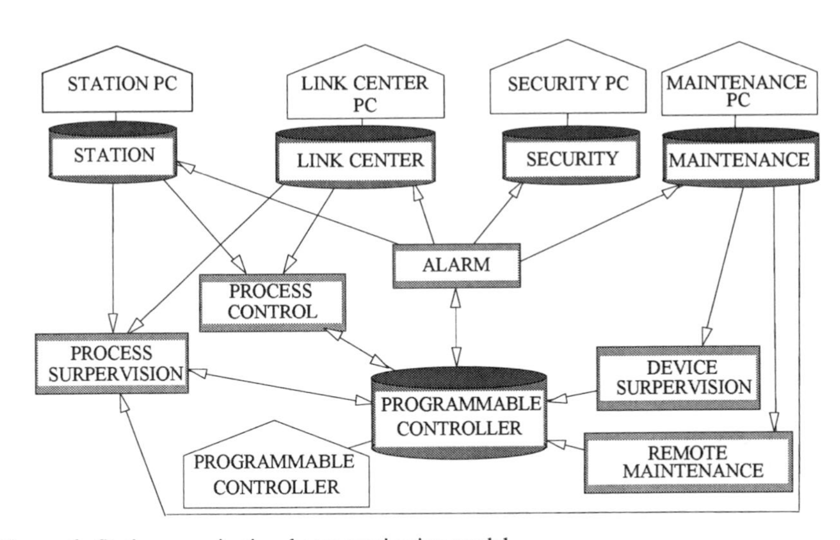

Figure 6 Station organisational communication model

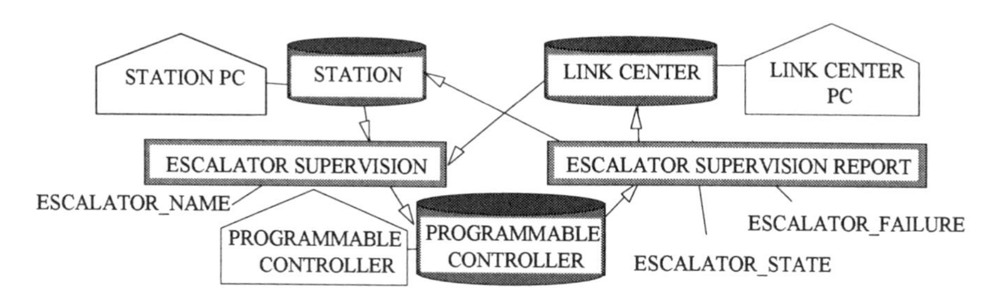

Figure 7 Escalator supervising organisational model

5.6 Conclusion

According to the term of these analyses, the system is now functionally and informationally well controlled. Their exploitation has yet to be carried out, with regards to expected objectives : modularity, genericity and control re-usability on different heterogeneous devices. Their virtualization allows to guarantee these objectives and to define standardized communications by the use of MMS.

6 THE MMS SITES CONFIGURATION

6.1 The VMD construction

The VMD definition, for each RATP device, is an important step of the study because it allows to standardise their access, and hence to establish a uniform profile of control applications. Thus, the VMD must not be modified because a complete revision of control programs should be required. From organisational communication models, which describe the relationships between functional and physical entities, five VMD families can be listed : maintenance, security, Link Center, station and programmable controller. In this paragraph, we will only present a partial view of the programmable controller VMD according to the informational model described on figure 5. Process control functions such as escalator management, video, etc... correspond to MMS domains. This allows to gather information which are characteristic of a specific activity, increasing in this way applications modularity and evolutivity. For example, adding an escalator only induces to specify a new escalator domain instance on the programmable controller VMD which controls it. This domain is defined as following :

Object : Domain
Key Attribute : Domain name (=Name)
Attribute : Control
Attribute : Order
Attribute : Status
Attribute : Failure
Attribute : Camera

All objects are defined as named variable class MMS objects whose types are simple or complex. For example, Escalator_Failure attribute is composed of six boolean fields as described in paragraph 5.4. So, a VMD elaboration consists in defining its attributes, possibly gathering them into a domain, then typing them. Thus the key attribute of an escalator domain always respects the EM_stationname_index syntax, where *EM* represents the escalator initials, *stationname* mentions the station where it is standing, and *index* is a number which differentiates several escalators located in a same station.

This job finished, VMD modelling could still be improved, using other MMS objects as semaphores. They could be applied in order to manage escalator read-write access which can be controlled either by a station or a Link Center. But taking these objects into account is at present impossible, because they are not implemented by all constructors.

6.2 MMS service identification

Apart from context management services (*Initiate, Conclude, ...*) which are essential to communicate in an MMS environment, required MMS services identification is specified from the organisational communication model shown on figure 6. Each message defined in this model induces the use of one or several MMS services we list in following chart.

Messages	Actions	MMS Services	Clients	Servers
Process Control	To refresh state graphs receptivities	Write	Station, LC[1], Security, PC[2]	PC[2]
Process Supervision	To read input/output of state graphs	Read	Station, LC[1], Security, PC[2], Maintenance	PC[2]
Alarm	To diffuse failures	Information Report	Station, LC[1], Security, PC[2], Maintenance	PC[2]
Device Supervision	To identify manufacturing equipment	Identify	Maintenance	PC[2]
	To observe manufacturing equipment status	Status	Maintenance	PC[2]
Remote maintenance	To clear manufacturing equipment	Start, Stop, Reset	Maintenance	PC[2]

1 : Link Center, 2 : Programmable Controller

6.3 Experimental Validation

The second step of the CRAN/RATP contract consists in testing all the devices in an real context (figure 8). So, an experimental platform has been developed and implemented in underground stations. This platform gathers :

• a PC microcomputer (located in the "Goncourt" station) which monitors with an Intouch Supervisor a Siemens programmable controller whose main function is to control an escalator of the same station;

• a second PC microcomputer (in the "République" station) which corresponds to a link center of the line 11. It means that the link center can control both "Goncourt" and "République" station escalators. (Those which belong "République" are managed by a Télémécanique programmable controller);
• then, a third PC microcomputer located in Noisy-le-Grand (Operational headquarters) supervises all escalators and calculates statistics about breakdowns which may occur, using Excel histograms.

A part from environmental problems, we have also validated inter-station communications which are supported by optic fibbers, and the RAMEAU (FDDI RATP network) routability in order to be able to have stations information available at headquarter level.

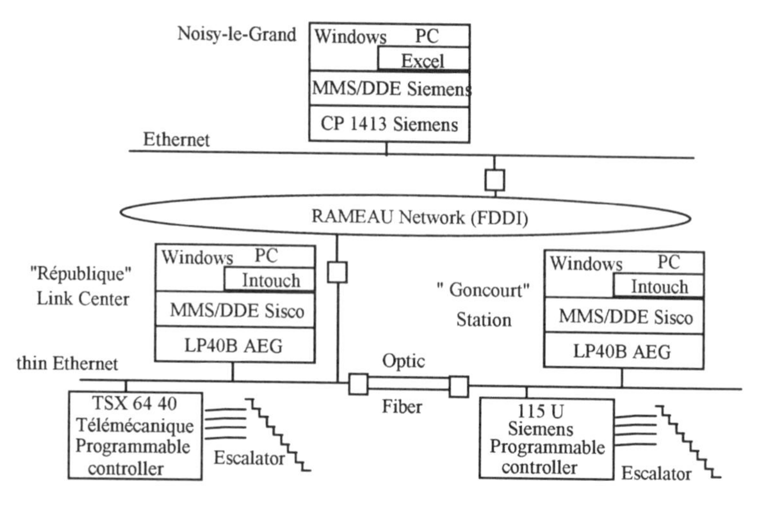

Figure 8 RATP platform

7 CONCLUSION

We have modelled for the RATP company the station functioning and the information which have to be exchanged. According to the CIM-OSA concept, and without choosing any software or hardware, MMS has been presented and proposed in order to harmonise the communications between a link center, a maintenance center, a station and its components. An experimental platform has been physically integrated. The objectives are reached, the control generecity is clearly demonstrated. This application was constrained by technological limitations which are inherent to the newness of the MMS implementations. That will be quickly resolved by the manufacturing device constructors. Today, when RATP invites tenders, they take into account this approach and all station equipment models that we have defined. Our collaboration goes on about both fieldbuses implementation and network management.

8 REFERENCES

Brill M., Gramm U. (1991) MMS : MAP application services for manufacturing industry. *Computer Networks and ISDN systems,* North Holland Publishers, vol 21, n°21, pp 357-380.
Hou W., Trauboth H. (1993) An approach to the development of the machine front end services in a CIM-OSA environment. *Flexible Automation and Integrated Manufacturing.* Proceeding of the Third International FAIM Conference. University of Limerick (Ireland), pp 115-124.
Jorysz H.R., Vernadat F.B. (1990) CIM-OSA Part 1 : total enterprise modelling and the function view. *International Journal Integrated Manufacturing*, vol 3, n°3 and 4, pp 144-156.
Jorysz H.R., Vernadat F.B. (1990) CIM-OSA Part 2 : information view. *International Journal Integrated Manufacturing*, vol 3, n° 3 and 4, pages 157-167.
Jorysz H.R., Vernadat F.B. (1991) Defining CIM Enterprise Requirements using CIM-OSA. *Computer Applications in Production and Engineering*, North Holland, pp 723-731.
Klittich M. (1990) CIM-OSA Part 3 : CIM-OSA integrating infrastruc-ture - the operational basis for integrated manufacturing systems. *International Journal Integrated Manufacturing*, vol 3, n° 3 and 4, pp 168-180.
Kosanke K. (1990) CIM-OSA : its role in manufacturing control. *Proceedings of the 11th triennal world congress of the International Federation of Automatic Control*, vol 5, pp 309-313.
Kosanke K. (1992) Open System Architecture (CIM-OSA), an european pre-normative development. *Cars & fof, 8th international conference on CAD/CAM, Robotics and Factories of the Future*, Metz, France, pp 486-498.
Querenet B. (1991) The CIM-OSA integrating infrastructure, *Computing & Control Engineering Journal*, vol 2, n° 3, pp 118-125.
Russel P.J. (1991) Modelling with CIM-OSA, *Computing & Control Engineering Journal*, vol 2, n°3, pp 109-117.
Rekoff M.G.(1985) On Reverse Engineering. *IEEE Transactions on systems, man, and cybernetics*, Vol. smc-15, n°2, pp. 244-252.
Rondeau E., Divoux T., Lepage F. (1993), *Définition et intégration expérimentale d'équipements virtuels dans un contexte MMS pour l'architecture N2S de la RATP.* Rapport de fin de contrat CRAN/RATP. Convention de Recherche C 93 U 003 du 1er juin 1993. Novembre.
Rondeau E., Divoux T., Lepage F., Veron M. (1994) *Creation of Virtual Manufacturing Devices by means of information system models.* Towards World Class Manufacturing 1993. Phoenix (Arizona). Elsevier Science B.V. (North Holland), IFIP, pp. 259-273.
Tabourier Y. (1986) *De l'autre coté de MERISE systèmes d'information et modèles d'entreprise.* Les éditions d'organisation.
Tardieu H., Rochefeld A., Coletti R. (1983) *La méthode MERISE, tome 1 : principes et outils.* Editions d'organisation.
Vernadat F.B., Deliva A., Giolito P.(1989) Organization and information system design of manufacturing environments : the new M* approach. *Computer Integrated Manufacturing Systems*, vol 2, n°2, pp 69-81.
Vernadat F.B. (1990) Modelling and analysis of enterprise information systems with CIM-OSA. *Computer Integrated Manufacturing,* proceedings of the sixth CIM-Europe annual conference. Lisbon, Portugal, pp 16-27.

Vliestra J. (1991) The architectural framework and models of CIM-OSA, *Advances in Production Management Systems*, proceedings of the 4th IFIP, North Holland, pp 15-26.

9 BIOGRAPHY

Eric Rondeau (31) is Associate Professor at the Henri Poincaré - Nancy 1 University (France). He teaches at the Mechanical Engineering Department of the "Nancy - Brabois Institut Universitaire de Technologies". He undertakes its researches at CRAN (Centre de Recherche en Automatique de Nancy) within the team " Industrial Local Area Networks". He has obtained its Doctorate of the Henri Poincaré University in 1993 by presenting its results on the integration of the MMS industrial communications by the information system study. These works are concretized in the framework of Research contracts with the RATP (Paris subway).

Thierry Divoux (32) is Associate Professor at the Henri Poincaré - Nancy 1 University (France). He teaches at the Telecommunications and Networks Department of the "Nancy - Brabois Institut Universitaire de Technologies". He co-manages the team " Industrial Local Area Networks" of CRAN (Centre de Recherche en Automatique de Nancy), and especially activities relevant to "Distributed System Engineering". He has obtained its Doctorate of the Henri Poincaré University in 1987 at the end of its works on heterogeneous communication integration in a CIM context. He continues today these researches by integrating multimedia communications, structured from EXPRESS modelling.

Francis Lepage (43) is Professor at the Henri Poincaré - Nancy 1 University (France). He is Chief of the Telecommunications and Networks Department of the "Nancy - Brabois Institut Universitaire de Technologies". He manages the team " Industrial Local Area Networks" of CRAN (Centre de Recherche en Automatique de Nancy), which especially works on the Distributed System Engineering and the Industrial System Communication Management. The team is essentially interested in the processing of the information in the communication and especially uses tools and methods of information system modelling.

Michel Véron (53) is Professor at the Henri Poincaré - Nancy 1 University (France). He is director of the CRAN (Centre de Recherche en Automatique de Nancy). He is member of IFIP, CIRP and several other international scientific organisations. He has contributed to the development of CIM both in industry and in university.

PART FOUR

Business Process Modelling and Reengineering

Process-Oriented Order Processing -A New Method for Business Process Re-engineering

Prof. Dr.-Ing. Dr. h. c. Dipl.-Wirt. Ing. W. Eversheim,
Dipl.-Ing. Th. Heuser
Laboratory for Machine Tools and Production Engineering,
Steinbachstr. 53 B, 52056 Aachen, Germany, phone: ++49/241/ 80-7404, fax: ++49/241/ 88 88-293, e-mail: heu@wzl-ps1.wzl.rwth-aachen.de

Abstract

Todays world is continuously changing. High quality and flexibility as well as low cost are the main objectives in production. Therefore it is necessary to have a process-oriented organisation of customer-led order processing from order clarification to shipping. In the following a model will be described to analyse and depict business processes as carried out in a company, to identify organisational problems, and to simulate lead time optimisation. In addition this model can be used to evaluate business processes according to the cause of cost under consideration of resource consumption.

Keywords

Process-oriented order processing, business process reengineering, activity based costing, organisation

1 INTRODUCTION

During the processing of customer orders, large problems appear especially in small and mid-sized companies due to the current economic situation. Despite the decreasing order volume, an increasing volume of work must be accomplished because of a growing number of special requests at lower prices for the product. Short lead-times and the keeping of guaranteed delivery dates based on good procedure organisation are the parameters which are decisive for competitiveness, whose optimisation for companies was until now an only unsatisfactorily solved problem. Even currently, many companies neglect an adjustment of their organization especially in respect to the production tasks which are becoming ever more complex (Eversheim, 1994-a). Through an organisation strongly characterised by job segmentation in manufacturing companies, the previously sensible functional-oriented structures in manufacturing have solidified themselves even in the ever growing area of administration (Eversheim, 1993-a).

Rationalisation measures, for instance, to influence lead-time and costs, have been increasingly taken over by the direct company departments manufacturing and assembly in the last few years. The rationalisation potential in the indirect company departments has not been taken fully into consideration. Companies indirect departments have a central meaning on efficiency of order processing. Especially there, however, the lead-times consist of 60-80% of total customer lead-time. 75% of the employees process information and not material. This leads to overhead costs which are about 50% of total cost [Groß, 1992; Eversheim, 1992; Traenckner, 1990). One of the main reasons is insufficient transparency. Hence, many important questions which occur in daily business remain unanswered (Figure 1) (Eversheim, 1994-b).

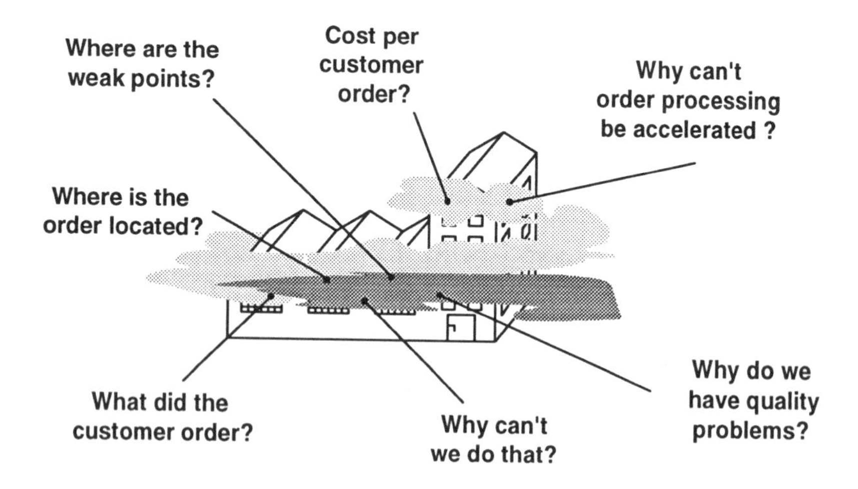

Figure 1 Insufficient transparency in order processing

Only through a total view of the order processing, aimed at realising an interconnecting, universal process, can single solutions be replaced by solutions which are department-wide with suitable requirements and can the processing of orders be placed at the center of the examination (Figure 2) (Striening, 1988; Gaitanides, 1983).

Within the framework of the organisational design, the perspective of the manufacturing costs of the resource consumption is linked to the order completion. This makes possible a differentiated view of resource consumption since it is related to the order measurably and manipulatively. The activities usually booked and hidden in the so-called overhead costs can now be analysed and examined in order to realise goals such as lean production.

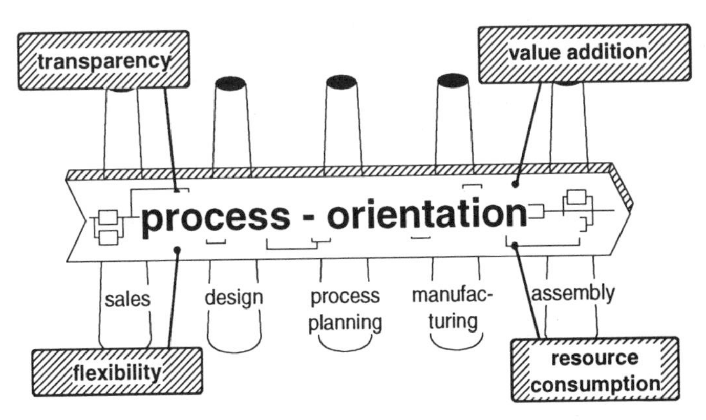

Figure 2 Process-Orientation-Precondition for improved competitiveness

2 FUNDAMENTALS OF PROCESS-ORIENTATION IN ORDER PROCESSING

Considering these problems, researchers of the Laboratory for Machine Tools (WZL) at the Aachen University of Technology developed a model of a process-oriented organisation of order processing. This model takes all necessary aspects into consideration, such as methods to implement customer-oriented organisation structures and methods for order scheduling and control. The model provides a method for busines process re-engineering.

The simultaneous optimisation of lead-time, costs and quality is mostly linked with a conflict of objectives since, for example, a minimisation of the lead-time carries with it high costs and possible sacrifices in quality. The application of the model process-oriented order processing allows to optimise the above mentioned objectives either individually or in the overall context (Figure 3).

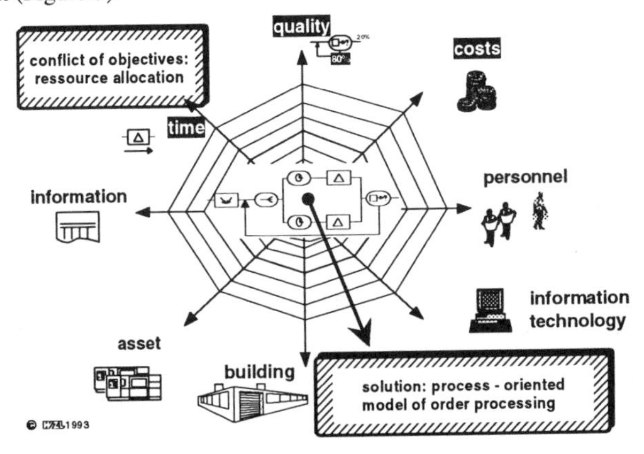

Figure 3 Integrated optimisation of time, quality and costs

A standardised modelling language was developed to depict the various mechanics of business order processing. Based on the modelling language, it is possible to reorganise and evaluate order processing. To make the evaluation practicable each element of the process is linked with resources, such as personnel, capital, asset etc. In this manner it can be evaluated in how far improvement measures can help in achieving the objectives time, costs and quality. Moreover it is possible to have a look at the changes e. g. in time and costs if you only optimise quality.

For the overall depiction of order processing, 14 basic "process elements" were developed (Figure 4). These elements compose the modelling language for order processing depiction from order placement to shipping. They are classified into direct and indirect element types. Indirect elements such as linkage, decide, and communicate etc., describe activities with an indirekt contribution to value addition of order processing. Direct process elements describe the activities causing a direct accretion to an order, such as designing, process planning, or manufacturing.

The method of business process re-engineering provides a general approach to apply the elements to their use in practice. It consists of the steps: building up the process-oriented model of order processing, calculate average lead-times or costs, identify weaknesses in order processing, and evaluate measures to be taken.

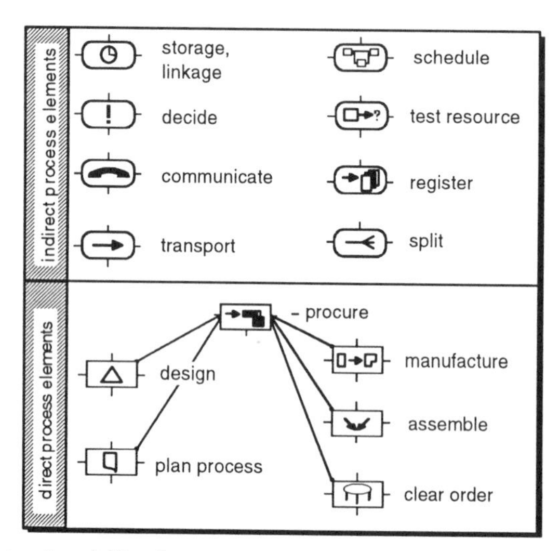

Figure 4 Elements of modelling language

For the application of the modelling language the elements are connected according to the sequence the order is processed. For that purpose, the elements provide one inlet on the left side and two resectively three outlets on the top, bottom, and the right side. The outlet to the right side represents the normal outlet for a trouble-free executed process. The bottom line shows the outlet for an interrupted process in case the alternative processes are known (e.g. test completeness of information during design process => procure lacking information from sales department). An order flows through the top outlet of an element when a decision about the following processes can not be made. The complete model is depicted in the so-called business process sequence plan.

3 APPLICATION OF THE MODEL

The implementation of analysis and reorganisation begins with the creation of a team of employees since order processing will be recorded based on interviews. The operative employees who have experience in order processing will be questioned. The analysis extends from offer and order processing in sales to product delivery through shipment or bringing into operational use for the customer. Employees from respective departments report in what sequence the incoming orders are processed (Figure 5). The analysis team presents the processes and their description in the business process sequence plan (BPSP). After completion of the BPSP, the depiction must be verified by the process owner (the employee who carries out the respective processes). Then, the corrections will be worked into the procedure depiction. This cycle will be implemented until the BPSP correctly reflects the procedures in the company (Eversheim, 1993-b).

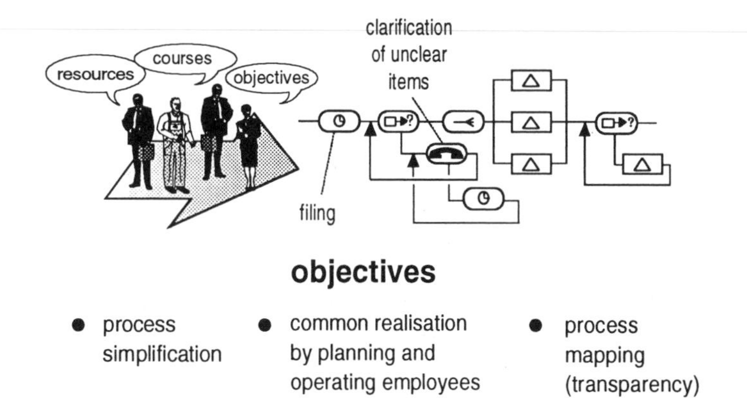

objectives

- process simplification
- common realisation by planning and operating employees
- process mapping (transparency)

Figure 5 Complexity reduction is based on process analysis (Eversheim, 1994-c)

The quantification phase follows the creation of the BPSP. For this, two kinds of data per process must be incorporated in the throughput reorganisation explained. First of all, the average length of time for the implementation of a process is to be determined, and second, the probability recorded with which an order will leave a process through the respective exit.

It is important that the employees who carry out each process are asked about the average lead-times per process. These values have proved themselves to be exact enough in practice.Processing time on the scale of minutes in a total processing time of several months, or conversion probability values after the decimal point supply in this case no extra information and overtax the employees. The incorporated times serve only the order processing time calculation and do not allow any conclusions about the capacity utilisation of an employee.

The result of the analysis steps "Inclusion and Quantification of Order Completion Processes" will be depicted graphically. As an example, the BPSP shown in Figure 6 is presented which indicates a section of the total process chain. In this case the procedures for the processing of individual components of an order is depicted.

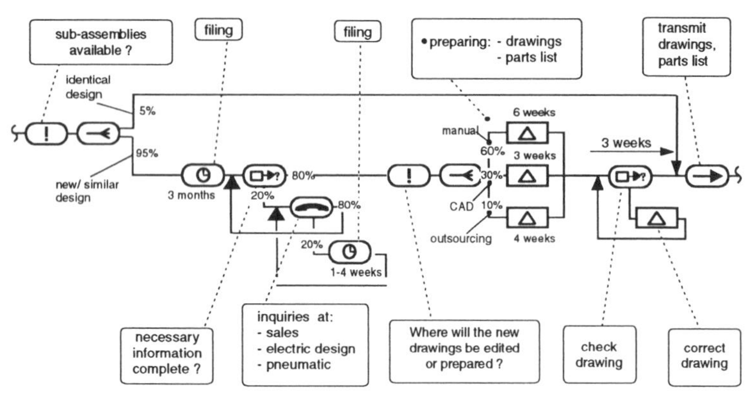

Figure 6 Business process sequence plan BPSP (example)

To support the approach an IT-Tool was developed. By applying the tool, the most important information about the processes carried out are recorded quickly and simply (Figure 7). To make the software accessible for industrial companies, as well, the system was lead to market maturity within the framework of a development cooperation between WZL, a software firm and 8 industrial companies. The analysis tool is operative on IBM compatible PCs under WINDOWS.

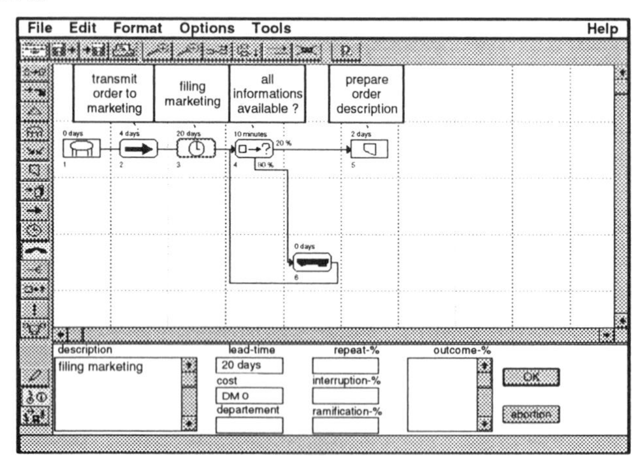

Figure 7 Proplan-IT-Tool for business process re-engineering

It is especially meaningful that the tools to be used, such as an EDP support, be simple and fast to employ, so that the employees are not held up from their daily tasks too long while carrying out the improvement process. Furthermore, the continual care and use of the BPSP serves as a continuous control of the organisation improvements made so far. The application of PROPLAN reduces the total efforts of business re-engineering up to 75%.

The BPSP contains all necessary information to calculate the average lead-time of an order regarding the considered process chain. For this, specially developed formulas are used. In these formulas, it is taken into consideration that time consumption based in feedback loops (e.g. inquiries based on missing information) is also based on the fact that some orders can run through it several times (Müller, 1992).

After the lead-time calculation regarding the current situation, the actual weakness analysis will be carried out. The weaknesses in order processing can be localised using various criteria. For example, processes with long lead-times in relation to the total lead-time will be examined. They could be value adding processes or processes which, for example, are characterised by idle times.

4 ACTIVITY BASED COSTING

In addition to a reduction of lead-time it is also important to reduce the costs. Efforts for costs reduction require a differentiated product and process cost consideration in all company departments. Especially in the up to now often neglected indirect company departments cost transparency is necessary. A charging of rendering services according to the cause of costs leads to an improved control of overhead costs. To share out overhead cost in a correct way, the WZL developed a method for a differentiated cost evaluation (Figure 8). A linkage of business processes and resource consumption is the basis of this method.

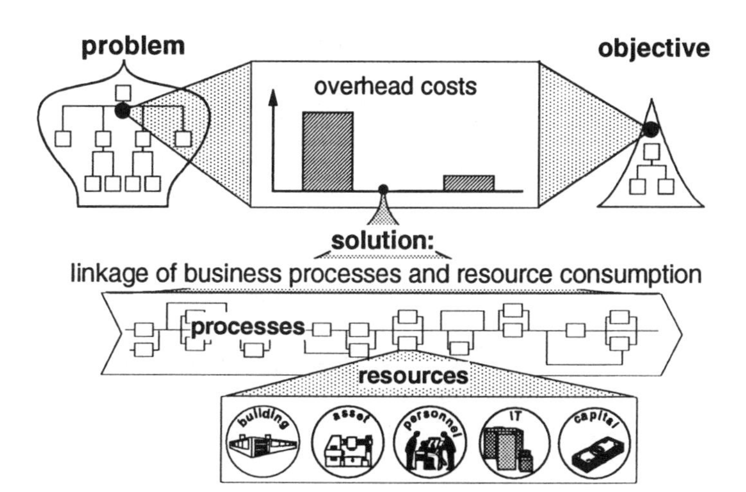

Figure 8 Overhead cost reduction

The resource consumption is described by the resource model (Figure 9). This model serves to evaluate business process comprehensively concerning the objectives lead-time, costs and quality. Therefore company resources personnel, information technology, asset, building, capital, material, information and time will be differentiated fundamentally.

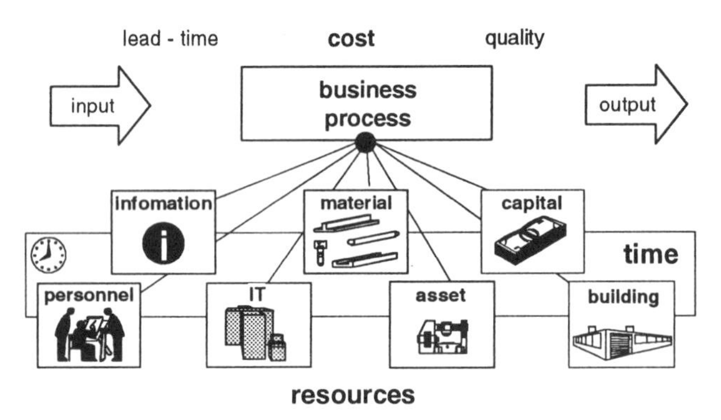

Figure 9 Elements of the resource model

The resource model has in view to describe the depreciation of examined business processes dependent on cost drivers (Figure 10). The depreciation of a business process will be determined by a relation between cost driver and resource consumption.

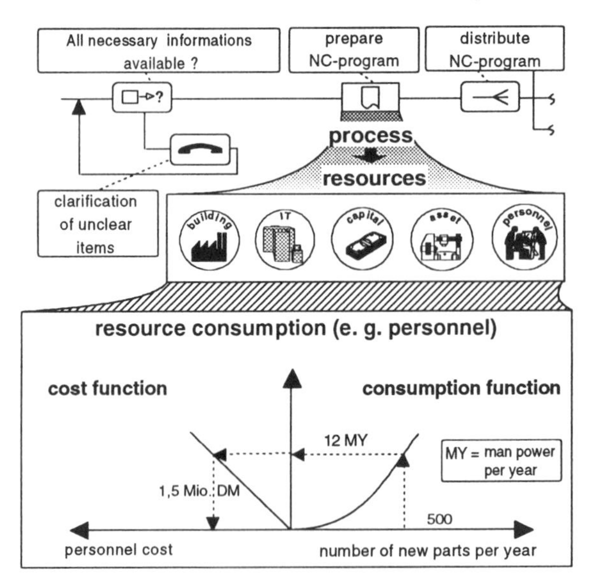

Figure 10 Evaluation according to the cause of costs

In general, cost drivers are technological and organisational oriented parameters (e.g. number of orders to be processed, number of new self-produced articles per year). With the specific cost rate the consumption function of a company resource is linked to a cost function (Eversheim, 1993-c)

The connection between these two functions represents the resource model. The application of the resource model integrates engineering and financial tasks (Schuh, 1988). Proved fields of application are optimisation of resource consumption during order processing, calculation of products according to the cause of costs, evaluation of variants, and the design of manufacturing and assembly lay-outs.

5 APPLICATION IN SME'S

The successes which WZL has had with the employment of business process re-engineering in industry show the meaningfulness of offering help toward selfhelp. The wide range of branches runs from special machine tool manufacture to series and large-series manufacture (Figure 11). The potential that was realised, for example through the reduction of lead-time, lies between 20%-50% of total lead-time.

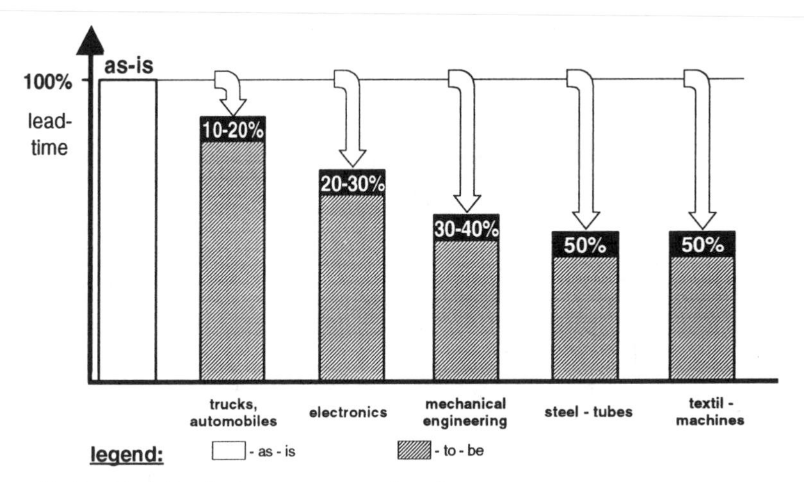

Figure 11 Results of business process re-engineering

By implementing the methods the following was achieved:

- transparency in the current sequences;
- exact localisation of the main weaknesses;
- concrete improvement measures;
- evaluation of all measures relating to their effect on shortening total lead-time time or cost reduction and
- setting up project priorities to realise the measures.

The experience gathered in practice shows that the method and the tool were well accepted by operative employees in direct and especially in indirect company departments. Due to the transparency produced, the preparedness for fundamental change in the company has drasticly increased.

6 SUMMARY AND OUTLOOK

The employment of the method leads to transparency of processes, resource consumption and costs. The accurate evaluation of reorganisation measures is made possible. An IT-Tool was developed, to reduce the expenditure of process analysis distinctly.

Following a project for continuous improvement with WZL some companies adopted the method of process analysis for continuous improvement. The continuous employment and improvement put up here an essential contribution to overcome the "operational blindness" and the increase of competitiveness particularly in today's difficult economic situation.

BPSP's which were constructed for the current and target condition are able to be used extensively after a project since they represent the company´s know-how in broad range. They represent, for instance, a suitable work foundation for the implementation planning and control. That means the continuous IT-supported documentation of process change is a constant measurement for the realisation of the reorganisation measures (Figure 12). The documented procedures, such as WZL´s experience has shown, furthermore can be used within the framework of a certification according to ISO 9001-4.

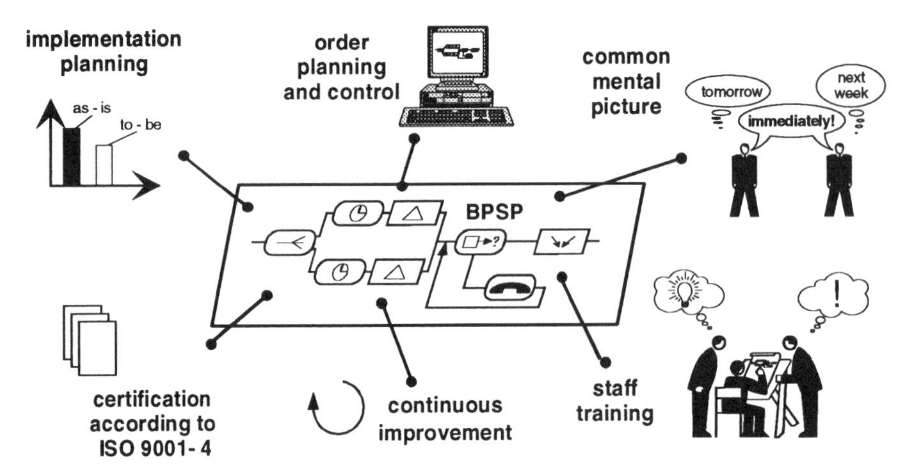

BPSP = business process sequence plan

Figure 12 Further fields of BPSP application

Further areas of application can be found in employee training. The transparent and easy-to-understand presentation of the company procedures simplifies training especially in view of new employees and leads to better understanding on all sides. A common mental picture for all employees is an important requirement for "mental fitness" for the consistent realisation of process organisation in a company.

7 REFERENCES

Eversheim, W.; König, W.; Pfeifer, T.; Weck, M. (1994-a) Manufacturing Excellence. The copetitive edge, Chapman & Hall, London

Eversheim, W.; Krumm, St.; Heuser, Th. (1994-b) Ablauf- und Kostentransparenz; CIM Management 10, Nr. 1, S. 57-59

Eversheim, W.; Heuser, Th.; Kümper, R. (1994-c) Verringerung und Beherrschung der Komplexität stärkt die Wettbewerbsfähigkeit; Referat zum Münchner Kolloquium; Hrsg.: Milberg, Springer Verlag, Berlin

Eversheim, W.; Krumm, St., Heuser, Th.; Müller, Th. (1993-a) Process - Oriented Order Pocessing. A New Method to Meet Customer Demands, In: Annals of CIRP, Vol. 42/1, S. 569-571

Eversheim, W.; Krumm, St.; Heuser, Th., Popp, W. (1993-b) Prozeßorientierte Reorganisation der Auftragsabwicklung; VDI-Z 135, Nr. 11/12, S. 119-122

Eversheim, W.; Krumm, St.; Heuser, Th. (1993-c) Prozeßorientierte Auftragsabwicklung; VDI-Z 135, Nr. 10, S. 48-51

Eversheim, W.; Müller, St.; Heuser, Th. (1992) "Schlanke" Informationsflüsse schaffen; VDI-Z 134, Nr. 11, S. 66-69

Gaitanides, M. (1983) Prozeßorganisation, Entwicklung, Ansätze und Programme prozeßorientierter Organisationsgestaltung. Verlag Franz Vahlen, München 1983

Groß, M.; Müller, St.; Heuser, Th. (1992) Schlank und rank; Zeitschrift Fabrik 2000, Nr. 2, S. 21-25

Müller, St. (1992) Entwicklung einer Methode zur prozeßorientierten Reorganisation der technischen Auftragsabwicklung komplexer Produkte. Dissertation RWTH Aachen

Schuh, G. (1988) Gestaltung und Bewertung von Produktvarianten, Dissertation RWTH Aachen

Striening, H.-D. (1988) Prozeßmanagement: Versuch eines integrierten Konzeptes situationsadäquater Gestaltung von Verwaltungsprozessen in einem multinationalen Unternehmen. Dissertation, Universität Kaiserslautern

Traenckner, J.-H.(1990) Entwicklung eines prozeß- und elementorientierten Modells zur Analyse und Gestaltung der technischen Auftragsabwicklung von komplexen Produkten; Dissertation, RWTH Aachen

8 BIOGRAPHY

Prof. Dr.-Ing. Dr. h.c. Dipl.-Wirt. Ing. *Walter Eversheim* born in 1937, study of mechanical and industrial engineering, PhD in 1965 at Aachen University of Technology. From 1969 to 1973 leading functions in several industrial companies. Since 1973 director of Laboratory for Machine Tools and Production Engineering at Aachen University of Technology, holder (head) of the chair for Production

Engineering. Since 1980 director of the Fraunhofer-Institute for Production Technology in Aachen. Since 1989 director of the Institute for Technology-Management at the University of Saint Gallen, Switzerland. Since 1990 director of Research-Institute for Rationalisation, Aachen University of Technology. In 1992, award of Dr. techinicae honoris causa from University of Trondheim, Norway.

Dipl.-Ing. *Thomas Heuser*, born in 1963, study of mechanical engineering at Aachen University of Technology. Since 1991 scientific employee at Laboratory for Machine Tools and Production Engineering at Aachen University of Technology. Since 1994 head of the section "assembly". Field of activity: business-process re-engineering.

8

Object-oriented modelling and analysis of business processes

K. Mertins, H. Edeler, R. Jochem, J. Hofmann
Fraunhofer Institute of Production Systems and Design Technology (IPK) Berlin
Pascalstraße 8-9, D-10587 Berlin, Germany,
Phone: ++49/(0)30/39 006 234, Fax: ++49/(0)30/39 11 037
joerg.hofmann@ipk.fhg.de

Abstract

Many problems within enterprises appear as a consequence of both organizational and technological issues. The integration of processes regarding aspects of dynamics and concurrency during decision making is a key element for achieving flexibility. Changed tasks and timeframes have to be reflected by restructured process chains.

To improve competitiveness, all efforts are traditionally concentrated on optimization of single functions - the enterprise is subdivided into a number of separate functions, which are easier to overview and control. This introduces a number of "interface" problems in organization and optimization of single functions at the expense of the manufacturing process and the organization as a whole.

The integration of separated functions and the optimization of business processes require a higher degree of transparency within the organization. In consideration of the complex relationships - looking on the manufacturing enterprise as a network of functions - modelling methods have to be applied, to support, to ease, and to systematize planning and integration of functions to business processes and to describe the related organizational structure. Suitable methods secure a common understanding of business processes and provide mechanisms for structuring the required information about processes and organization.

The authors describe a methodology for integrated modelling of business processes, related organsational structures, and information based on an object-oriented approach which is in discussion at ISO TC184/SC5/WG1 and CEN TC310/WG1 for standardisation. Examples of industrial application for different areas and a supporting modelling tool prototype are presented.

Keywords

Integrated enterprise modelling, object-oriented modelling, business process, standardisation, modelling tool

1 INTRODUCTION

In the nineties enterprises face a higher pressure of time and pressure to succeed at global markets, increasing competition, shorter product life cycles and, related to this, a higher flexibility in all areas. The market demands additional product differentiation exactly meeting customer needs, advance of technological level of products, shortening of both product development and order troughput times, advance of delivery time, and advance of product quality. To fulfil these requirements the enterprise of the future is characterized by:

- flexibility of production resources;
- automation of manufacturing and assembly;
- interconnection of all manufacturing systems, organizational processes, and decision centres.

Therefore, a higher transparency of enterprise organization and processes is required. Flexible automation presupposes a clear structure of the organization of information and material flows. Commercial just as organizational and technical matters have to be considered. The common understanding of all participants about the objectives and the core processes of the enterprise as well as about project goals is essential for the efficiency of both planning and development of business process organization.

In the following, the method of Integrated Enterprise Modelling (IEM) is presented. IEM uses the object-oriented modelling technique for modelling business processes, related organizational structures and required information systems as well. It provides a model for planning and optimizing the processes and organizational structures within the enterprise.

Models developed according to the IEM method give a transparent representation of planning information and therefore, are the basis for discussion between project participants. For evaluating the variety of planning information and description requierements, it allows different views on one consistent model.

IEM models provide the means to precisely assign the value of planning goals, like improvements in time, cost, or quality, to each business process and resource and, by that, to optimize the process organization.

2 OBJECT-ORIENTED MODELLING

The Approach

Object-oriented techniques are broadly used for the development of applications in various areas. The main advantage of this approach is the entirety of data and functions operating on these data. Provided with the powerful inheritance mechanism it yields models which are more stable and easier to maintain than those based on other modelling approaches (Coad and Yourdon, 1990). However, methods providing an entire, object-oriented approach for enterprise modelling are not seen so far.

In order to utilize it's advantages and to provide a comprehensive and extendable enterprise model, the IEM method uses the object-oriented modelling approach, thus allowing the integration of different views on an enterprise in one consistent model and the easy adaptation of the model to changes within the enterprise (Coad and Yourdon, 1990; Mertins and Jochem, 1993; Süssenguth, 1991).

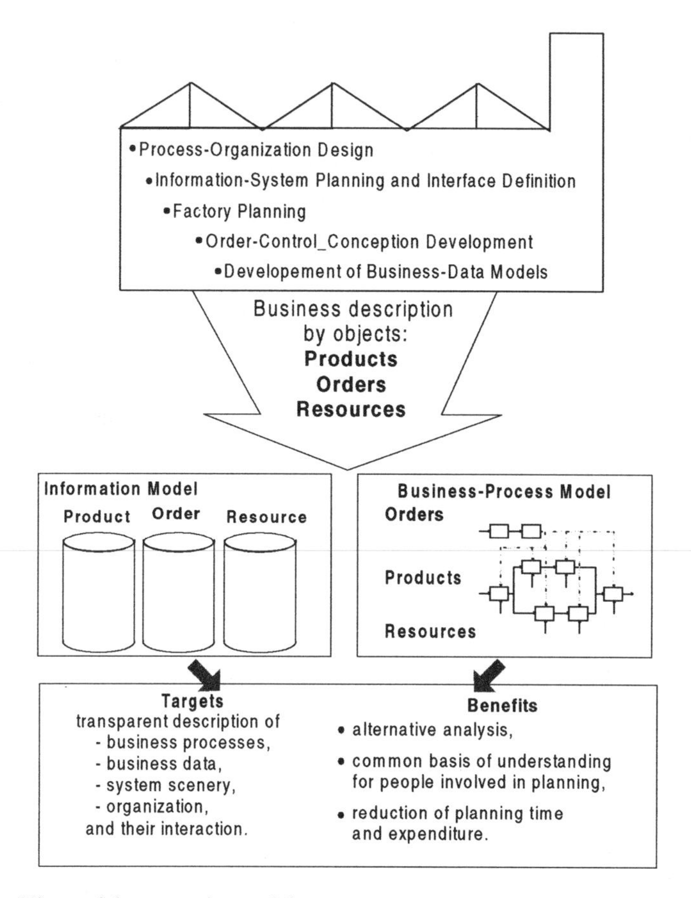

Figure 1 Views of the enterprise model.

Generic classes of objects

The generic classes Product, Resource, and Order form the basis of Integrated Enterprise Modelling for developing models from the user's point of view. They will be specialized according to the specifics of an individual enterprise (ISO TC 184/SC5 N 148; Süssenguth, Jochem, Rabe and Bals, 1998; Mertins, Süssenguth and Jochem, 1991). Each generic class prescribes a specific generic attribute structure, thus defining a frame for describing the properties and the behaviour of objects of it's subclasses (cf. Figure 4). Real enterprise objects will be modelled as objects of these subclasses.

Required enterprise data and the business processes, i.e. the tasks referring to objects, are structured in accordance to the object classes (see below). Furthermore, the relations between objects are determined. The result is a complete description of tasks, business processes, enterprise data, production equipment, and information systems of the enterprise at any level of detail (Spur, Mertins and Jochem, 1993; Mertins, Süssenguth and Jochem, 1994).

The model kernel comprises two main views. The tasks, which are to be executed on objects, and the business processes are the focal point of the Process Model View, whereas the Information Model View primarily regards the object describing data (Figure 1). Thus, the kernel of the enterprise model consists of the data and process representations of classes of objects. The views are interlinked by referring to the same objects and activities, although they represent them in different ways, levels of detail and context. Any view on the model can be derived from this standardized model kernel. Additional features can be tied to the kernel if necessary.

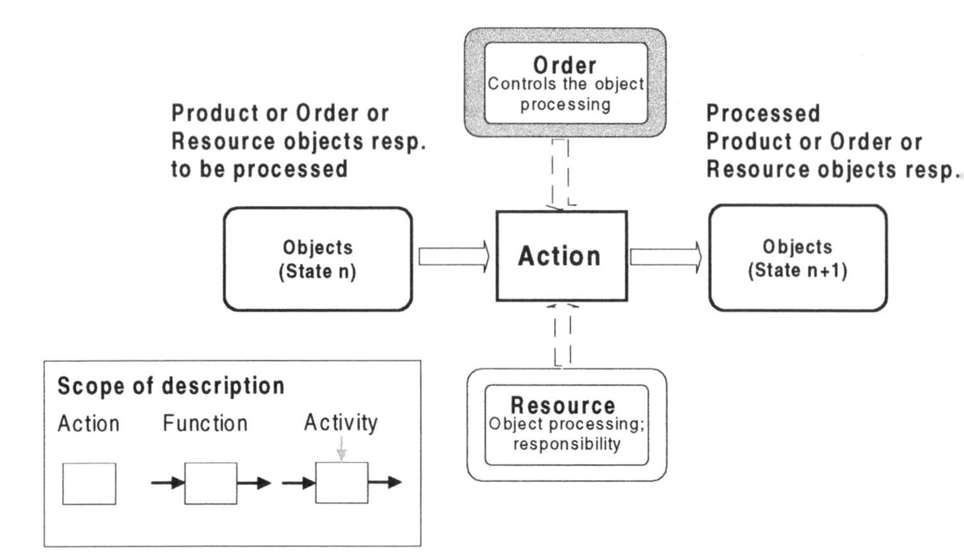

Figure 2 Interactions of Objects described by Generic Activity Model.

Business Processes as Interactions of Objects

Everything that happens in a manufacturing enterprise as part of some business process can be described by activities. In general, activities process and modify objects which were classified above as Products, Orders, and Resources. The execution of any activity requires direct or indirect planning and scheduling, it is executed by Resources owing the needed capability. The IEM method suggests three levels of describing the essentials of an activity.

- The **Action** is an object-independent description of some work or business, a verbal description of some task, process step, or procedure.
- The **Function** describes the processing of objects as a transformation from one determined (beginning) state to another determined (ending) state.
- The **Activity** specifies the Order controlling the execution of the Function and the Resource(s) being in charge of executing the Function.

Figure 2 graphically represents the Generic Activity Model. The beginning and ending states are connected with the action rectangle by arrows from left to right. The controlling of the activity is represented by an Order state description and a dashed vertical arrow from top; the required or actually assigned capability for executing the function is represented by a Resource state description and a dashed vertical arrow from bottom.

The Generic Activity Model represents the processing of objects of Product or Order or Resource classes respectively indicating the object interactions at processing. The related organizational structure is described by specific Resource classes along with their interrelations (see Chapter 5).

Using special concatenating constructs (cf. Figure 3), Actions, Functions, and Activities are combined to represent business processes. The decomposition and aggregation of processes is supported as well. The IEM modelling constructs of the Process Model View are shown in Figure 3.

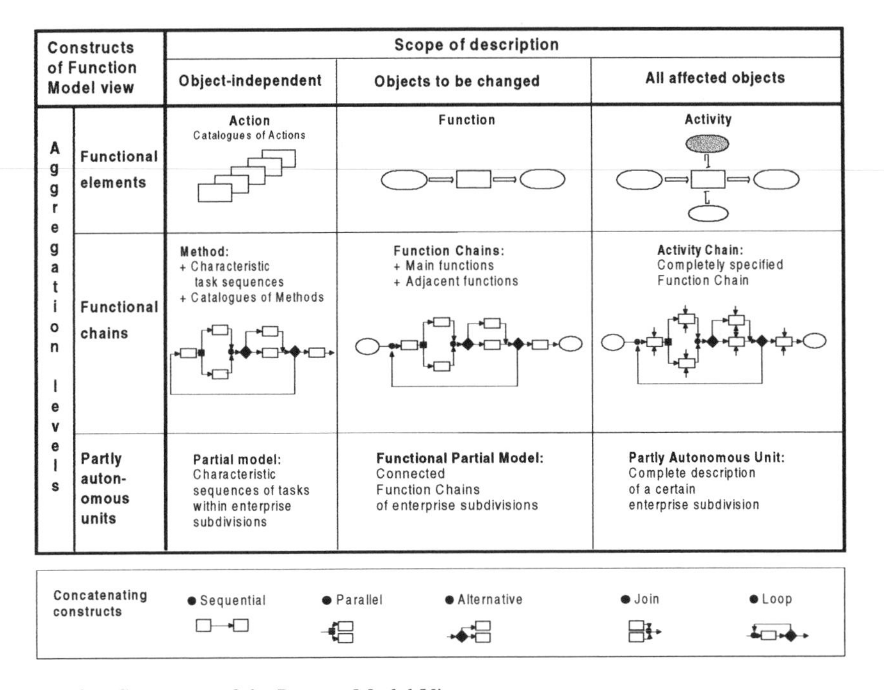

Figure 3 Constructs of the Process Model View.

Product classes represent the main results of the whole enterprise process - the products. Resource classes represent all means, including organizational units, being necessary for carrying out any activity in the enterprise. Order classes represent planning and control information. Figure 4 represents the attribute schemes of the generic classes (Spur, Mertins and Jochem, 1993; Mertins, Süssenguth and Jochem, 1994; Mertins and Jochem, 1992).

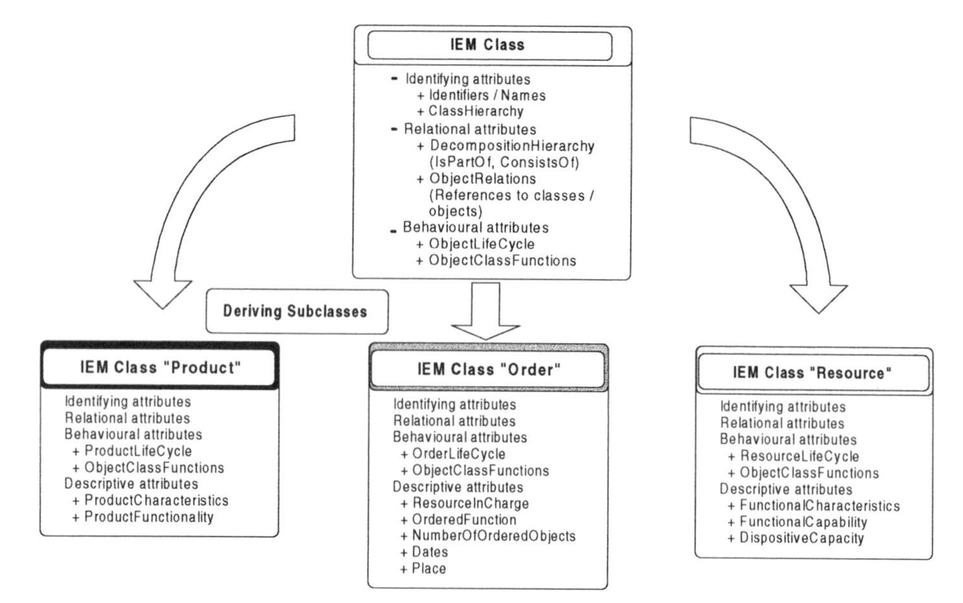

Figure 4 IEM object class attributes.

3 MODELLING PROCESS

Business Process Modelling

The description of enterprise processes starts with the analysis of the actual situation, normally applying a top-down procedure. Business processes along with the respective classes of objects to be processed are the starting point for modelling a certain part of the enterprise. This part is delimited in a first step concerning e.g. Products and respective "ordering" with regard to the main task, required resources and the interfaces to the environment as well. The main task and objects of the application area are described by that (Mertins and Jochem, 1992 and 1993; Süssenguth, 1991).

The products have to be identified as subclass instances of the IEM class Product and the business processes have to be modelled according to these products, independently of organizational structures (Figure 5).

Next, resources and controlling orders have to be identified for each function of the defined enterprise processes.

To obtain more detail, business processes of generation, processing and supply of orders and resources as well as of processing of sub-objects of the products should be modelled. The network and the interdependencies of the business processes are described with the concatenating constructs of the IEM (Mertins, Süssenguth and Jochem, 1991, Mertins and Jochem, 1992).

The development of particular business process models has to be extended by the order and resource flow, the analysis of concurrency of business processes and their mutual influence. For this purpose simulation and other methods should be applied.

Information Modeling

The collection and structuring of the data of all objects which were identified at modelling lead to a particular enterprise information model. For this purpose, a structuring frame for representing the relevant data is required.

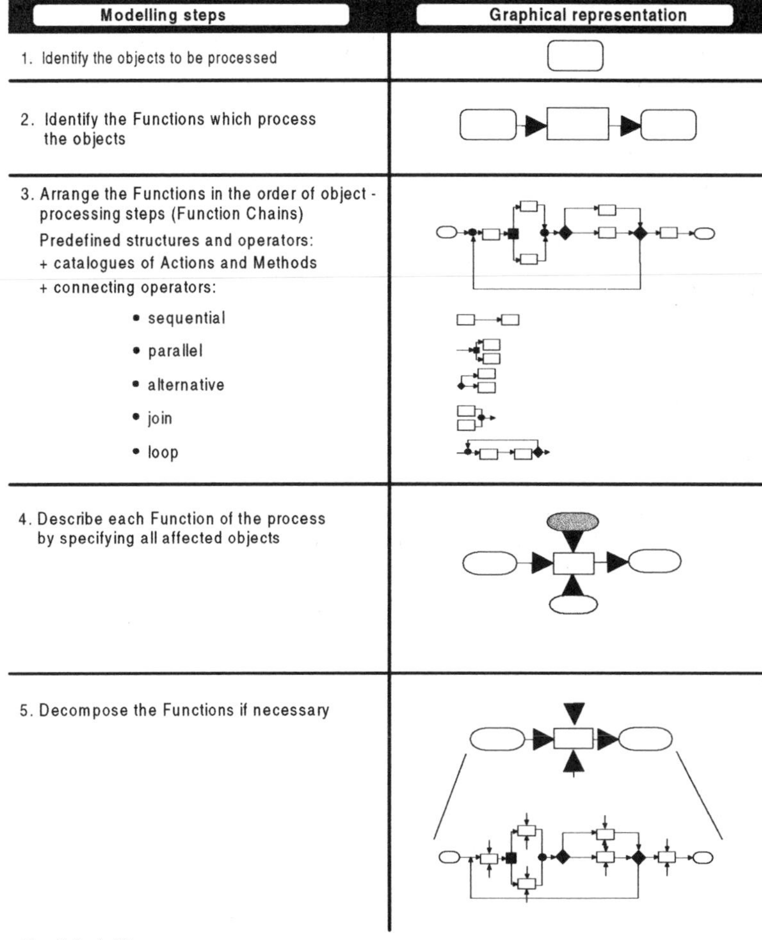

Figure 5 Modelling process.

The differentiation of the generic classes of objects and their internal structure define the structure of the enterprise information model. Three interconnected submodels, the Product, Control and Resource models are defined. The internal structure of the submodels is represented by layers which enable a grouping of object data by different criteria.

Beside the aid for data structuring, the IEM data modelling provides the advantage of close relation to the application area by considering real world objects with regard to the generic classes. The independence of data of a specific information processing system secures the extendability and interchangeability of data between several systems. A particular enterprise information model will provide the preconditions for a general use of data bases and support the recognition of priorities at data exchange (Mertins and Jochem, 1993).

The first step for the development of a particular information model of an enterprise is the cataloguing of objects, their descriptive data and object relations in data dictionaries. Different kinds of lists were defined in the layers of the enterprise information model.

4 INTEGRATION OF ORGANIZATIONAL STRUCTURES AND INFORMATION SYSTEM SUPPORT BY ADDITIONAL VIEWS

Further aspects of modelling related to special purposes can be integrated as additional views on the model. Examples of such views are special representations of control mechanisms, organizational units and costs. The relevant properties of the additional views can be represented by deriving specific subclasses of the generic classes:

- Determination of class specific attributes,
- determination of respective attribute values.

An example for the integration of additional views related to the special purpose of planning and introduction of CIM systems is shown by Figure 6.

The model kernel is the basis for developing application-oriented modelling constructs, views and partial models. Existing application-oriented classes, constructs, views, and models can be traced back to the main views of an enterprise model. Therefore, changes in business processes or in organization and their impact on information system support can be evaluated.

This example shows how the model kernel can be used and extended by representing additional subjects and views. In the same principle way, several other models can be integrated into an entire enterprise model (Süssenguth, 1991).

The scheme comprises eight layers, which represent the interlinked fields of design within the enterprise.

The layers "Functions" and "Data" represent the business process and the required information independently of technical solutions, i.e. the model kernel as presented above.

The layers "Application systems", "Data storage", "Network" and "Hardware" are the technical fields of information system planning related to the first two layers.

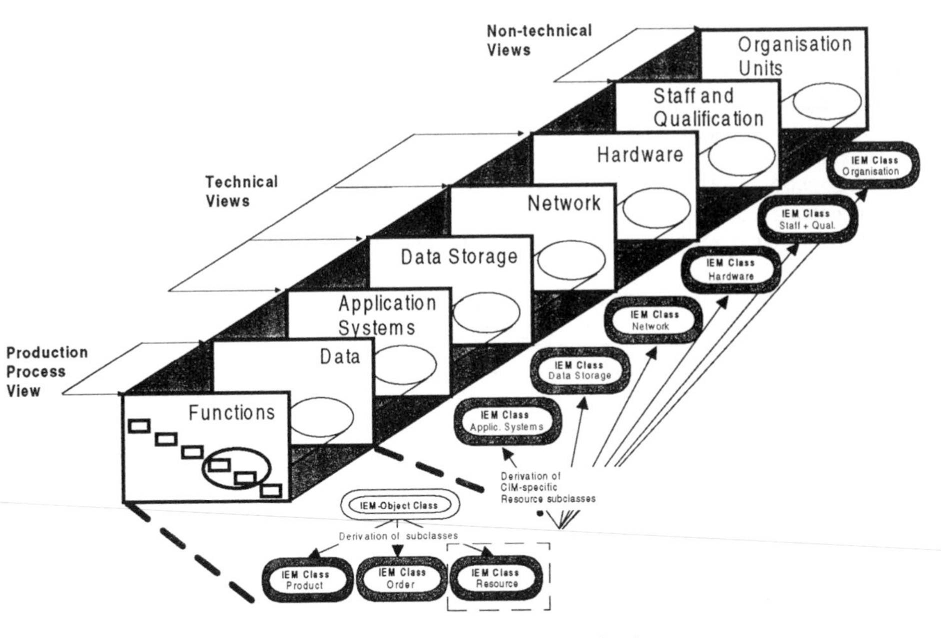

Figure 6 Views of Information System Support and Organization.

The layers "Organizational units" and "Staff and Qualification" are non-technical fields of design, which have to be considered simultaneously to the others. The tube across the layers illustrates their interrelations incorporated in the functions and data of the respective classes and objects.

Specific Resource subclasses have to be defined to represent additional views, like organization, with the respective properties (cf. Figure 4), (Mertins and Jochem, 1992).

The right choice of the level of detail is important for the modelling effort and benefit. For the task of information system planning, an overall, not too detailed modelling of a number of enterprise areas should be preferred.

The modelling of sub-classes leads to data, function and system interfaces, which are indicators of obstructions for integration. Interfaces in that sense have to be avoided. Therefore, alternatives within each view have to be worked out and evaluated against their potential benefit (Süssenguth, 1991).

5 APPLICATION AREAS

The described method of Integrated Enterprise Modelling is suitable for various planning and structuring measures in enterprises. It covers the aspects of material and information flows as well. In addition to the systematization of planning in a certain project, the benefits of applying the method are obtained from the reusability of the models for further projects with different tasks (Spur, Mertins and Jochem, 1993). Examples for the application of the IEM as an accepted description method in industrial projects are the following:

- Ascertainment of kernel processes in distributed enterprise structures of rail vehicle manufacturers;
- analysis and ascertainment of layout structures for the organization of production preparation and execution in distributed enterprises of rail vehicle manufacturers;
- presentation of potentials for saving time and costs by improved information system at order execution (Figure 7);
- development of a CIM architecture and an order control concept in a mechanical engineering plant (Figure 8);
- concept of interfaces between head office and decentralized production plants in distributed company structures;
- reorganization and development of a modern rail vehicle production.

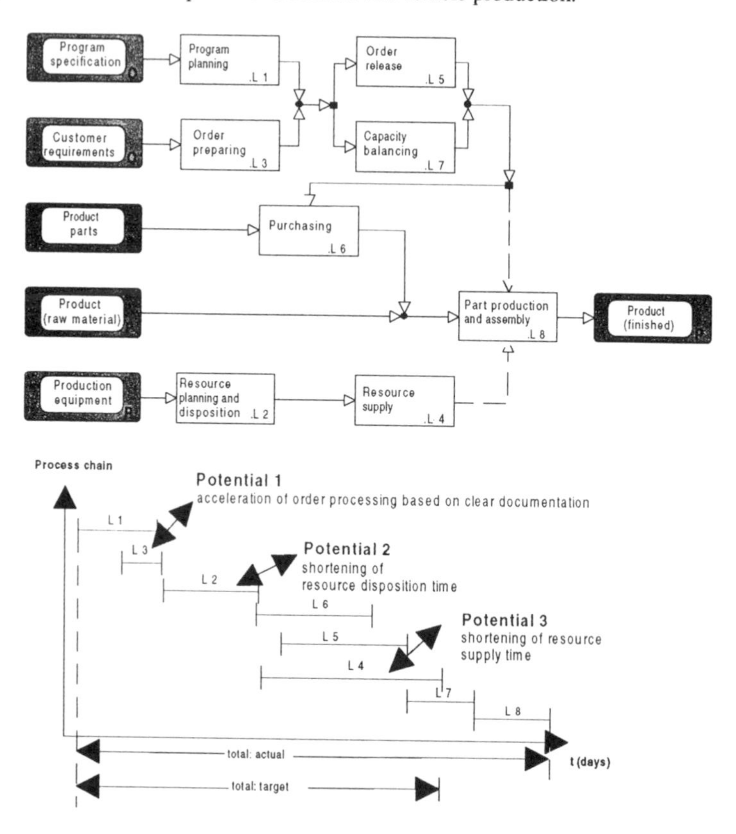

Figure 7 Derivation of potentials to save time at order processing.

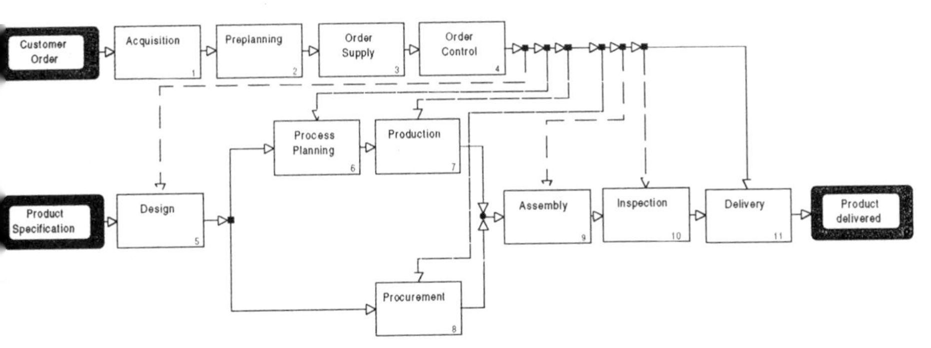

Figure 8 Entire enterprise process.

6 MODELLING TOOL

The use of the described method for enterprise modelling is only efficient in connection with a modelling tool. A PC-based verison, which supports the method, is available (Figure 9).

It is based on Windows 3.1 platform and uses Windows standard. The development of the graphical notations (diagram technique) of the user-interface was supported by a special commercial software package. The logics and the application rules of the method are implemented in C++ independently of the supporting software via a special programming interface (Spur, Mertins and Jochem, 1993). The following modelling tasks are supported by the functionality of the tool:

- Classification of products, orders and resources of the enterprise;
- hierachy of business processes;
- description of product, order and resource structures;
- generation of documents;
- rapid model construction by predefined model structures and navigation facilities;
- support of different user views.

The Tool is used in industrial projects of IPK as well as in internal projects of industrial and consulting companies.

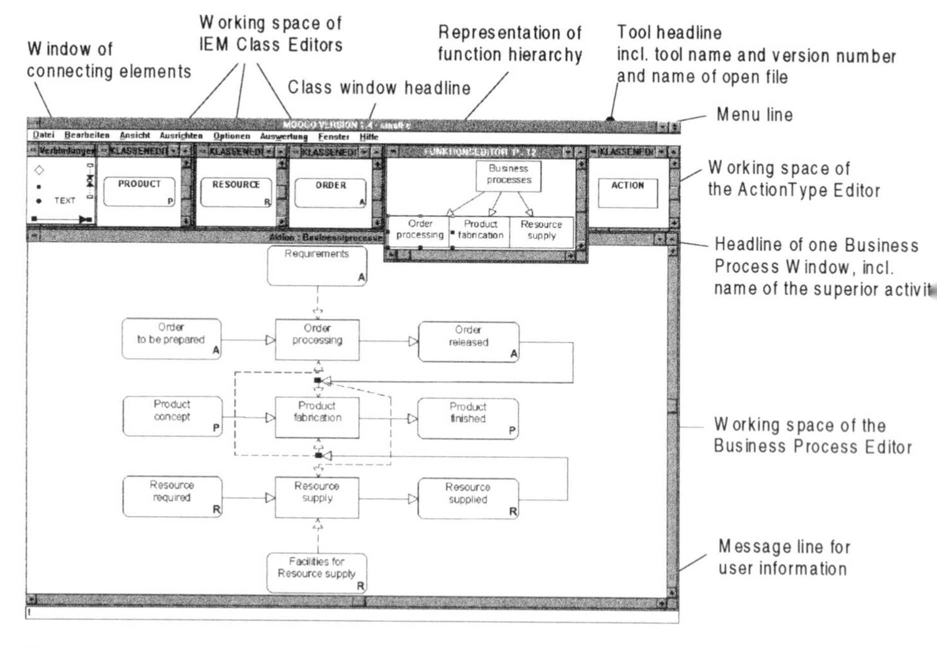

Figure 9 Tool user interface.

7 RELATION TO OTHER MODELLING METHODS

There exist a number of methodologies for enterprise modelling with their own purposes and advantages. In (Mertins, Süssenguth and Jochem, 1994) a detailed analysis of several modelling methods was conducted. Here we mention only ICAM/IDEF and SADT as examples of function-oriented methods. Generally speaking, the main topic of these methods are data processing functions, data and functions are represented in separate models loosely tied via the names of data entities.

Besides the well-known difficulties at systems design, implementation and maintenance, the application of these methods for business process re-engineering faces the problem that, starting with function analysis, one rather "copies" the existing organizational structure than to analyze the (object-related) nature of processes. Thus, for the task of process re-engineering these methods are useful but not optimal.

CIMOSA is another major methodology for enterprise modelling. CIMOSA models, based on a process modelling concept, support the complete enterprise life cycle of enterprise engeneering and operation. CIMOSA provides a reference architecture for particular enterprise architectures, supporting the complete enterprise life cycle and providing different views on the model.

The method of Integrated Enterprise Modelling (IEM) views an enterprise starting from the production process in it's entirety and the resulting requirements. The aim of IEM is to build a comprehensive (conceptual) enterprise model as an aid for analyzing and re-engineering business processes, and as a platform for developing and integrating various CIM applications independently of a particular implementation environment.

Recently, representatives of CIMOSA Association and Working Group Quality-oriented Enterprise Modelling (part of German QCIM Project) elaborated a comparison paper of CIMOSA and IEM modelling constructs in order to assist European standardization of enterprise modelling constructs at CEN TC310/WG1 (CEN TC310/WG1 (Architecture) N41).

This comparison demonstrates transferability (on the basis of constructs) and complementarity (considering details related to intended use) of the two approaches. Transferability means that it is possible to construct a transformation which maps a CIMOSA model to an IEM model and vice versa. Complementarity expresses the fact that, for each method, some parts are elaborated with more detail related to the intended use.

So, the common efforts of CIMOSA and IEM hopefully will pave the way towards an useful standard of modelling constructs being conducive to enterprise integration and to meet the challenges of recent economical development.

8 REFERENCES

CEN TC310/WG1 (Architecture) N41, *Comparison CIMOSA - IEM Modelling Constructs.* June 1994.

Coad, P. and Yourdon, E.: *Object Oriented Analysis.* Yourdon Press/Prentice Hall, Englewood Cliffs, New York, 1990.

Flatau, U.: *Digital's CIM-Architecture*, Rev. 1.1. Digital Equipment Corporation, Marlboro, MA U.S.A., April 1986.

Harrington, J.R.: *Understanding the Manufacturing Process. Key to Succesful CAD/CAM Implementation.* Marcel Dekker, inc: 1984.

ISO TC 184/SC5 N 148, *Technical Report: Reference Model for Shop Floor Production*, Part 1.

Mertins, K. and Jochem, R.: *An Object-oriented Method for Integrated Enterprise modelling as a Basis for Enterprise Coordination*, International Conference on Enterprise Integration Modeling Technology (ICEIMT), Hilton Head (South Carolina), US Air Force-Integration Technology Division, June '92.

Mertins, K. and Jochem, R.. *Planning of Enterprise-Related CIM-Structures.* In: Proceedings of 8th International Conference CARS and FOF. Metz; France, 17.-19. August 1992.

Mertins, K. and Jochem, R.: *Integrierte Unternehmensmodellierung - Basis für die Unternehmensplanung.* DIN-Tagung. April 1993.

Mertins, K., Süssenguth, W..and Jochem, R.: *Integration Information Modelling.* In. Proceedings of Fourth IFIP Conference on Computer Applications in Production and Engineering (CAPE '91). Bordeaux, France. Elsevier Science Publisher B.V. (North Holland).

Mertins, K., Süssenguth, W. and Jochem, R.: *Modellierungsmethoden für rechnerintegrierte Produktionsprozesse* (Hrsg.: G. Spur). Carl Hanser Verlag. München, Wien. 1994.

Spur, G., Mertins, K. and Jochem, R.: *Integrierte Unternehmensmodellierung*. Beuth Verlag. Berlin. 1993.

Süssenguth, W., Jochem, R., Rabe, M. and Bals, B.: *An Object-oriented Analysis and Design Methodology for Computer Integrated Manufacturing Systems*. Proceedings Tools '89, November 13-15, 1989, CNIT Paris/France.

Süssenguth, W.: *Methoden zur Planung rechnerintegrierter Produktionsprozesse*. Dissertation. Berlin 1991.

9 BIOGRAPHY

Dr.-Ing. Kai Mertins,

Born in 1947, Education in Electro-Mechanic, Study of Electrical Engineering at the Engineering School of Hamburg. Several Industrial Experiences as Electrical Engineer. Study of Economical Engineering at TU Berlin. 1984 Doctoral Thesis at TU Berlin. Since 1982 Head of Department and since 1988 Director of the Division Systems Planning at the Fraunhofer-Institute for Production Systems and Design Technology (IPK Berlin).

Dipl.-Soz. Hermann Edeler,

Born in 1955, Study of Social Sciences and Economics at University of Bielefeld and Free University of Berlin. Since 1986 researcher at IPK and since 1992 Head of Department Production Management, Division Systems Planning at the Fraunhofer-Institute for Production Systems and Design Technology (IPK Berlin).

Dipl.-Ing. Roland Jochem,

Born in 1962, Study of Mechanical Engineering at TU Berlin. Industrial Experiences as Mechanical Engineer. Since 1988 researcher and since 1991 Group Leader for Business Process Modelling at the Fraunhofer-Institute for Production Systems and Design Technology (IPK Berlin), Division Systems Planning.

Dipl.-Math. Jörg Hofmann,

Born in 1952, Study of Mathematics at the Donetsk State University (Ukraina). Experiences in basic and applied research. Since 1992 researcher at the Fraunhofer Institute of Production Systems and Design Technology IPK Berlin, Division Systems Planning.

PART FIVE

Manufacturing System Specification

9

Using a Formal Declarative Language for Specifying Requirements Modelled in CIMOSA

Eric Dubois, Michaël Petit
Facultés Universitaires de Namur, Institut d'Informatique
Rue Grandgagnage 21, B-5000 Namur (Belgium)
{edu, mpe} @ info.fundp.ac.be

Abstract

Requirements Engineering is more and more considered as a central phase in the development and implementation of computer systems. Within the context of CIM, the CIMOSA project proposes a set of models based on adequate concepts for expressing requirements. In this paper, we suggest how these models can be supported by the use of a fully formal requirements specification language called ALBERT and based on an agent-oriented real-time temporal logic framework.

Keywords

Requirements Engineering, CIMOSA, ALBERT, real-time temporal logic, agent-oriented framework.

1 INTRODUCTION

It is now widely recognised that the implementation and the maintenance of successful and adequate CIM infrastructures can only be achieved through the adoption of a rigourous development process made of a number of well-defined activities covering the whole lifecycle of a CIM infrastructure.

Among these activities, requirements analysis (or *requirements engineering* (RE)) is an activity which appears as crucial since it is in charge of eliciting and capturing customers wishes and goals for the CIM infrastructure to be settled in an enterprise. From recent research trends related to the RE activity, we can learn that a key issue relies on the use of an adequate language for modelling the requirements expressed by customers. Two specific qualities are expected from such a language:

- On the one hand, the language should be *expressive* enough so that customers requirements can be modelled in a natural way, without the introduction of any overspecification. To ensure this traceability property, a suitable requirements language will be based on an *ontology*

of concepts large enough so that it permits a straightforward modelling of a large variety of requirements belonging to the considered application domain.
- On the other hand, the language should be a *formal* language in order to support powerful verification and validation checks, i.e. a language equipped with an adequate mathematical/logical semantics made of (i) rules of interpretation which will guarantee the absence of ambiguities in a requirements document and (ii) rules of deduction which will permit to reason on a specification document in order to discover potential incompletenesses and/or inconsistencies.

Within the CIM context, languages have been proposed for the purpose of modelling requirements. Examples include, e.g., SADT (Ross, 1977), IDEF-0 (CAM.I., 1980), ... However, such languages are not really formal languages since they are mainly based on a set of 'boxes and arrows' notations with only a poor underlying semantics. Another problem is that such languages have been originally designed for the purpose of modelling requirements inherent to business information systems and, thereby, it is not proved that their underlying ontology of concepts is rich enough for capturing the whole complexity of a CIM enterprise modelling. Among some initiatives at the level of the identification of an adequate ontology of concepts, one may quote the CIMOSA project (ESPRIT Consortium AMICE, 1993), (Vernadat, 1993) which will be heavily referred throughout the rest of this paper.

At the level of formality, there are formal languages proposed for specifying CIM applications. Some examples include (Bastide, 1991), (Jaulent, 1990) ,(Zerhouni, 1990) and are based on Petri-Nets and/or on an operational semantics. However, such languages are basically *design specification* languages and not *requirements specification* languages. This means that there are more adequate for expressing the solution to a problem rather than to model the problem itself. In most cases, such languages encourage a procedural style of specification while a more declarative (logical) style is required at the RE level.

Some recent examples of formal requirements specification languages include RML (Greenspan, 1986), GIST (Feather, 1987), MAL (Finkelstein, 1987) and ERAE (Dubois, 1991). In the specific CIM context, as examples, one may quote the on-going researches of Bussler (Bussler, 1993) related to the use of workflows and the work of Fox (Fox, 1994) relying on a specific logical framework. In this paper, we propose to evaluate the ALBERT language, a language that has been designed recently with the purpose of modelling requirements inherent to distributed real-time safety-critical systems. In order to proceed to this evaluation, the expressiveness of ALBERT will be judged through its capacity of modelling concepts presented in the CIMOSA RE framework (viz. domains, processes, inputs and outputs, resources, etc).

All along this paper, we will illustrate the use of ALBERT by considering a small case study, originally introduced in (Lutherer, 1994) for the purpose of illustrating the application of the CIMOSA concepts at the requirements and design levels.

The example concerns the automation of a concrete batching and mixing plant. The goal of the installation is to produce several types of concrete corresponding to the customer's demand. For that purpose, the plant is composed of:

- several bins to store the cements, the aggregates and the water;
- a mixer, for the preparation of the concrete;
- two scales, to weight the ingredients;
- two conveyor belts, to carry the ingredients from the bins to the mixer.

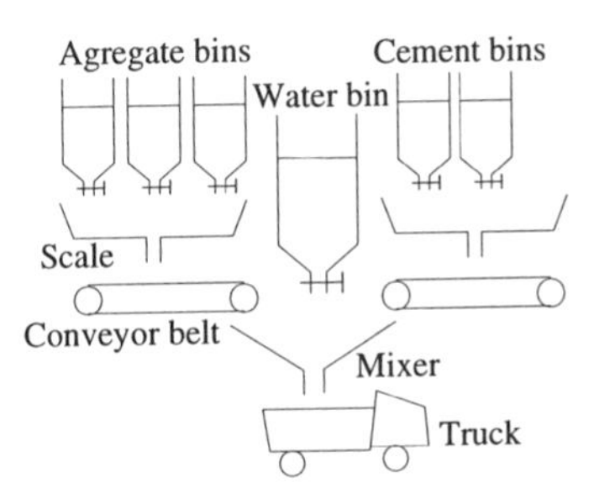

Figure 1 Schema of the plant.

The objective of the system is to produce automatically the concrete corresponding in composition, quantity and delay to the customer order. The system is depicted on Figure 1.

In Sect.2, an in-depth presentation of the ALBERT language is provided together with its illustration through fragments of the case study presented above. In Sect. 3, we carefully review the ontology of concepts made available in CIMOSA for the modelling purposes and we compare each of them with an equivalent construct available in ALBERT. Finally, Sect. 4 concludes by providing a brief overview of our recent researches in the CIM context.

2 THE ALBERT LANGUAGE

In this paper, we do not provide an in-depth presentation of the ALBERT language (such presentation can be found in (Dubois, 1994b)). Our aim is just to illustrate some basic features of the language through the handling of the concrete plant case study.

Basically, ALBERT is based on a variant of *real-time temporal logic*, a mathematical language particularly suited for describing histories (i.e. sequences of states) and expressing performances constraints (like, e.g. "this property holds for at least 3 minutes"). This logic is itself an extension of multi-sorted first order logic, still based on the concepts of variables, predicates and functions. The language introduces mainly three extensions:

1. the introduction of **actions**. Actions are associated with changes that may alter states in histories. Using actions in ALBERT makes possible to overcome the well-known *frame* problem, a typical problem resulting from the use of a declarative specification language;
2. the introduction of **agents**. An agent, which can be seen as a specialization of the object concept, is characterized through (i) its **internal state** recording the knowledge maintained by the agent and the "visibility" that it offers to the other agents of the environment, (ii) its **perception** of what is happening in its environment and (iii) its **responsibility** with respect to actions having some effects on its state or on states of other agents.
3. the identification of **typical patterns of constraints**. The use of a logical formal language may be compared to the use of an assembly programming language. A set of basic constructs are available in the language but there is a lack of support for the analyst in writing complex and consistent statements. To overcome this problem, a number of typical patterns of

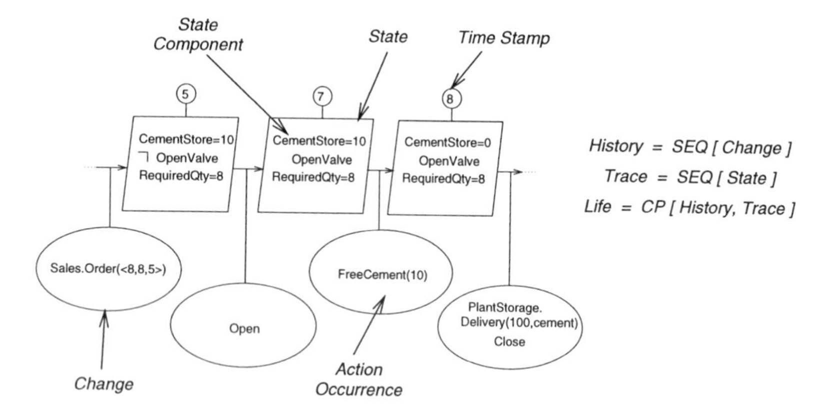

Figure 2 A graphical representation of a possible life of a *CementBin* agent (excerpt).

formulas have been predefined in ALBERT (in the past, a similar need for typical patterns was identified in RML (Greenspan, 1986)).

Using the language involves two activities: (i) writing *declarations* introducing the vocabulary of the considered application and (ii) expressing *constraints*, i.e. logical statements which identify possible behaviours of the different agents and exclude unwanted ones. A graphical syntax (with a textual counterpart) is used to introduce *declarations* and to express some static properties. The expression of the other constraints is purely textual.

Before to present the language constructs in Sect.2.2 and 2.3, we first provide a brief insight of the mathematical model underlying an ALBERT specification.

2.1 Models of a Specification

The purpose of our requirements language is to define admissible behaviours of the system to be developed. A specification language is best characterized by the structure of *models* it is meant to describe.

In order to master their complexity, models of a specification are derived at two levels:

- at the agent level: a set of possible behaviours is associated with each agent without any regard to the behaviour of the other agents;
- at the society (group of agents) level: interactions between agents are taken into account and lead to additional restrictions on each individual agent behaviour.

The specification describes an agent by defining a set of possible *lives* modelling all its possible behaviours. A life is an (in)finite alternate sequence of *changes* and *states*; each state is labelled by a time value which increases all along the life. Figure 2 illustrates, in a graphical way, the concept of life by giving an excerpt of a possible life of the *CementBin* agent.

The term "history" refers to the sequence of changes which occur in a possible life of the agent. A change is composed of several occurrences of simultaneous *actions* (the absence of action is

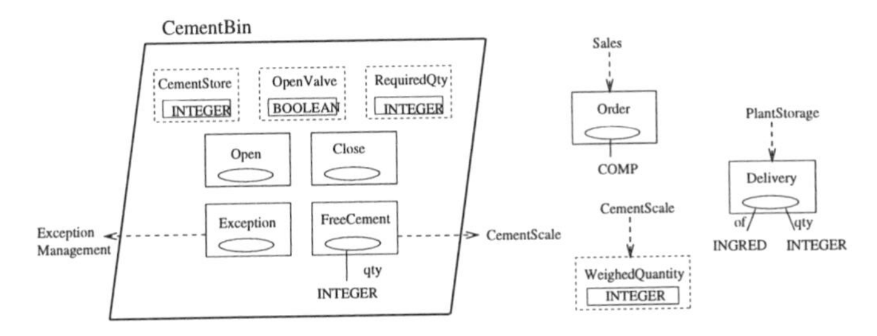

Figure 3 Declarations of the *CementBin* Agent.

also considered as a change). The term "trace" refers to a sequence of states being part of a possible life of the agent. A state is structured according to the information handled in the considered application in terms of *state components*. Notice that the value of a state at a given time in a certain life can always be derived from the initial state and the sub-history containing the changes occurred so far.

2.2 Declarations

Declaration of Agents

The declaration part of an agent consists in the description of its states structure and the list of the actions its history can be made of. Importation and exportation links between agents are also graphically described.

Since agents are considered as *specialized* objects, our modelling of a state structure is largely inspired by recent results in O-O conceptual modelling (see, e.g., OBLOG (Sernadas, 1989) and O* (Brunet, 1991)).

Agents include a key mechanism that allows the identification of the different instances. A type is automatically associated to each class of agent. For instance, each *CementBin* agent has an identifier of type *CEMENTBIN*.

The state is defined by its components which can be *individuals* or *populations*. Usually populations are *sets* of individuals but they can also be structured in *sequences* or *tables*. Elements of components are typed using (i) predefined elementary data types (like, *STRING, BOOLEAN, INTEGER,...*), (ii) user-defined elementary types (for which no structure is given), (iii) user-defined constructed types built using predefined type constructors like,e.g. Cartesian product, sequence, union, enumerated type, etc (e.g. *INGRED*, in our example, is defined as an enumeration of three possible values, namely *Cement, Aggregate* and *Water*), or (iv) types corresponding to agent identifiers.

Figure 3 proposes the graphical diagram associated with the declaration of the *CementBin* agent. It can be read that:

- *CementStore* is an instance of type *INTEGER* indicating the current amount of cement in the bin;
- *FreeCement* is an action which may be issued by the *CementBin*. Actions can have arguments; for example, each occurrence of a *FreeCement* action has an instance of type *INTEGER* as argument which indicates the quantity of cement released by the bin when the valve is open.

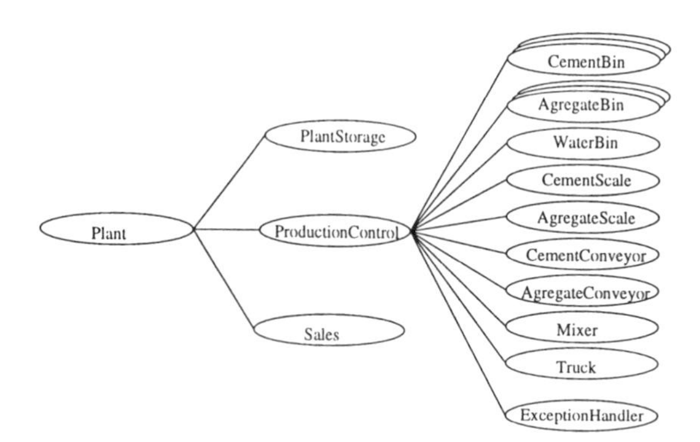

Figure 4 Declarations of the resources hierarchy of the concrete plant.

The diagrams also includes graphical notations used to express the visibility relationships linking the agent to the outside (*Importation* and *Exportation* mechanisms). Boxes without arrow denote information (state components or actions) which is not visible from the outside while boxes with arrow denote information which is exported to the outside. From the declaration of the *CementBin* agent, it can be read, for example, that the *Exception Management* department may have knowledge of *Exception* occurrences inside the bin and that the *CementScale* informs the bin of the currently *WeighedQuantity*.

Importation and *Exportation* are static properties; *Perception* and *Information* are their dynamic counterparts and provide the analyst with a finer way of controlling how agents can see information inside each other (perception and information constraints will be discussed in Sect. 2.3).

Declaration of a Society

Agents are grouped into societies. Societies themselves can be grouped together to form larger societies. In fact, a specification consists in a hierarchy of agents (a tree-like structure). Figure 4 shows the hierarchy associated with the concrete plant.

The existing hierarchy among agents is expressed in term of two combinators: cartesian product and set. In our specific case, e.g. the *ProductionControl* agent is an aggregate of one *Mixer*, several *CementBins*, one *CementConveyor*, one *CementScale*, a *Truck*, ...

2.3 Constraints

Constraints are used for pruning the (usually) infinite set of possible lives (histories) of an agent. Unlike usual design specification languages, the ALBERT semantics is not operational. A life must be extensively considered before it can be classified as possible or not, i.e. adding new states and changes at the end of a possible life does not necessarily result in a possible life.

Figure 5 introduces the specification associated with the behaviour of the *CementBin* agent and refers to the graphical declaration introduced in Figure 3. Formal constraints are expressed in terms of the different *patterns* available in the language. Rather than describing in details those patterns (which are extensively presented in (Dubois, 1994b), we will only refer to the informal comments written on the figure in order to exhibit the expressiveness and the non-operational style of an ALBERT specification.

Basically, properties are classified under 8 different headings grouped into two families: **Local Constraints** precise the responsibility of an agent with respect to changes that may alter its state of knowledge, and **Cooperation Constraints** specify how the agent interacts with the other agents of its environment. Finally, it should be also noted the use of **Basic Constraints** to define the initial values of the system.

Local Constraints

Local constraints are related to the internal behaviour of the agent. These constraints prescribe the sequence of actions that may/must happen through the whole life of the agent as well as the sequence of states which results from action occurrences. Local constraints are classified under four headings: **State Behaviour**, **Effects of Actions**, **Causality** and **Capabilities**.

CementBin

BASIC CONSTRAINTS
INITIAL VALUATION

CementStore=0 $\wedge$ *OpenValve=*FALSE $\wedge$ *RequiredQty=0*

LOCAL CONSTRAINTS
STATE BEHAVIOUR

$\neg$ *(CementStore=0* $\wedge$ *OpenValve)* $\wedge$ $\neg$ $(\Box_{>5min}$ *OpenValve)*

```
The valve of an empty bin is always closed. In order to avoid technical problems,
bin valves can not stay open for more than 5 minutes.
```

EFFECTS OF ACTIONS

Sales.Order($< qc, \_, \_ >$)*: RequiredQty=qc*
PlantStorage.Delivery(q,_): CementStore=CementStore+q
*Close: OpenValve=*FALSE $\wedge$ *RequiredQty=0*
*Open: OpenValve=*TRUE
FreeCement(q): CementStore=CementStore-q

CAUSALITY

Sales.Order($< qc, \_, \_ >$) $\overset{\Diamond<10sec}{\longrightarrow}$ *Open*

CAPABILITY

$\mathcal{F}$ (*FreeCement(q) / OpenValve=*FALSE $\vee$ *CementStore*$< q \vee q \leq 0 \vee q > 100$)

```
The bin empties only when the valve is open and the quantity of cement freed
in one time unit is less than 100
```

$\mathcal{XO}$ (*FreeCement(_) / OpenValve=*TRUE $\wedge$ *CementStore*$\neq 0$)

```
A non empty bin always empties when the valve is open.
```

$\mathcal{XO}$ (*Close / CementScale.WeighedQuantity* $\geq$ *RequiredQty*)

```
As soon as the required cement quantity has been weighed, the valve must be closed.
```

$\mathcal{XO}$ (*Exception / CementScale.WeighedQuantity* $>$ *RequiredQty*)

```
An exception message is generated if the weighed quantity outnumbers the required one.
```

COOPERATION CONSTRAINTS
ACTION PERCEPTION

$\mathcal{XK}$ (*PlantStorage.Delivery(_,i) / i=cement*)

```
Only an ingredient of type cement can be delivered in a cement bin
```

$\mathcal{XO}$ (*Sales.Order*($< qc, \_, \_ >$)*/ CementStore* $\geq$ *qc* $\wedge$ *RequiredQty=0*)

```
Sales orders are taken into account if sufficient stock exists for producing them
and if no other order is already in production.
```

STATE PERCEPTION

$\mathcal{XK}$ (*CementScale.WeighedQuantity* / TRUE)
```
The cement bin has always access to the dial of the scale.
```

ACTION INFORMATION

$\mathcal{XK}$ (*FreeCement(_).CementScale* / TRUE)
```
Everything freed by the cement bin goes into the cement scale
```
$\mathcal{XK}$ (*Exception.ExceptionManagement* / TRUE)
```
Each time an exception occurs, the Exception Management is warned.
```

STATE INFORMATION

$\mathcal{XK}$ (*CementStore.Controller* / TRUE)
```
The bin always lets the controller know its content.
```

Figure 5 Constraints on the CementBin Agent.

Under the heading **Effects of Actions** are described the effects of the different actions happening. The effect of an action is expressed in terms of a property characterizing the state which follows the occurrence of an action (see examples in Figure 5). Under the heading **Capability** is described the responsibility of the agent with respect to the occurrences of its own actions. The default rule is that all actions are permitted whatever the situation but specific constraints can be added for making possible to express circumstances under which obligations and preventions are associated with actions occurrences (see for example the prevention constrain on the *CementBin* agent stating that no *FreeCement* action can occur when the bin valve is closed).

A specification of requirements only written with **Effects of Actions** and **Capability** constraints leads the analyst to adopt a rather *operational* style of specification where are described states transitions and conditions under which these transitions may/must occur (such a style of specification is advocated in, e.g., MAL (Finkelstein, 1987). Adopting this style may lead to the introduction of extra information in the agent state at the risk of over-specifications.

ALBERT provides more freedom to the analyst by letting him/her to adopt a more declarative style of specification. Under the **Causality** heading are described the causality relationships existing among action occurrences. In our case study, an example of causality exists among e.g. an *Order* request issued by the *Sales* department and a *Open* action under the control of the cement bin. It relies upon the necessity of having one unique occurrence of the *Open* action in response to each occurrence of the *Order* action. Finally, under the **State Behaviour** heading are described two kinds of constraints which must hold from the admissible history of an agent:

- static constraints are constraints which are true in all states (usually referred as *invariants*);
- dynamic constraints are constraints on the evolution of the state. They are expressed using temporal connectives. See for example the constrains on the *CementBin* agent.

Cooperation Constraints

Under the **Action Perception** and **State Perception** headings is described the responsibility of an agent with respect to the perception that it has to guarantee (i) for the actions happening in other agents and (ii) for state information made visible by other agents. In any case, using the appropriate pattern, the perception guaranteed by an agent may vary with time and depend on circumstances (see, the example of an **Action Perception** constraint in Figure 5 stating that *Order* requests issued by the *Sales* department are ignored if another order is currently being processed ($RequiredQty \neq 0$) or if not enough cement is present in the bin.

Under the **Action Information** and **State Information** headings is described the responsibility of an agent with respect to the visibility that it may offer to some other agents (i) for some of its performed actions and (ii) for parts of its state. See, in Figure 5, the example of an **Action Information** constraint in the specification of the *CementBin* agent stating that the *FreeCement* action is always showed to the *CementScale* agent. In any case, using the appropriate pattern, the visibility offered by an agent may vary with time, depend on circumstances and be restricted to some agents.

The use of *perception* and *information* constraints makes possible to describe various cooperation protocols (reliable or not) that may take place among agents. Again, ALBERT permits to express the different protocols at a high level of abstraction and without any regard to implementation details (like, e.g., the use of acknowledgment messages, messages storage buffers).

3 FROM CIMOSA CONCEPTS TO ALBERT CONSTRUCTS

The CIMOSA proposal (ESPRIT Consortium AMICE, 1993), (Vernadat, 1993) relies upon a framework made of different integrated models (with associated concepts and supporting formalisms) used for the purpose of covering the different phases (i.e. requirements definition, design and implementation) of the engineering activity of CIM systems.

In this section, we will concentrate on the relationship existing between ALBERT and CIMOSA at the requirements and design engineering levels. At the requirements level, CIMOSA concentrates on a logical view of the different functionalities to be performed in the CIM system without any regard to the resources required for their execution. These resources are taken into account at the design level where elementary functions identified in the functionalities are taken in charge by the resources.

Throughout this section, we will refer to the terminology used in CIMOSA and illustrate the mapping between CIMOSA and ALBERT through the case study already introduced and completely handled in (Lutherer, 1994).

3.1 Requirements Definition

In CIMOSA, the *Requirements Definition Modelling Level* concentrates on collecting the end-user needs and on structuring them in terms of a hierarchy of purely logical processes where:

- At the top level, processes correspond to the *Domains Processes* (DP) belonging to the *Domains* establishing the scope of the studied problem. DP's are further decomposed into *Business Processes*.
- At an intermediate level, processes correspond to *Business Processes* (BP). BP's identify the logical activities as well as the required sequencing of them needed for solving the manufacturing system problem. BP's can be themselves further decomposed into finer BP's and/or into *Enterprise Activities*. Within the context of our case study, Figure 6 shows an example of such decomposition.

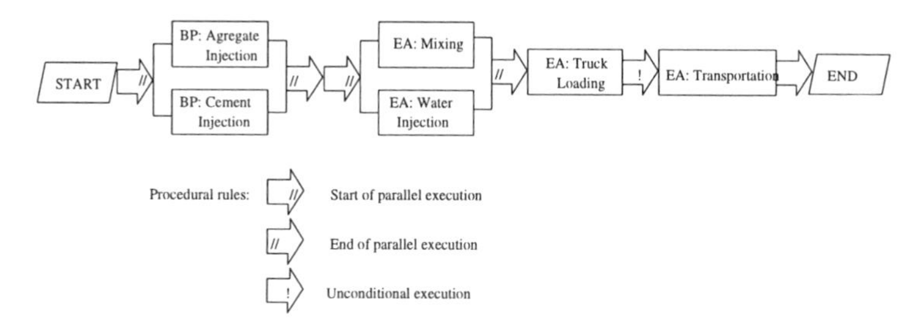

Figure 6 Decomposition and behaviour of the *Production* BP.

- At the lower level, processes correspond to *Enterprise Activities* (EA). EA's are terminal nodes in the hierarchy and are associated with the description of the ultimate functionalities (inputs/outputs transformations) guaranteeing the effectiveness of the solution to the manufacturing problem. Within the context of our case study, Figure 7 shows an example of such decomposition.

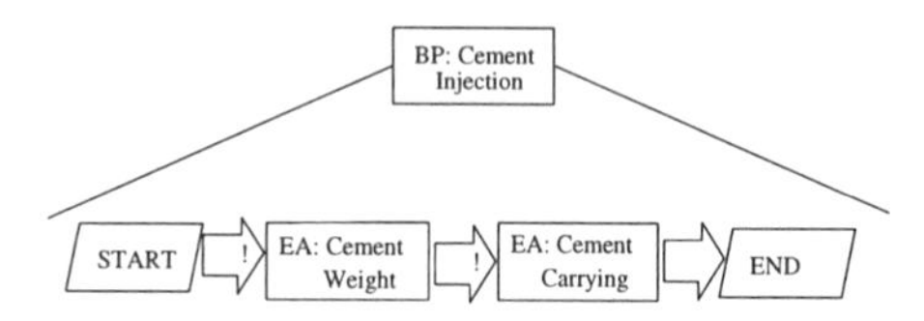

Figure 7 Decomposition and behaviour of the *CementWeight* BP.

Processes and their behaviour

Using ALBERT, it is straightforward to map the results of the *Requirements Definition Modelling Level* through the elaboration of an equivalent hierarchy where:

- societies (group of agents) are introduced at the higher and intermediate levels for representing DP's and BP's;
- agents are introduced at the lower levels for representing EA's.

Within the context of our case study, such hierarchy is presented on Figure 8.

In CIMOSA, at the *Behaviour Analysis Level*, *Procedural Rules* (PR) are used for defining the logical sequence of BP's and EA's inside a specific DP/BP (see Figure 6). Such rules have their counterpart in ALBERT where *causalities* (see Local Constraints in Sect. 2.3) can be used to express the unconditional execution, the conditional execution, the parallel execution and the rendezvous of different processes.

Besides the dynamical aspect, in CIMOSA, at the *Operational Analysis* level, ultimate elements introduced in the different functionalities are characterised by *Inputs* and *Outputs*. Inputs are of three types:

- *function* inputs are a set of object views (information) representing information and physical objects processed by the EA;

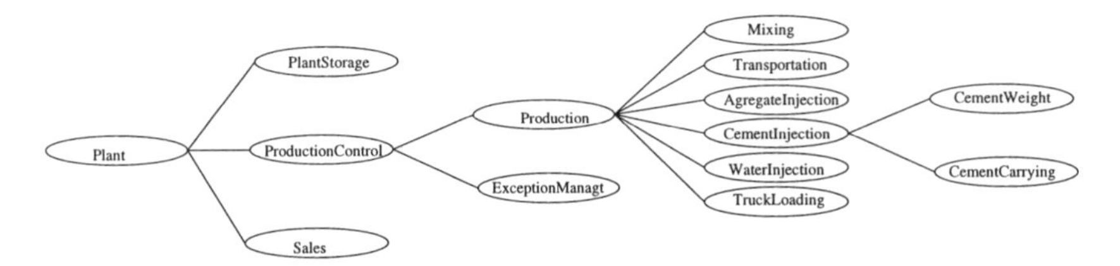

Figure 8 Declarations of the process hierarchy of the Production Control Domain.

- *control* inputs represent information used to control or to constrain the execution of the EA but which are not processed by the EA. A typical example is a NC-Programme;
- *resource* inputs are either the current resources used for executing the EA or a set of required capabilities of candidate resources. Required Capabilities may describe abilities of machines as well as of humans.

Outputs of an EA are also of three types:

- *function* outputs are, similarly to function inputs, a set of object views (information) representing information and physical objects produced by the EA;
- *control* outputs are a set of events generated by the EA and which may cause the triggering of a domain process;
- *resource* outputs report the information to be recorded about the usage of resources after execution of the EA (e.g. tool usage times, ...).

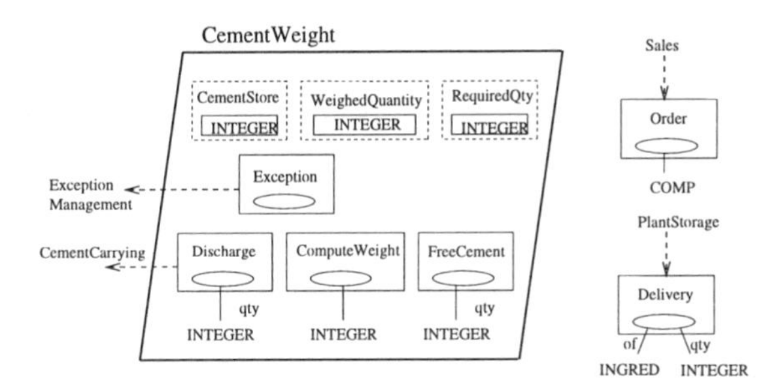

Figure 9 ALBERT declarations associated with the *CementWeight* EA.

In Figure 9, we provide the specification of the EA *CementWeight* where we have identified:

- *RequiredQty* as a control input recording the quantity of the cement to be weighed;
- *CementStore* as a function input attached with the stock of cement;
- *Exception* as a control output associated with the occurrence of an exception and the request issued to the *Exception Management* Domain Process to handle it;
- *Weighed Quantity* as a function output attached with the cement weighed before its delivery.

All the constraints attached with these different inputs/outputs are presented in Figure 10.

Information view

In parallel with the specification of the different processes, CIMOSA also recommends to perform an *Information Analysis* based on an object oriented paradigm. The concepts used are *Object Views* (perception of objects by users or applications), *Information Elements* (data items composing object views), *Enterprise Objects* (EOs) (entities of the enterprise on which views are defined), *Object Abstraction Mechanisms* (generalisation and aggregation between objects), *Object Relationships* (connecting pairs of objects), and *Integrity Rules* (constraints on information).

The information modelling in ALBERT is based on *data types*. Every information in a specification (be an action argument or a state component) is typed. Information Elements (IEs) correspond to simple (basic or predefined) data types. These are the basic elements (attributes) with are used to define more complex objects. Enterprise objects correspond to complex data types (defined by applying constructors –like Cartesian Product, Set, Sequence, ...– on simple and other complex data types (just like EOs are constructed from information elements and EOs). They represent the objects (physical and informational) which are manipulated in the enterprise.

ALBERT does not provide any explicit representation of object views. An object view would correspond in ALBERT to a particular data type which is a sub-sort of another. For example, using this mechanism makes possible to define a cartesian product of three components as a subset of a cartesian product of five components.

ALBERT provides usual abstraction mechanisms associated with abstract data types (viz genericity and specialisation) and thereby can be used for the definition of generic objects (encompassing several IEs or EOs) and the definition of specialisation of them.

Objects Relationships can be expressed by two means: cartesian products or tables. Both solutions can equally be used. The choice of a solution rather than the other depends on constraints and accesses that have to be expressed on the relation. The expression may be easier with a table in some cases, and more easier with a Cartesian product in others.

Integrity rules are easily expressed in ALBERT where, under the *state behaviour heading* (see Sect. 2.3), any temporal logic formula can be used. This enables the expression of constraints expressing derivation of values of components from values of others, invariants and constraints on the evolution of states.

3.2 The Design Model

In CIMOSA, the *Design Specification Modelling Level* is concerned with the technological choices to be made for the implementation of the CIM system. At that level, the major design decision is concerned with the identification of the actual resources (called *Functional Entities* in CIMOSA) that will be able to carry out the different elementary functions identified during the *Requirements Definition* level. In other words, at the design level, we need to transform the idealized (process-driven) view of the manufacturing system into a degraded view of the system where constraints are superimposed due to the nature and the capabilities of the different resources which are used.

In CIMOSA, the resulting design model presents the different functional entities together with the behaviour of the activities that they perform. In ALBERT, the hierarchical structure of the resources is reflected in the graphical diagram presented in Figure 4 where the agents correspond to the different functional entities. In the particular case of the *CementBin* resource, we have presented its complete specification in Figure 3.

Cement Weight

BASIC CONSTRAINTS

INITIAL VALUATION

CementStore=0 $\wedge$ *RequiredQty=0* $\wedge$ *WeighedQuantity=0*

LOCAL CONSTRAINTS

EFFECTS OF ACTIONS

PlantStorage.Delivery(_,q): CementStore=CementStore+q
FreeCement(q): CementStore=CementStore-q
Sales.Order(< qc, _, _ >): RequiredQty=qc
*Discharge(_): Compositon=*UNDEF
ComputeWeight(q): WeighedQuantity=q

CAUSALITY

Sales.Order(< qc, _, _ >) $\xrightarrow{\Diamond}$ *Exception* $\oplus$ *((FreeCement(_))* ; Exception)* $\oplus$ *((FreeCement(_))* ; Discharge(qc))*

    Exceptions may occur during cement weighing: either at the beginning or after
    a certain quantity (below the required quantity) of cement has been weighed.

CAPABILITY

$\mathcal{XO}$ (*ComputeWeight(q) / RequiredQty* $\neq$ *0* $\wedge$ *WeighedQuantity < RequiredQty*)

    Weight computing occurs continiously from the order receipt until the required
    quantity has been weighed.

$\mathcal{F}$ (*FreeCement(q) / WeighedQuantity* $\geq$ *RequiredQty* $\vee$ *q > CementStore*)

    No more cement than what is required, nor than what is available, can be weighed.

COOPERATION CONSTRAINTS

ACTION PERCEPTION

$\mathcal{XO}$ (*Sales.Order(< qc, _, _ >) / CementStore* $\geq$ *qc* $\wedge$ *RequiredQty=0*)

    Sales orders are taken into account if sufficient stock exists for producing them
    and if no other order is already in production.

$\mathcal{XO}$ (*PlantStorage.Delivery(_,_)* / TRUE)

    Deliveries are always possible (capacities are infinite!)

ACTION INFORMATION

$\mathcal{O}$ (*Exception.ExceptionManagement* / TRUE)

    Exception Management is always informed of occurences of exceptions.

$\mathcal{O}$ (*Discharge(_).CementCarrying* / TRUE)

    Cement Carrying is always informed that the cement weighing activity is over.

Figure 10 ALBERT constraints associated with the *CementWeight* EA.

4 CONCLUSION

In this paper, we have used the ALBERT language, a formal requirements for the specification of real-time composite systems, in the context of CIM infrastructures modelling. In order to evaluate the degree of expressiveness offered by ALBERT in this context, through the handling of a case study, we have compared it with respect to the modelling constructs proposed in CIMOSA for the requirements and design levels. From that experiment, we can draw two conclusions:

- The ALBERT language offers enough expressivity for capturing the different informal and semi-formal descriptions provided in the CIMOSA models;
- Thanks to the underlying formal semantics, ALBERT helps the analysts in detecting desired properties like e.g., the absence of deadlocks, the completeness of the specification, ...

Finally, we would like to conclude this paper by providing some insights about our researches in the CIM context. These researches are going in two directions:

1. On the one hand, we are working on the specification of CIM requirements fragments for the purpose of defining a library of reusable components for CIM applications. Such reusable components can be composed through the use of well-defined structuring mechanisms available in ALBERT but not fully detailed in this paper. Preliminary result is the identification and the specification of a general CIM *framework* presented in (Dubois, 1993).
2. On the other hand, we are working on the elaboration of a general architecture in terms of which *functional* as well *non functional* requirements can be reconciled within a unique formal requirements document. In CIMOSA, these non functional requirements are related, for example, to the *Objectives* and *Constraints* attached to *Domains*. Preliminary results are reported in (Dubois, 1994a).

Acknowledgements

Part of this work was supported by the Belgian Walloon Region under Project 2337 of the FIRST program and by the European Community under Project 8319 (MODELAGE) of the ESPRIT program.

REFERENCES

ESPRIT Consortium AMICE, editor. (1993) *CIMOSA: Open System Architecture for CIM*, volume 1 of *Research Reports ESPRIT, Project 688/5288 AMICE.* Springer-Verlag, 2nd revised and extended edition.

R. Bastide and C. Sibertin-Blanc. (1991) Modelling a flexible manufacturing system by means of cooperating objects. In G. Doumeingts, J. Browne, and M. Tomljanovich, editors, *Computer Applications in Production and Engineering CAPE'91*, pages 593–600, Bordeaux, September 10-12. IFIP, Elsevier Science Publishers B.V.

J. Brunet. (1991) Modelling the world with semantic objects. In *Proc. of the working conference on the object-oriented approach in information systems*, Québec (Canada).

Christoph Bussler. (1993) Enterprise process integration model and infrastructure. In H. Yoshikawa and J. Goosenaerts, editors, *Preprints of JSPE-IFIP WG 5.3 Workshop on the Design of Information Infrastructure Systems for Manufacturing – DIISM'93*, pages 415–426, Tokyo (Japan), November 8-10. Japan Society for Precision Engineering.

CAM.I. (1980) Architect's manual: ICAM definition method IDEF0. Technical Report DR-80-ATPC-01, CAM.I., April.

Eric Dubois, Philippe Du Bois, Frédéric Dubru, and Michaël Petit. (1994b) Agent-oriented requirements engineering: A case study using the albert language. In A. Verbraeck, H.G. Sol, and P.W.G. Bots, editors, *Proc. of the Fourth International Working Conference on Dynamic Modelling and Information System – DYNMOD-IV*, Noordwijkerhoud (The Netherlands), September 28-30. Delft University Press.

Eric Dubois, Philippe Du Bois, and Michaël Petit. (1993) Elicitating and formalising requirements for CIM information systems. In C. Rolland, F. Bodart, and C. Cauvet, editors, *Proc. of the 5th conference on advanced information systems engineering – CAiSE'93*, pages 252–274, Paris (France), June 8-11. LNCS 685, Springer-Verlag.

Eric Dubois, Jacques Hagelstein, and André Rifaut. (1991) A formal language for the requirements engineering of computer systems. In André Thayse, editor, *From natural language processing to logic for expert systems*, chapter 6. Wiley.

Eric Dubois and Michael Petit. (1994a) The formal requirements engineering of manufacturing systems. In S.M. Deen, editor, *Proc. of the Second International Working Conference on Cooperative Knowledge Based Systems – CKBS'94*, pages 67–82, Keele (UK), June 14-17.

Martin S. Feather. (1987) Language support for the specification and development of composite systems. *ACM Transactions on Programming Languages and Systems*, 9(2):198–234, April.

Anthony Finkelstein and Colin Potts. (1987) Building formal specifications using "structured common sense". In *Proc. of the 4th International Workshop on Software Specification and Design – IWSSD'87*, pages 108–113, Monterey CA, April 3-4. IEEE, CS Press.

Sol J. Greenspan, Alexander Borgida, and John Mylopoulos. (1986) A requirements modelling language. *Information Systems*, 11(1):9–23.

Patrick Jaulent. (1990) Flexible manufacturing system specification with object-oriented method sys-p-otm. In *Proc. of the International Conference CIM'90: Integration Aspects*, pages 297–306, Bordeaux (France), June 12-14. Productic-A, Teknea.

Eric Lutherer, Soumeya Ghroud, Michel Martinez, and Joël Favrel. (1994) Modelling with CIMOSA: A case study. In *Proc. of the IFIP WG5.7 Working Conference on Evaluation of Production Management Methods*, pages 233–241, Gramado (Brazil), March 21-24.

Douglas T. Ross. (1977) Structured analysis (sa): a language for communicating ideas. *IEEE Transactions on software engineering*, SE-3(1):16–34, January.

A. Sernadas, C. Sernadas, and H.-D. Ehrich. (1989) Abstract object types: a temporal perspective. In B. Banieqbal, H. Barringer, and A. Pnueli, editors, *Proc. of the colloquium on temporal logic and specification*, pages 324–350. LNCS 398, Springer-Verlag.

Mark S.Fox and Michael Gruninger. Ontologies for enterprise integration. In Michael Brodie, Mathias Jarke, and Michael Papazoglou, editors, *Proc. of the Second International Conference on Cooperative Information Systems – CoopIS-94*, pages 82–89, Toronto (Canada), May 17-20, 1994.

Francois Vernadat. (1993) CIMOSA: Enterprise modelling and entegration using a process based approach. In H. Yoshikawa and J. Goosenaerts, editors, *Preprints of JSPE-IFIP WG 5.3 Workshop on the Design of Information Infrastructure Systems for Manufacturing – DI-ISM'93*, pages 161–175, Tokyo (Japan), November 8-10. Japan Society for Precision Engineering.

Noureddine Zerhouni and Hassane Alla. (1990) Continuous petri nets: a tool for production systems dynamic analysis. In *Proc. of the International Conferefence CIM'90: Integration Aspects*, pages 417–424, Bordeaux (France), June 12-14. Productic-A, Teknea.

10

Specification Environment for Multi-agent Systems Based on Anonymous Communications in the CIM Context

A. ATTOUI, A. HASBANI, A. MAOUCHE
Laboratoire d'Informatique/ ISIMA
B.P. 125
63173 AUBIERE CEDEX, FRANCE
Tel: 73 40 74 40 Fax: 73 26 88 29
mail: ammar@sp.isima.fr

Abstract

In this paper, we propose a formal specification method for manufacturing systems software development. Our approach is based on rewriting logic and multi-agent paradigm. It proposes a methodology for the analysis and the structuration of the command part of manufacturing systems, in terms of cooperative and specialized agents. Rewriting logic constitutes the formal framework. The approach may be seen under two complementary aspects; the first one consists, in general, in the distributed system conception and the second concerns the use of IAD principles for the representation, the distribution and the knowledge co-operation through a system of cognitive, autonomous and co-operating agents. The method supports modularity and abstraction, follows the great principles of a multi-agent systems approach and supports real time applications.

Keywords

Specification, Validation, Formal methods, Manufacturing Systems, Multi-Agent Systems, , Methodology.

1 INTRODUCTION

It is widely recognized that methodologies and tools used in analysis requirements and specification stages determine directly the quality of the development of software systems. For real-time systems, these methodologies and tools must additionally provide concepts, formalisms, and mechanisms to express at a high level of abstraction the concurrency of multiple computation threads, the synchronization and the communication between these threads and the handling of internal and external events. Inconsistency and incompleteness of the specifications must be detected as soon as possible in order to avoid costly readjustments in the design and development stages. Manufacturing systems belong to the more general class of distributed and real time systems. The intelligent systems designers (Ayel, 1991]) are not out of the way of this evolution; They use both systems with blackboard architecture (Laasri,1989), (Bouzouane,1993), where the contribution to solve a problem goes by a common data structure, and Multi-agent systems, where communications are done solely by message passing. We can distinguish two distribution levels: running distribution (parallel inferences) and data distribution (knowledge) which favours the performance improvement (sequential or parallel execution) (Occello, 1993). Implementation of such systems requires two different stages. The first stage consists in designing a conceptual unambiguous model for specifying the requirements analysis. The second stage consists in deriving an executable model for evaluation and validation of the system. The information provided by this model are very useful for the final definition of the control part of the system.

Numerous methods are available to specify and design such systems. Among these methods we can mention SA/RT, SA/DT, OMT(Hatley,1990), (Fayad,1994). These methods are adequate to capture and describe the behavior of complex systems. But most of them do not offer a formalized framework facilitating the elimination of any inconsistency and incompleteness. Indeed, these methods use multiple different and incompatible models with the same specification. For example, SA/RT uses data flow diagrams, control flow diagrams (finite state machines) and a process activation table. Timing constraints are specified separately in the timing constraints table. In the other hand, Petri nets and their extensions (Bruno,1986), (Peterson,1981) can be efficient when the studied system is not too complex.

Formal methods and techniques have been suggested over the last several years to prove properties about specifications. CCS, Z, VDM, ESTELLE, LOTOS, SDL (ISO,1989), (Courtiat,1991), (Sijelmassi,1991), (Binding,1991), (Busttard,1992), (IS8807,1988), (Vigder,1991), (Coelho,1992), (Hoare,1985), (Lightfoot,1991), are the most prominent ones. In a general way, these formal methods have not been widely used in industrial software development environment for several reasons (Fraser,1994). Among these reasons, we can mention: a formal specification language provides a notation (syntactic domain), a univers of objects (semantic domain), and a precise rule defining which satisfy each specification. This makes them an inappropriate tool for communicating with the end user and requires that software engineers, designers, and implementors master the notation and the conceptual grammar of the language

In this paper, we present an approach based on a unique formalism: the rewriting logic. This approach is dedicated to support the distributed system design in general, and in particular, to the design of multi-agent systems (SIC,1992). In this domain, our contribution is to palliate to some difficulties (of communication, concurrence, and real-times) proper to the cognitive approach by message passing (Bouront,1993). This approach, has not yet been formalized as blackboard approach, nevertheless, it seems avoid some limits of the blackboard approach. Its main features are the following:

- It takes into account an important variety of systems based on a sequential or a concurrent execution model.
- It uses the object model to implement multi-agent systems as in (Cardozo,1993), (Stinckwich,1993).
- It supports modularity and abstraction.
- It offers the possibility for validating the specifications and for generating the code automatically, according to a predefined distributed architecture.

This approach does not necessitate the mastering of the concepts and the foundations of the rewriting logic. It also integrates architectural considerations and is able to support the design of complex systems.

2 FORMAL MODEL FOR MULTI-AGENT SYSTEMS

Our model is based on a formalism that is closed to the one of the MAUDE language (Meseguer,1990). It takes into account the complexity of the information processed within an enterprise, it includes a description of the various entities manipulated, the actions that these entities may undergo, the temporal constraints and the tracability of information. The model is independent of the target language.

An agent will be associated to a module representing both the cognitive body and the resolution strategy. It is made up of rewriting rules basis expressing production rules, facts basis and a communication interface.

Rules and fact basis encapsulate the agent knowledge and transcribe its resolution strategy in regard with its specification. The communication interface allows, to a basic agent, to share its results with others agents of the same abstraction level .

2.1 The formal modules (Description of an agent)

A formal module is a quadruplet (Σ, a, R, S). Σ is the set of symbols for the functions of the module. It permits the formal description of the static part of an agent. Symbols can be simple (i.e. characters) or complex syntactical units. a is the set of structural axioms necessary to achieve the rewritings in a concurrent manner modulo the axioms.

The doublet (Σ, a) is called the signature of the agent. R is a set of rewriting rules. It permits the formal description of the dynamical part of a system. A rule has the form [t] ==> [t'] where t and t' are terms constructed from Σ. The notation [t] is used to indicate that t represents an element of the class of terms modulo the axioms. We will use three structural axioms called ACI : associativity, commutativity and identity. A rule indicates that the current state of the agent corresponding to the configuration t, becomes a new state corresponding by the configuration t'.

A configuration is defined in terms of agents and messages. It is represented through a sentence in the language of the corresponding formal module. More precisely, a configuration at a given time is composed of:

- identified agents. Each agent possesses intrinsic properties and is in a particular defined state.
- messages. Messages are generated by external excitations or internal interruptions. A same message can be sent to different agents.

During the specification stage, each agent can be considered if necessary as composed of other agents. A hierarchical decomposition is thus naturally introduced. An agent can use for its specification other agents. We will use the notions of level and visibility. An agent A is at a level immediately lower of the one of the agent B, if B is directly used in the composition of A. Usually, agent A knows only attributes of agents of its level. In some situations it is necessary that an agent see attributes of agents of lower levels. Such attributes are called visible.

2.2 The signature

For (Σ,a) we use a general formulation (figure 1) which is closed to the one proposed by (Meseguer,1990). This general form of signature integrates different operators. Op < _: _ / _> is the constructor of agents. Op _ = _ permits to affect a value to an attribute. Op _, _ is the syntactical constructor of attribute lists. These lists are used for the designation of the attributes of an agent. Op _ _ is necessary for the construction of distributed configurations. It permits the specification of any configuration. A configuration is composed of agents and messages. This operator has been declared modulo the ACI axioms. Thus, the order with which agents and messages are declared has no influence on the reduction process used further.

```
/*Alphabet of the system*/
Type Agent, Attribute, Attributes, Msg, Configuration, Value, AgentId, ClassId,
AttributeId;
/*Hierarchies and structural relations between the agents*/
Subtype AgentId, ClassId, AttributeId < Value;
Subtype Attribute < Attributes;
Subtype Agent, Msg < Configuration;
/*Operators for constructing the words and the sentences of a formal module*/
Op <_: _/ _> : AgentId Value -> Object;
Op _ = _ : AttributeId Value -> Attribute;
Op _, _ : Attributes Attributes -> Attributes [ Assoc, Com, Id = Nul];
Op _ _ : Configuration Configuration -> Configuration [Assoc, Com, Id = Nul].
```

Figure 1 The general signature of a formal module

The description of a real multi-agent system consists to instantiate the metatypes of Σ (agent, Attributes, Msg, ...).

2.3 The rewriting rules

Actions of messages on an agent are described through the rules. An agent can receive messages from agents in the same level or in the upper levels of its hierarchy. An agent can send messages to agents in the same level or in the lower levels of its hierarchy. A message can also be intercepted through a rule and routed to any level. A rule signals the occurrence of a communication in which n messages and n agents are involved. All the agents participating in a rule are at the same level. The general form of a rule is given by figure 2.

```
/*Syntax*/
M1M2..Mp<AG1: C1/ listeAt1>...<AGi: Ci/listAti>
<Aj: Cj/ listAtj>...<Ak: Ck/ listAtk> ==>
<Aj: Cj/ listAtj>...<Ak: Ck/ listAtk>
<AGm: Cm/ listAtm>...<AGn: Cn/ listAtn>Mq ... Mr [T]
```

Figure 2 Syntax and effects of a rewriting rule

Effects:
The messages M1M2..Mp are deleted after the execution of the rule.
The states of the agents Aj,..., Ak are modified.
Agents A1,..., Ai which appear only in the left part of the rule, are deleted.
New Agents AGm,...,AGn defined in the right part, are created.
New messages Mq, ..., Mr are created.
[T] is a temporal constraint. It can take:
1: every (T, msg) : to each time interval T, send the message msg;
2: within (T, msg) : after the time T, send the message msg;
3: AT (T, msg) : at the time T, send the message msg;
4:before(T, msg) : before the time T is elapsed, send the message msg.

In this syntax, an agent is represented by the term <idAgent:C/list At>. idAgent is the agent identifier. C is the agent class. listAt is a list of conditions on the attributes of the agent. Attributes which are modified must be visible at this level. At least, T expresses the service time for the execution of the rule. T can be a random law.
This general form permits to specify, at a high level of abstraction, the different conditions of co-operation and synchronisation between agents.

A rule indicates that the system goes from the configuration defined by the left part to a new configuration defined by the right part. A rule can be activated when all the messages of the left part are present and when all the conditions on the attributes of the left part are satisfied. Rules describe the actions of the events associated to the messages. They permit to reason on the state changes of the system and to draw valid conclusions about its evolution. They constitute the formal description of the dynamic aspects of the system.

2.4 Synchronous rules and asynchronous rules

A rule is synchronous if several agents are simultaneously modified through an atomic action. Otherwise it is asynchronous.

An example of synchronous rule is given by the following rule which specifies the carriage of a part by a an AGV (Automated Guided Vehicul) from a turning unit to a control unit:

(Carry by V Part P from T to C) < T: TurningUnit / CurrentOperation: finish>
< C: ControlUnit/ (queue : N) <= 10 > <V: AGV/ state: free>
==>
< T:TurningUnit / CurrentOperation: O> < C: ControlUnit/ queue : N+1 >
<V: AGV/ state: S>

In this rule, the carriage of P from T to C is an atomic operation. The trigger of this rule is:

(Carry Part P from T to C by V) and (T.currentoperation= finish) and (C.queue <=10)

The same problem can be formulated through an asynchronous rule:

(Carry Part P from T to C by V) ==>(TurningUnit T exit Part P)
(AGV V accept part P) (ControlUnit C accept Part P)

This rule deletes the message (Carry Part P from T to C by V) from the configuration and produces three new messages. The rules which intercept these messages can then fire in parallel.

Thus, formulation of a rule in an asynchronous way expresses possibilities of parallel processing.

Internal working rules

These rules have the following form:

Ai ==> Ai, Am...An, Mq...Mr[T]

They permit to express that state changes result from an internal working not visible at this level. They generally concern an agent which can evolve according to a self working. Through the parameter [T], which can be any random law it is possible to express random behaviour.

These rules permit specifications very similar to those proposed by (AGHA,1987) for actors.

3 DEVELOPMENT METHODOLOGY FOR MULTI-AGENT SYSTEMS

The method covers almost the entire life cycle of a distributed system in general, and of a multi-agent system in particular.

Specifications use concepts and notion of concurrent rewriting logic which leads to a coherent description of the system. The automatisation of the passage from one phase to the next permits to kept the coherence and to ensure a uniform life cycle.

3.1 Analysis

This step concerns the definition of agents which will compose the future command part of the studied manufacturing system according to needs and objectives.
The automatic passage from this phase to the specification phase remains very difficult. Consequently (ever though our specification method lightly overlaps the analysis phase), our methodology does not integrate a method of analysis which is already intrinsic to it. But it use may prove efficient, if it is situated after an analysis of needs made with E/R model, SADT or the diagram of data flows.

3.2 Specification methodology

The method is based on a systemic approach for the analysis and the decomposition of the studied systems. Indeed, we are interested in complex reactive systems which can be decomposed in three sub-systems:

- **the physical and logistic sub-system**. This is the part of the studied system composed of physical resources. According to the type of the studied system, these resources can be engines, machines, hardware systems, software systems, etc.
- **the decisional or monitoring sub-system**. It is the set of rules or functions, when applied to the physical sub-system, permits to reach the fixed goals: the decisions, the regulation, etc.
- **the information sub-system**. Its main feature is to establish the connection between the two other sub-systems. It intercept the data flows from the physical sub-system, if necessary processes them, and sends the information to the decisional sub-system (figure 3).

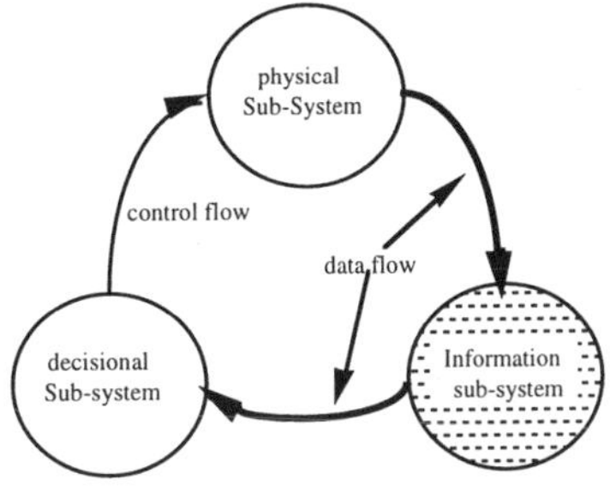

Figure 3 Complex system decomposition in basic sub-systems

The analysis and the design of a complex system can be done with respect of the two following different specification stages which are strongly coupled.

First stage: identification of the physical sub-system resources

Five steps are used:

1- Identification of the physical sub-system objects or resources. It consists to highlight the different components of the studied system. This identification concerns the available objects if the studied system is an existing system, or the resources or objects necessary to build a new system. In this step, the Entity Association model or any other simple formalism can be used for the analysis of the physical part or the static part of the system.

2- For each identified object or resource, precise its interface (cooperation and communication protocol):

- Input data flow messages,
- Input control flow signals or events,
- Output data flow messages,
- Output control flow signals or events,

3- For each object, identify, if they exist, the state variables or visible attributes. These attributes are necessary to write the control or the decisional rules. In general, these state variables are used in the system global synchronisation and monitoring rules (machine state: On, Off, Occupied, etc.).

4- For each resource decide if it is an active resource type or a passive resource type. Active resources, generally, concern an object which can evolve according to a self working (their states result from an internal working not visible at this level.) and realize one or several tasks in accordance with the received control command (robot, workstation, a software active program (server), etc.). These reactive objects interact with their environment by messages exchanges. They are complex systems as well as the studied system.

Passive resources are objects which perform a particular task, but they have not an internal logic which allows them to evolve or to perform actions in an autonomous manner (pallet, machining tools, sensor, captor, database, etc.). The distinction between these two kind of objects is of a nature to facilitate the analysis and the specification of the decisional sub-system in the following stage. At this step, it is important to have an accurate vision of the nature and the type of each component of the sub-system, because, this will determine the structuration of the decisional sub-system as it will be shown in the following stage.

5- For some cases passive resources are data storage means. They are also components of the information sub-system.

second stage: Hierarchical decomposition by level abstraction of the decisional sub-system

This stage uses, also, five steps.

1- associate a ***monitoring agent*** for each active resource. The interface of this agent will be made of input and output control and data flow of its resource (machine command and utilization protocol, software system invocation interface, etc.). The visible attributes or state variables values must be integrated in the agent interface as input or output messages. The agent is the only entity qualified to retrieve or to give the contents of these kind of attributes (encapsulation principle) to the other agents of the same or the upper levels.

2- For each passive resource or object necessary for the implementation of the decisional sub-system, associate an access ***manager agent***. Its interface have to integrate the resource access protocol massages (Database access protocol, captor or sensor access commands, etc.).

3- Identify the input:output control flows of the decisional sub-system.

4- For each input control flow (signal) or input data flow (message) of the studied system, associate an interception rewriting rule ("Handler"). This rule may implicate several monitoring agents and manager agents (synchronous rewriting rules). It can also use visible state variables of the physical sub-system via its associated agents. If an interception rule of a given event or signal have to use a complex logic in plus of the simple synchronisation and the control of the implicated monitoring agents and manager agents, it's advised to associate to this signal or event a decisional functional object which have to be decomposed in next level of the hierarchical decomposition process of the decisional sub-system. This abstract object is called ***expert agent.*** Indeed, to process such events to take a decision, a complex logic must be used. This logic can use some expertise in a particular domain (scheduling algorithms, production planing, etc.). The expert agents use their specific and private data and passive resources. Make these resources or data visible at this level is of a the nature to compromise the readiness and the comprehension of the decision logic of this level.

5- For each expert agent highlighted in the four step, precise:

- the knowledge on its environment in order to complete its interface and, above all, to identify its resources during its decomposition process.
- its expertise.

Then, apply to it this second stage of the method: hierarchical decomposition by level abstraction, only, if this agent has to be created.

third stage: top-down decomposition of the physical sub-system

This stage presents an interest only if some passive or active physical resources have to be defined. In this case, each resource of this kind identified in the first stage, must be considered as a new system to be studied and we apply to it the three stages of the method. For more details, see Attoui(95a)

3.4 Inter-agent Communications

One of the most important factors in the inter-agents communication protocols is the designation of agents implicated in this kind of interaction. To which agent a message will be sent? from which agent a message will be received?

With the variety of messages which transit between an application parallel entities and the diversity of sources, it becomes important and possible to have automatic methods of information filtering. An agent needs a fraction of messages. To have access to these messages without knowing, beforehand, their sources is not easy to make use for the designers of multi-agents system.

In this context, we have defined and made use of a mechanism based on anonymous communication by message passing. No designation of agents implicated in the communication is necessary; whatever it may be, explicit or implicit, direct or indirect (Attoui,1994b). In order to enssure communication according to message content and requested services, we have used a filtering process of messages emitted by anonymous agents. All messages are filtered and oriented towards their addressee.

The filtering is a process integrated to anonymous communication mechanism. The information filtering is strongly linked to information retrieval because they have a common objective: retrieve an information requested by an agent.

Nevertheless, there is a main difference. Information filtering is applied to an incoming data-flow whereas information retrieving is applied on an existing database which may evolve in time. In multi-agents system, the data flow is constituted of messages produced by parallel entities.

In our environment, the inter-agent communication is modelled by a channel. A channel is an abstraction of a physical communication network, so it provides a communication path between several agents. It may be seen as a communication "software bus". At the time of channel declaration, we permit only its later utilisation by the agents. The agents identities and the direction of transfers are not indicated. A channel does not have associated data types. This reduces the number of channels required if data of different types are transmitted between agents.

Figure 4 highlights the logical architecture of a multi-agent application. Each agent is bound to a channel to carry out inter-agents communication. A channel is seen by agents as a logically shared variable. A nested agent Aij can use a communication channel Cij. Two agents situated on different levels of the nested structure can use any common communication channel. For instance, Ax may communicate with Aik (k<>j) by using Ci channel. Cp channel can also be used. This principle applied is identical to the use of variables by nested procedures. Then the scope of the channel for agents is the same as the scope of variables for procedures in the classical programming languages.

Anonymous communication mechanism allows solving problems of the agents designation and the messages description. Rewriting rules enssure one of "languages acts" characteristics namely the correspondence between messages and actions to do.

In the specification level of multi-agents system behaviour, messages specified in the left part of rules represent, in reality, the profiles of these messages. No designation, explicit or implicit, of agents is specified in these profiles. Any message respecting a profile will be received by an agent independently of the sender. A complete anonymity on sender identity and a total transparency on its localisation in the communication network are ensured by the anonymous communication mechanism.

The anonymity is also applied in emission. Indeed, the right part of a rule specifies the actions to execute when conditions described in the left part are satisfied. These actions include the sending of messages.

With this anonymous communication mechanism, interrogation, request and supply messages may be exchanged without any designation of the concerned agents. More details on this macanism can be found in (Attoui,1994b), (Maouche,1994)

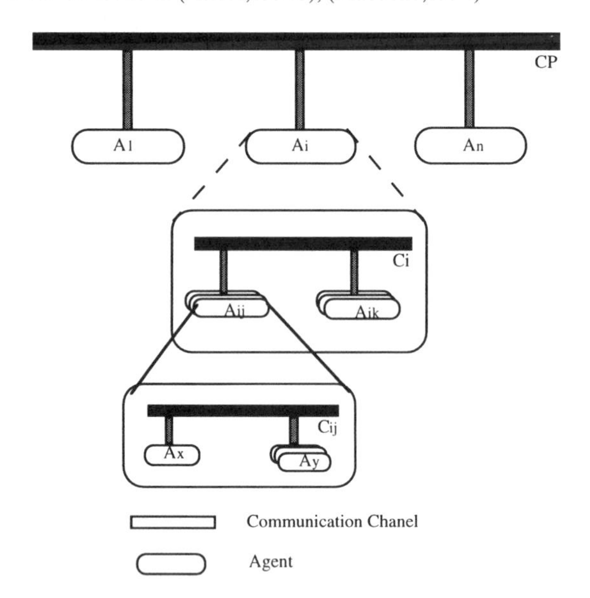

Figure 4 Hierarchical multi-agent architecture

4 Verification and validation

The inference engine of VALID environment allows the verification of the specification previously constructed with the graphical editor.

The user has to submit to the reduction engine the system description using objects and inference rules. Each test session is based on the notion of "scenario". A scenario is a foreseeable evolution of the system behavior from an initial state to a final state. (a state is defined by attribute values of the instanciated objects that constitute the real system).

The user specifies the state in which must be the system either at the beginning of the scenario (initial state) and at the end of the scenario (final state). If it is possible he can also specify several undesirable states of the system for this scenario.

The inference engine has the relevant information to perform a syntactical reduction on the specified objects. This reduction will respect the principle of locality and visible levels between objects. We remind that a message sent by an object can be intercepted by its father, its sons or its brothers in the specified hierarchy.

We can make either a global verification of the hole specified system or a partial incremental verification by respective levels. So we have to define for the engine the object that we consider as "root" (beginning object) and the scenario (initial and final state) associated to this object. This verification can detect:

- deadlock situations : the reduction process reaches an intermediate configuration not foreseen by the user and can no more evolve.
- undesirable boundless cycles : the system reduces in cyclical manner the same set of rules; so the engine reaches the maximal inference number specified by the user.
- impossibility to reach the final state indicated by the user.
- undesirable state (if specified)

This automatic detection is performed in an interactive way with the user. Each time a problem is detected, the system can show the historic of the behavior from the initial state. This historic contains the list of the explored configurations an the rules triggered since the beginning of the verification and the possible errors messages. The step by step inference mode is also available, it is very useful for a detailed supervision of a complex system reduction.

The temporal constraints in the simulation process, and especially the integration of the various temporal constraints primitives defined by the VALID syntax, produce time references in the historic file.

5 APPLICATION TO A CONCRETE INDUSTRIAL WELDING LINE

As an illustration of our approach, we consider the example of a welding line for electrical motors (figure 5). The configuration which is considered corresponds to the one of a real line installed by a French automobile manufacturer (Kellert,1990).

There is four similar welding stations for welding the motors and two types of conveyors for accessing it: a main conveyor to permit the routing of the motors along the line and several bi-directional conveyors to permit access to the welding stations (one for each station). Different workings of the line can be considered and several policies can be studied. The objective is to maximise the putting through of the line. We only give the description of a policy called “ordered policy”.

Unwelded motors must access to one of the four stations to be welded. They circulate on the main conveyor and they can take the elevator and the bi-directional conveyor leading to a station if there is space available in the queue of this station; otherwise they continue their routing on the main conveyor.

At the output of a unit, a welded motor must go down the bi-directional conveyor in order to output the line through the main conveyor. In any case, there can be only one part in an hashed section (a bi-directional conveyor and the two elevators situated at each of its extremities). An untreated motor can reach the end of the line. In this case it is recycled at the line input through the recycling conveyor. Figure 5 shows the schema of this line.

The system can be described through three levels (figure6). An agent type called "Unit" has been introduced to characterise one section of the line. A section is composed of a welding station and the corresponding means for accessing it: a section of the main conveyor and the critical section composed of the bi-directional conveyor and the elevators at its extremities.

Each level of rules represents the part of the decisional subsystem dealing with this level. This approach permits a hierarchical decomposition of the decisional subsystem as well as the physical subsystem. The locality and the encapsulation principles are naturally respected.

For instance, rule "R0" of the Welding Line level is used to routing parts from the last unit "Unit4" to the terminal Unit because the decomposition leads to a structure of the terminal unit which is different from that of the other unit. On the other hand, the rule "R3" of the Unit level is used to propagate the internal message to the upper level.

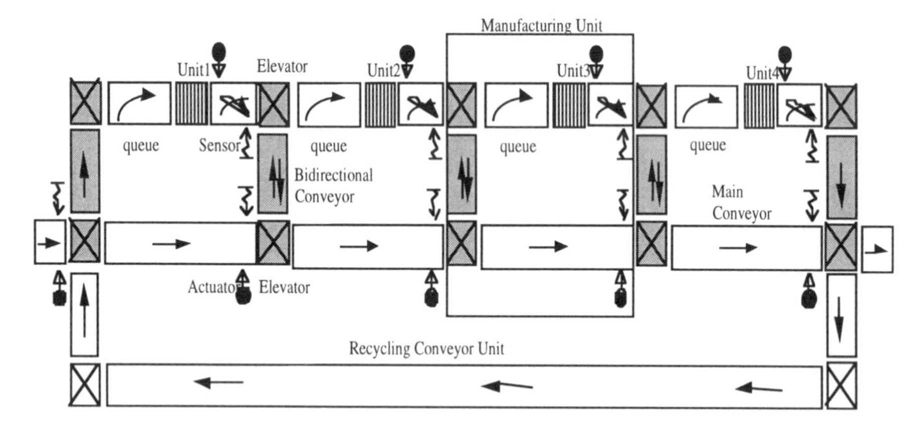

Figure 5 Schema of the Welding Line

It is important to note that when a free variable [I] is used in a message, this variable must be instanciated to all the possible values of the corresponding parameter. For instance, (Elevator [I] is free) represents both (Elevator E1 is free) and (Elevator E2 is free) messages.

A set of abort messages (Missfunctionning CS, Missfunctionning ML, ...) are used at each level. They allow agents to perform recovery operations which are the most critical operations in this kind of real time systems. Temporal constraints ([t1: (Missfunctionning CS)]) can be used to express that the execution of rules must not exceed the given deadline value, otherwise, the following message is generated. Finally, rules R0 and R1 of the Unit level specify the ordered policy mentioned above.

The figure 7 give a VALID session for the description of the first level of this system.

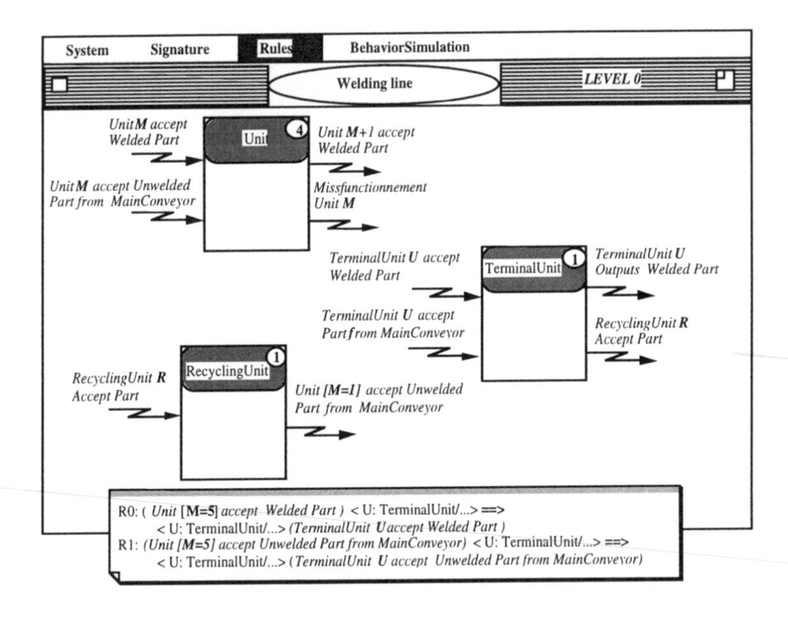

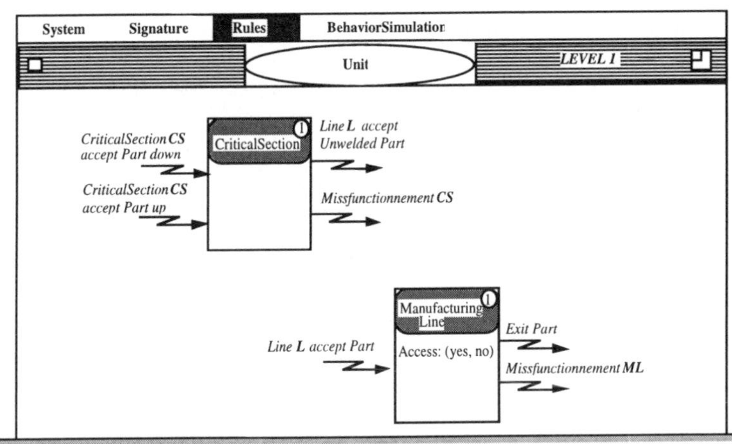

R0: (Unit [M>1] accept Welded Part) <CS: CriticalSection> ==> <CS: CriticalSection> (CriticalSection CS accept Part down)
R1: (Unit M accept Unwelded Part from MainConveyor) <CS: CriticalSection> < L: ManufacturingLine/ Access: yes:>==>
<CS: CriticalSection> < L: ManufacturingLine/...>(CriticalSection CS accept Part up)
R2: (Exit Part) <M: Unit/...> ==> <M: Unit/...> (Unit M+1 accept welded Part)
R3: (Missfunctionnement CS) <CS: CriticalSection/...><M: Unit/...>==>
<CS: CriticalSection> <M: Unit/...>(Missfunctionnement Unit M)
R4: (Missfunctionnement ML)<ML: ManufacturingLine/...><M: Unit/...>==>
<ML: ManufacturingLine/...><M: Unit/...>(Missfunctionnement Unit M)

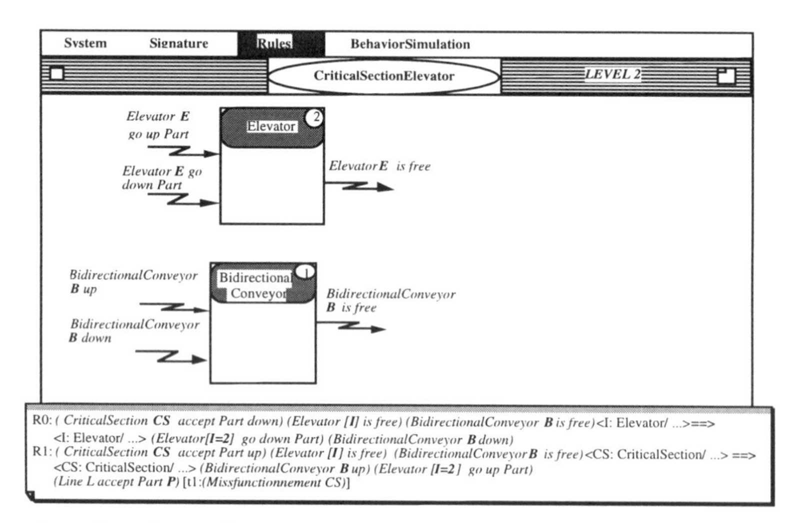

Figure 6 Description of the welding line

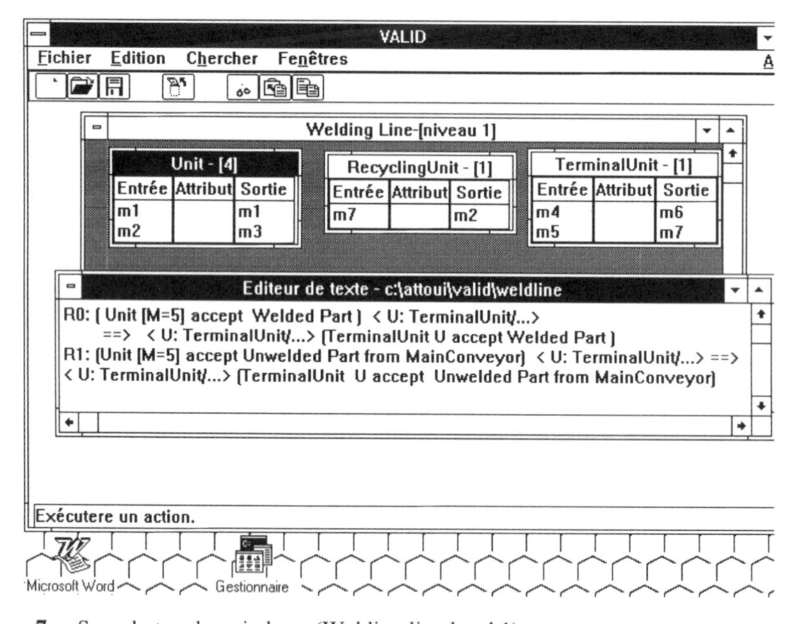

Figure 7 Snapshot under windows (Welding line level 1)

6 Conclusion

The specification and validation approach for manufacturing systems software development presented in this paper has the advantage to enhance the insight into and understanding of software requirements, helps clarify the customer's requirements by revealing or avoiding contradiction of specifications and ambiguities in the specifications, enables rigorous verification of specifications and their software implementation.

Verification of specifications would increase specification quality there by reducing life cycle costs. This approach is based on rewriting logic and multi-agent paradigm.

The development process has two main stages:

-The first one is a specification stage with a specific methodology for the analysis and the structuration of the command part of manufacturing systems, in terms of co-operative and specialized agents. The result is a conceptual model. Rewriting logic constitutes the formal framework.

-The second stage is a verification and validation stage based on a syntactical reduction engine.

After this two main stage it's possible to obtain the translation of the conceptual model into an executable model which can be directly used as a prototype. The executable model is generated into a target programming language (C++, ADA, VHDL...).

The environment for this approach is composed of a graphical editor for agents and rewriting rules description, a distributed inference engine for rules activation and a generator for the automatic translation of the formal specifications into a target programming language.

The approach and the environment have been used to develop a multi-agent application for a specific manufacturing system. We have used it for the specification of different kinds of applications and systems including manufacturing systems (Attoui,1994a) and transactional information systems (Attoui,1994b).

Although our environment constitutes a step towards a multi-agent architecture, it remains far from true multi-agent systems such as they are theoretically defined. Its evolution depends in effect on an integration of the modelisation of phenomenon like intentionality, rationality, commitment and representation of beliefs. Those phenomena have not yet attained the stage of concretisations.

References

AGHA G., HEWITT C. (1987) Concurrent Programming Using Actors: OOCP87, Yonezawa, MIT Press, .

ATTOUI A., SCHNEIDER M (1994a) Valid: An Environment Based on Rewriting Logic for the Formal Modelling of Manufacturing Systems: CIMPRO'94, Rudgers' Conference on Computer Integrated Manufacturing in the Preocess Industries, New Jersey, USA, April 25-26.

ATTOUI A., SCHNEIDER M (1994b) A Formal Approach for Prototyping Distributed Information Systems: IEEE International Workshop on Rapid System Prototyping, Grenoble, France, 21-23 June.

AYEL J. (1991) CIMES, un système d'intelligence artificielle distribuée pour la supervision en continu des activités de gestion de production: Thèse de Doctorat, Université de Savoie.

BINDING C., SARIA H., NIRSCHI H.(1991) Mixing LOTOS and SDL Specifications: FORTE'91, Sydney, 12-22 Nov.

BOURON T.(1993) Structure de communication et d'organisation pour la coopération dans un univers multi-agents: Thèse de Doctorat, Université de Paris VI, LAFORIA 93.04.

BOUZOUANE A.(1993) Un modèle multi-agent basé sur le tableau noir: application au pilotage d'une délégation d'assurances: Thèse de Doctorat, Ecole Centrale de Lyon.
BRUNO G., MARCHETTO G.(1986) Process translatable Petri nets for the rapid prototyping of process control systems", IEEE Transactions on Software Engineering, SE-12, February 1986.
BUDKOWSKI S.(1992) Estelle Development Tooltest (EDT): Computer Network and ISDN Systems, Special Issues on FDT Concepts and Tools, Vol.25, N°1.
BUSTTARD D.W., NORRIS M.T., ORR R.A., WINSTANLEY A.C.(1992) An Exercise in Formalizing the Description of Concurrent Systems: Software Practice & Experience, Vol 22, N° 12, Dec.
CARDOZO E.(1993) Using the object model to ilplement multi-agent systems: IEEE International Conference, Boston.
COELHO DA COSTA R.J., COURTIAT J.P.(1992) A True Concurrency Semantics for LOTOS: FORTE'92, Lannion (France), 13-16 Oct.
COURTIAT J.P., DIAZ M., MAZZOLA V.B., DE SAQUI-SANNES A.(1991) Description formelle de protocoles et de services OSI en Estelle et Estelle*- Expérience et méthodologie: CFIP' 91.
SIC 5TIMC/IMAG, Grenoble (1992) Modèles de connaissances et systèmes multi-agents: Journée Systèmes Multi-Agents du PRC-IA, 18 Déc. 1992, Nancy.
KELLERT P, FORCE C. (1992) Knowledge Model Building of Manufacturing Systems with SADT: 8th International Conference on CAD/CAM Robotics and Factories of the Future, Metz (France), 17-19 Août.
ISO 9074 (1989) Information Processing systems- OSI Estelle: a Formal Description Technique Based on an Extended State Transition Model.
FAYAD M. E.(1994) Objects Modeling Technique (OMT): Experience report: Journal of Object-Oriented Programming, Nov-Dec.
FRASER Martin D.(1994) Strategies for Incorporating Formal Specification in Software Development: communications of the ACM, N°10, Vol.37, October.
LAASRI H., MAITRE B.(1989) Coopération dans un univers multi-agents basée sur le modèle du blackboard: étude et réalisation: thèse de Doctorat, Université de Nancy 1.
LIGHTFOOT DAVID (1991) Formal Specification Using Z': The Macmillan Press.
IS 8807 (1988) LOTOS, a formal description technique based on the temporal ordering of observational behavior, December 88.
MAOUCHE A. and ATTOUI A.(1994) A programming environment for distributed applications: Proceedings of the 5th international training equipment conference and exibition, The Hague, The Netherlands, April 26-28.
MESEGUER J.,(1990) A Logical Theory of Concurrent Objects: Concur 90 Conference, Springer Verlag, Amsterdam, August 1990.
OCCELLO M.(1993) Blackboards distribués et parallèles: application au contrôle de systèmes dynamiques en robotique et en informatique musicale: Thèse de Doctorat, Université de Nice, Sophia-Antipolis 93.01.
PETERSON J.L.(1981) Petri Nets Theory and the Modelling of Systems: Prentice-Hall, Englewood Cliffs NJ.
SIJELMASSI R. and STRAUSSER B. (1991) NIST Integrated Tool Set For Estelle: Formal Description Techniques, Quemada (ed), North-Holland.
STINCKWICH S. (1993) Modèle et environnement objet dédié aux systèmes milti-agents: Premières journées IAD & SMA, Toulouse, Avril.
VIGDER M. (1991) Using LOTOS in a Design Environment: Proceedings FORTE'91, Sydney, 12-22 Nov.

11

Interflow systems for manufacturing: concepts and a construction

J. Goossenaerts and D. Bjørner
United Nations University, International Institute of Software Technology
P.O. Box 3058, Macau

Abstract

A conceptual framework for modelling, simulation and control in manufacturing industry is proposed. The framework builds on three organizing principles: a *decomposition principle*, the *life cycle principle*, and the *interflow principle*. The links of these principles with the provision of information and command infrastructure services for manufacturing enterprises are indicated. A terminology is proposed for the construction of interflow models of manufacturing enterprises and industry and a generic plant interflow model is constructed. Issues such as the particularization of models and the interfacing of enterprise models to manufacturing phenomena are also touched upon.

Keywords

Enterprise models, industry models, model execution services, formal methods

1 INTRODUCTION

A *manufacturing industry information and command infrastructure* (abbreviated: MI^2CI) is an information processing and activity control and monitoring system which provides information and control services in support of the operations and projects of manufacturing enterprises. Sophisticated such services may enable industries in moving in a relatively short period of time from an early stage of development towards a lean/agile supply-based sustainable system. An industrial system is lean when it is capable of achieving results without using superfluous resources (e.g., equipment, workers, investments, stock), it is agile when it is capable of deciding quickly and intelligently and accordingly adapting operations quickly and easily. An industry is called supply-based when a large number of the parts in products are procured from (a large number of) suppliers. An industry is sustainable when products are designed, produced, distributed and disposed with minimal (or none) environmental and occupational health damages, and with minimal use of resources (materials and energy) (Alting and Jorgensen, 1993).

Information and command infrastructure services that rely on model execution services and interflow with manufacturing phenomena, are expected to become the enabling technology for lean operations and agile projects of extended enterprises. Without an integrated problem domain understanding, and without inter-operable services built on such, non-productive concerns threathen to inflate costs and complexity due to the increasing number of specialized enterprises and other actors involved in engineering projects and manufacturing operations. A rigourous understanding of the interactive and synchronized dynamics which is possible between computational flows on networked computers and the life cycles of products, plants, enterprises, extended enterprises and markets, is instrumental for the definition and implementation of information and command infrastructure services and for integrated manufacturing systems engineering.

Overview and relationships to other work

Adequate tools and services in support of integrated manufacturing systems engineering should build on a proper conceptual understanding of manufacturing industry. Such conceptual understanding should be cast in a formal framework which affords construction, computation and formal reasoning.

The *decomposition principle* and *life cycle principle* proposed in section 2 complement the concepts of *generic*, *partial* and *particular* model consolidated in ENV 40 003 (1989). The decomposition principle has been elicited from the discussion about the relationship between an enterprise model (catering for operations) and the full life history of an enterprise (in which also improvements and innovations must be considered) (Williams, 1993); and that about the relationship between enterprise models and environment models (Goossenaerts, 1993). Decomposition is a device for classifying the wide range of concepts that must be catered for in enterprise and industry models. It reflects a manufacturing domain understanding which draws on those of CIM-OSA (Vernadat, 1993), ATLAS (ATLAS, 1993) and MS^2O (Matthiesen, 1994), all EU ESPRIT Projects. The *life cycle principle* reflects the concern with life cycle engineering in manufacturing. Indeed whether an industrial system is sustainable or not depends on the extent to which the life cycles of plants, products and materials affect the environment, during their creation, their operation or use, and after their disposal.

Section 3 proposes the *interflow principle* and links it to commonly used terms in enterprise modelling. The principle draws our attention to the interface between enterprise model driven computational flows and shop floor and engineering office activities. An interflow system is a system in which computational flow is kept in step with activities and changes in an application domain. The generic plant interflow model constructed in section 4 synthesizes the generic properties of plant systems into a single coherent system, and it articulates the distinction between computational and physical activities. The selection of the primitive templates for the bottom-up construction was made on the basis of work by Scheer (1989) and Atsumi (1993). The construction focuses exclusively on operations, as separated from innovations and improvements. A RSL specification (The RAISE Language Group, 1992) of the construction can be found as an annex to the report by Goossenaerts & Bjørner (1994).

2 INFORMATION & COMMAND SERVICES FOR MANUFACTURING INDUSTRY

Information and command infrastructure services for manufacturing industry have to be conceived in reference to full *plant life cycles* and full *product life cycles* seen in the context of extended enterprises embedded in business environments. Below the concepts of product and plant life cycle are defined and a decomposition of manufacturing industry is proposed. The latter is referred to in a summary of the information and command infrastructure services for manufacturing industry.

Plant and Product Life Cycles in the Business Environment

The *life cycle of a product* is concerned with: (a) Its *creation* in a (virtual) plant. (b) Its *usage* by consumers on the market. Usage of the product may include maintenance, repair, upgrading, etc. (c) Its *decomposition* – preferably with recycling or reuse of its composing materials – in a (virtual) plant.

The *life cycle of a plant* is concerned with: (a) *Projects* which phase *a plant and its products* into *a market*: the gradual introduction of a new plant and its products into a market (and environment). (b) *Operations*: a plant will respond to *orders* issued by customers, by exploding the orders into *plant programme steps*, by scheduling and carrying out the (production) steps, and by delivering the goods ordered. *Projects* during operations may concern: how to phase in&out *products*, *equipment* and *human resources* into a plant, and how to improve *plant operations*. (c) *Projects* which phase out *a plant and its products*: the gradual withdrawal from use of an old plant and its old products.

The Orthogonal Systems of Manufacturing Industry

The focus on (collaborative) labour in the product and plant life cycles and the requirement of orthogonality for the systems to describe these suggest: First, to make, in an enterprise, a distinction between "repetitive" *operations*, to be catered for in a *plant model*; and "one-of-a-kind" *improvements* and *innovations* to be catered for in *projects* carried out by *teams*. Secondly, when extending the viewpoint beyond the boundary of the enterprise to let *extended-enterprise operations* be carried out by so-called *virtual plants*, and to let *extended-enterprise projects* be carried out by *virtual teams*. A virtual enterprise (or extended enterprise (Browne *et al.*, 1994)) comprises a virtual team and a virtual plant. And thirdly, to consider *markets* as the contexts within which enterprises are embedded, and to use the term *industry* to denote this context when considering also the projects and virtual projects executed in the market (e.g. industrial policy projects).

The following domains (Table 1) are proposed to organize the objects, processes and issues in industry (see also Goossenaerts & Bjørner (1994)): (a) A *team* phases in and out a plant or a product in response to *goals* formulated by entrepreneurs or other agents of change. A team refines and enriches the goals into *project programmes* for changing or making a (virtual) plant, it schedules and carries out these programmes, finally, it delivers the new or improved plant, capable of producing goods, with the performance expressed in the goals. (b) A *plant* responds to *orders* issued by customers, by exploding the orders into *plant programme steps*, by scheduling and carrying out the (production) steps, and by delivering the goods ordered. (c) A *virtual plant* comprises several plants which synchronize their operations as if they were carried out by a single plant. (d) A *virtual team* comprises several teams which synchronize their work and decisions as if they would carry them out as a team within a single enterprise. (e) A *market* forms the context for the operations of plants and virtual plants. Market regulations (e.g. such as in Company Law and Tax Law) affect the operations of plants and virtual plants. (f) An *industry* forms the context for the projects of (virtual) teams. Regulations (e.g. such as in Environmental and Technical Regulations and international standards) affect projects.

The desirable conditions of operations (leanness) and projects (agility) must be realized in different contexts: in the – extended – enterprise and in the industry.

Projects are further classified. *Entrepreneurial Projects* deal with changes of enterprise level programmes during the enterprise life cycle (e.g. the introduction of new products, business process reengineering, investment appraisal). *Engineering Projects* are concerned with product life cycle and production processes. Usually they require innovation. *Quality Improvement Projects* are concerned with product and production process improvements. *Industrial Policy Projects* pertain to the industry. Virtual plants result from the successful implementation of projects by virtual teams. These require inter-organisational engineering – in the extended enterprise – and a (standardized) technology infrastructure.

Table 1 Domains for organizing objects, processes and issues in industry

activities / *context*	*operations* "repetitive"	*projects* "one-of-a-kind"
enterprise	plant / production	team / innovation
extended enterprise	virtual plant	virtual team
environment	market / trade	industry / (policy) innovation
desirable condition	leanness	agility

Information and Command Infrastructure Services
The provision of information and command infrastructure services requires (computer network supported) enterprise model execution services. See CEN/TC310/WG1 (1994a) for a statement of requirements for such services and CEN/TC310/WG1 (1994b) for an evaluation of some related initiatives. As to design services, relevant requirements can be sourced from the literature on computer aided design and project support environments.

Subsystems of an information and command infrastructure for enterprises and industry are classified in accordance with the domain decomposition in Table 1 (IMES stands for Interflow Model Execution Services and ISDACS stands for Interflow System Design and Construction Services). An enterprise information and command infrastructure comprises:
Plant IMES: a plant component which supports the repetitive operations of the enterprise (cf. enterprise wide information systems built around enterprise-wide data models (Scheer, 1989)).
Plant ISDACS: a team component which supports the innovations and improvements which the enterprise needs to compete in the market. Application packages in a Plant ISDACS may include an *entrepreneuring project coach*, an *engineering project coach* and a *quality project coach*.

An industry information and command infrastructure comprises:
Market IMES: a market component which supports the repetitive operations at the market. Application packages in a Market IMES may include *material flow monitors*, *employment monitors*, *(value added) tax collection systems*, *trade and transportation systems*, *an intellectual property monitor*, etc.
Market ISDACS: an industry (team) component which supports the one-of-a-kind (industrial policy) innovations and improvements which the industry needs to sustain its competitiveness in the global market. Application packages in a Market ISDACS may include an *industrial policy project coach*.

3 INTERFLOW MODELS FOR MANUFACTURING INDUSTRY

Considered are the interplay of phenomena flows and computational flows; assumptions and a basic terminology for interflow models; the use of enterprise (reference) models as tools for organizing and integrating information about enterprise processes and industry, with attention for generic models and how particularizations can be accommodated.

The Interplay of Phenomena Flows and Computational Flows
The division of labour – and the productivity gains and accelerated innovation enabled by it – has made manufacturing industry increasingly dependent on computational flows to support operations and projects. A manufacturing system has become a pair of a phenomena flow – on the shop floor and in the engineering office – and a computational flow. The former can

be described by mathematical models, the latter is defined over the mathematical model and implemented by some computing system. Both flows are linked by multiple interface channels which enable synchronized dynamic behaviour.

A better understanding of the interplay of computational flows and business/manufacturing processes is useful for a number of reasons. Costs and performance in manufacturing inherently depend on physical world phenomena. Consider for instance the case of a "hand over" – the passing on of a job to a fellow worker – in business processes (see Hammer & Champy (1993) for elaborate illustrations). Criteria as to whether or not a hand over (and a specialized work) is required are contingent on phenomena flow conditions and resources (both for material transformation and computation). But resulting decisions should be consolidated in an enterprise model, to scrutinize them as to their consistency with other decisions, to propagate changes due to new decisions, to support the automatic derivation of programmes for implementing decisions, etc. By focussing on the division of activities between phenomena flow and computational flow and by modelling the enterprise, one also can swiftly compare business and manufacturing processes which are more or less automated and support decisions about whether or not to model objects and activities in the enterprise (for instance prior to automation or reengineering) or in the business environment.

Assumptions and Terminology: Interflow Models
Mathematical models are built for a selection of objects, concepts, and changes in these, as designated and measured in an application domain. The utilization of mathematical models for modelling an application domain rests on two assumptions:

The **two worlds assumption**: We deal with two worlds: one created by means of mathematical symbols, and another one, the *physical system* (application domain) catering for the physical space & time, with the physical objects having life cycles in it.

The **measurement assumption**: There exists methods (*measurements* or *observations*) of assigning mathematical symbols to particular states or conditions of phenomena in the physical system. It is possible to model properties of phenomena by means of mathematical constructs.

These assumptions are of fundamental importance in physical science as succinctly expressed by Hertz (1894). They are also referred to in the General Design Theory (Yoshikawa, 1981).

Some terms for phenomena flows and computational flows: (1) An *application domain* is a part of the world to which we restrict our attention with the purpose to build a mathematical model, for the derivation of properties, for simulation or interflow. (2) *Phenomena Flow* over an application domain is the amalgam of observable changes of physical objects in the course of time and in a designated environment (space). An *object* can be a building, a part, a tool, a worker, a work sheet, etc., and *change* may result from work by man or machine (in which case we use the term *activity*), from the laws of physics, chemistry or biology, etc. (3) The *designation* of an application domain is by means of a *space-time-resource filter* which selects objects and changes and their properties (concepts for which to execute measurements, changes and the objects involved, the precision of measurements, the intervals between measurements) for consideration in the model. (4) A *template* is a formally defined collection of data which is used to reproduce a same thing (typically a representation of an object or an activity) many times. It comprises two orthogonal kinds of data. *Attributes* contain data about the *intrinsic* properties of the template's instances. *Linkages* contain data related to the *extrinsic* or *accidental* aspects of the template's instances. (4.a) Illustration. In manufacturing industry, geometric properties and material of a part are among its intrinsic properties. Extrinsic aspects of a part include the parent parts in which a part can be assembled, the activities in which it can be transformed, as well as the positions (on the shop floor) where a part can be found during the manufacturing operations.

Another example: an extrinsic property of an instance of an activity template, is the part on which it is invoked. (4.b) Intrinsic and extrinsic properties of templates are particularized in accordance with a *space-time-resource filter* which determines a *space-time interval* and a space-time-concept granularity. (4.c) In the constructions in section 4, the particularization attribute "τ" indicates an abstraction of the descriptions in templates as regards their intrinsic properties (granularity, formality, and focus (completeness)). When particularizing a particular model from a generic model, it is recommended when introducing, enriching or refining the templates of the particular model, to develop granularity, formality, and focus simultaneously. I.e. it is recommended that for instance geometric attributes (assumed under τ in the generic system), are simultaneously taken into consideration for all templates for which they are relevant in the particular system. (5) A mathematical model *models* ($\boxed{\models}$) an application domain if and only if it supports the derivation of (observable) properties of the corresponding phenomena. The mathematical model is called a *model* of the domain. (5.a) Illustration: The schematic of a house supports the derivation of properties (e.g. dimensions of rooms) of the house: schematic $\boxed{\models}$ house (one or more tangible buildings). (6) A computational flow over a mathematical model *simulates* ($\boxed{\wr\models}$) a phenomena flow in an application domain if and only if successive states of the computational flow describe successive states of affairs of the phenomena flow. Measurement methods are used to validate or deny "description". Successive states of the computational flow are mutually linked by transitions (symbol manipulation, computation). There is no interaction nor synchronization between the computational flow and the phenomena flow. The computational flow is called the *simulation model* of the phenomena flow.

The utilization of computational flows for interflowing a phenomena flow rests on one further assumption. The **interface assumption**: it is possible to establish *interface channels* which link particular objects and changes in a phenomena flow to particular instances in a computational flow. Some terms for interfaces: (7) An *interface* is the area between a phenomena flow and a computational flow , it consists of a number of *interface channels.* (8) An *interface channel* is a link between a phenomena flow - some objects and changes in it - and instances in a computational flow. Over a channel *tasks* and *signals* are transmitted. They are the elementary units of information exchange and synchronization. Tasks and signals which the computational flow receives are called *inputs*, those which it sends are called *outputs.* (9) A computational flow *interflows* ($\boxed{\aleph\models}$) a phenomena flow if and only if its behaviour simulates the phenomena flow and if it synchronizes with it along one or more interface channels. The computational flow and the phenomena flow proceed *synchronously* and *inter-actively* with tasks and signals flowing (simultaneously) along *interface channels.* The computational flow over the mathematical model is called the *interflow model* of the phenomena flow in the application domain. (10) An *interflow system* comprises the joint behaviour of the computational flow and the phenomena flow.

Enterprise and Industry Models; From Generic to Particular

The use of enterprise models and reference models as tools for organizing and integrating information about enterprise processes is well established. See for instance ENV 40 003 (1990), Scheer (1989), CIMOSA (1993), Spur *et al.* (1994). Goossenaerts & Bjørner (1994) introduce the related concept of industry model.

Enterprise Modelling

The distinction in the enterprise between repetitive operations and one-of-a-kind projects allows one to split an enterprise model into two components: a plant model and a team model. These models are related to each other and to phenomena flow. Projects, supported by the team model in execution, transform the plant model and the physical plant. Operations, controlled and supported by the plant model in execution, transform material inputs and produce the products.

Enterprise modelling is concerned with the construction of generic, partial or particular (as defined in ENV 40 003 (1990)) plant models and team models and their integration.

The Construction of a Generic Plant Interflow Model.
A construction of a generic plant interflow model is summarized in Table 2. The first three steps deal with statical structural properties of plant systems. The fourth step allows one to define computational flows over *instances* of the templates (cf. types) in a plant structure. Computational flows involving these instances can simulate plant operations. The fifth step deals with the interfaces between the computational flows and plant operations. Some remarks to justify the focus of the five steps in the construction: (**1**) Materials and the work to transform them are described *statically* by means of *part templates* describing units of material and *work templates* describing units of change to material. A *Part-Work Structure* integrates the information in a bill of materials and in process charts and describes the parts and works required for making products. (**2**) Cells and the orders they send and receive incorporate the division of labour and the coordination of results. *Cell Templates* describe cell(s), a cell is a unit capable of sustaining activities/work. Cf. resource model. *Order Templates* describe orders. An order is sent by one cell to another to request the delivery or transformation of a part. A *Cell-Order Structure* integrates the properties described in organization charts and in information flow charts. It is concerned with the division of labour and coordination of results. (**3**) A *Plant Structure* results from joining a part-work structure and a cell-order structure. It integrates the statical properties of a plant. (**4**) *Plant System.* Instances of cell templates and their responses to order instances can simulate the operations at a shop floor. The term *pulse* denotes the response by a cell instance to an order instance it receives. Cf. the order handling process. The term *flow* denotes the integration of pulses. A *plant system* controlled by a model execution system and interfaced to a suitable discrete event generator, can simulate shop floor operations. (**5**) *Plant Interflow Model. Channel templates* describing interface channels are added to the plant system. Interface channels are used to interface a computational flow to phenomena flow on a shop floor. They support synchronization through the input and output of particular tasks and signals.

Table 2 The bottom-up construction of a Plant Interflow Model

1	*Part-Work Structure* (*Part Template* $\models$ part(s), *Work Template* $\models$ work(s))
2	*Cell-Order Structure* (*Cell Template* $\models$ cell(s), *Order Template* $\models$ order(s))
3	*Plant Structure* $\models$ plant(s)
4	*Plant System* $\wr\models$ plant operations (*Pulse* $\wr\models$ response to order instance, *Flow* $\wr\models$ operations)
5	*Plant Interflow Model* $\aleph\models$ plant operations (*Channel Template* $\models$ interface channel) *Plant Interflow System, sync(hronize) cycle*

Team Interflow Models
A team working in an (virtual) enterprise aims for improvements or innovations in a (virtual) plant interflow system. In pursuit of a *goal* – the sequence of activities towards its accomplishment has not been secured yet – a team will first – during the design&drafting phase – transform or particularize a plant interflow model (templates, plant structure and plant system consolidate the deliverables during this phase) and next – during the construction phase – construct or

upgrade the plant interflow system, by transforming the physical plant and its computational flow. In the course of a project, the composition of the team is adapted according to changing project activities. In contrast, the cell structure of a (traditional) plant remains static during operations.

Team interflow models in execution support team activities. A generic team interflow model forms a basis for the implementation of an *Integrated Project Support Environment.*

Industry Modelling

The contemporary market is governed by a large and changing collection of regulations and constraints for the operations of plants, producers and consumers. Coping with the regulations and their changes requires high costs. More integrated and harmonized industry models and the corresponding market-wide information and command infrastructures, and proper model-driven interfaces between their enterprise infrastructure and the industry infrastructure are desirable.

Industry Modelling is concerned with the construction of generic, partial or particular market interflow models or industry interflow models. Such models enhance the understanding, coordination and harmonization of rules and other market information.

A Generic Market Interflow Model

The construction of a generic market (interflow) model parallels the construction of a generic plant (interflow) model. A significant difference is that the cell-order structure is replaced by a person-contract structure. Cells (in plants) communicate with each other in a master-slave relation, client-to-server, whereas legal persons (persons, companies and public bodies) in a market communicate (also) as equals, peer-to-peer. *Contracts* commit two or more legal persons to carry out, during a certain period, a number of exchanges of products or services (including labour, money, etc.). They may express terms of synchronization, obligations, etc.; e.g. to incorporate a company a suitable "legal person template" must be selected, this will determine reporting obligations, and – to some extent – the contracts the company can enter into. The legal personality and contracts may also be correlated with the products/services which the enterprise can provide on the market (e.g. banks must be registered in a special commission).

Templates of all legal persons and contracts that exist in a market are joined in a *person-contract structure.* This structure includes person templates for the public bodies that – because of statutory regulations – are involved in the registration and monitoring of (certain) contracts and exchanges. It also includes contract templates such as for incorporation contracts, labour contracts, etc. (Goossenaerts & Bjørner, 1994).

Industry Interflow Models

Industry-wide projects such as industrial policy and legislative projects aim at transformations of a market, industry or sector as a whole. Such transformations may be planned (or designed) in terms of particular market interflow models, prior to their implementation.

Our approach suggests to consolidate the deliverables of industry-wide projects as templates of products, exchanges, persons and contracts, (in) a market structure, (in) a market system, (in) a particular market interflow model, and eventually, as (in) a market interflow system.

Models and their Execution Services

Particular (virtual) plant models, (virtual) team models, market models and industry models will drive the information and command infrastructure services for manufacturing enterprises and industry.

Generic models, and the particularization operators for them, should ensure that particular models can easily be developed, and that they can be executed on commercial computer and communications hardware, with interfaces to shop floor and engineering office resources.

4 CONSTRUCTION OF A GENERIC PLANT INTERFLOW MODEL

A factory is the application domain. The space-time-resource filter selects: Space: the shop floor and the engineering offices of the factory; Time: the hours of operation of the factory; Resources: materials and parts, machines, workers and related concepts. The statical model of the factory should cater for bill of materials, process charts and resource models. The dynamics of a simulation model should be related to order handling (incoming orders require creation of new variables and transitions), work in progress records, production plans, work schedules, etc.

The construction which is summarized in Table 2 is explained.

4.1 Part-Work Structure

Terminology

1. A *Part* is one of the things that make up another thing. The concept of part includes: *end product*: a part which receives no further processing within the firm; *assembly*: a part which is constructed from other parts and used in the construction of further parts; and *component (material)*: a part that is not produced in the firm, but bought-in, including raw materials.

2. *(i) Work* is any activity that transforms a part. The concept of work includes: *transporting*: moving a part without changing its properties; *assembling*: constructing a part from other parts; *transforming*:adding or removing features to or from a part; *purchasing* and *selling*.

Notations

1. Π is the universe of part templates; Γ is the universe of work templates; $\subset$ or $\subseteq$ denote sub-set relationship; $\in$ denotes element relationship; ϵ denotes instance relationship; Γ^S denotes the powerset of Γ; Γ^ϵ $(\mathbf{G}^\epsilon)$ denotes the set of instances of elements of Γ $(\mathbf{G})$; $\emptyset$ denotes the empty set $\{\ \}$; $\mathbf{P}, \mathbf{Q}, \mathbf{R}, \ldots \subseteq \Pi$ and $\mathbf{G}, \mathbf{H}, \mathbf{I}, \ldots \subseteq \Gamma$; $\mathbf{p},\mathbf{q},\mathbf{r}, \ldots \in \Pi$ and $\mathbf{g}, \mathbf{h}, \mathbf{i}, \ldots \in \Gamma$; $p_1, p_2, p_3, \cdots \epsilon\ \mathbf{p}$ (p_1 $\boxed{\models}$ a particular physical part, $\mathbf{p}$ $\boxed{\models}$ a class of physical parts) and $g_1, g_2, g_3, \cdots \epsilon\ \mathbf{g}$.

2. The following functions are referred to in the definitions of Part and Work Templates:

(i) creation_works: κ: $\Pi \to \Gamma^S : \mathbf{p} \mapsto \kappa\ (\mathbf{p})$ (works that create a **p**-instance)
(ii) affecting_works: α: $\Pi \to \Gamma^S : \mathbf{p} \mapsto \alpha\ (\mathbf{p})$ (works that affect a **p**-instance)
(iii) usage_works: υ: $\Pi \to \Gamma^S : \mathbf{p} \mapsto \upsilon(\mathbf{p})$ (works that use a **p**-instance)
(iv) is_component: **p** is a *component iff* $\kappa(\mathbf{p}) = \emptyset$ (no work to create a **p** instance)
(v) is_product: **p** is a *product iff* $\upsilon(\mathbf{p}) = \emptyset$ (no work uses a **p** instance)
(vi) input_parts: ι: $\Gamma \to \Pi^S : \mathbf{g} \mapsto \iota(\mathbf{g})$
(vii) affected_parts: β: $\Gamma \to \Pi^S : \mathbf{g} \mapsto \beta(\mathbf{g})$
(viii) output_parts: ω: $\Gamma \to \Pi^S$: $\mathbf{g} \mapsto \omega(\mathbf{g})$
(ix) is_consumer: **g** is a *consumer iff* $\omega(\mathbf{g}) = \emptyset$ (a **g** activity has no output)
(x) is_producer: **g** is a *producer iff* $\iota(\mathbf{g}) = \emptyset$ (a **g** activity has no input)

Definitions

1. A *Part Template* describes (all relevant) attributes of a single part (for production, design, etc.). It is 4-tuple: $(\tau, \kappa : \{\mathbf{g}_1, \ldots, \mathbf{g}_m\}, \alpha : \{\mathbf{g}_{m+1}, \ldots, \mathbf{g}_{m+n}\}, \upsilon : \{\mathbf{g}_{m+n+1}, \ldots, \mathbf{g}_{m+n+o}\})$.

2. A *Work Template* contains (all relevant) attributes of a single work.
It is 4-tuple: $(\tau, \iota : \{\mathbf{p}_1, \ldots, \mathbf{p}_k\}, \beta : \{\mathbf{p}_{k+1}, \ldots, \mathbf{p}_{k+l}\}, \omega : \{\mathbf{p}_{k+l+1}, \ldots, \mathbf{p}_{k+l+m}\})$.

3. Let $\mathbf{P} \subseteq \Pi$ and $\mathbf{G} \subseteq \Gamma$. The pair ($\mathbf{P}$, $\mathbf{G}$) is a *Part-Work Structure iff (i)* for all $\mathbf{p} \in \mathbf{P}$: $\kappa(\mathbf{p}) \subset \mathbf{G}$, $\alpha(\mathbf{p}) \subset \mathbf{G}$ and $\upsilon(\mathbf{p}) \subset \mathbf{G}$; *(ii)* for all $\mathbf{g} \in \mathbf{G}$:$\iota(\mathbf{g}) \subset \mathbf{P}$, $\beta(\mathbf{g}) \subset \mathbf{P}$ and $\omega(\mathbf{g}) \subset \mathbf{P}$.
4. On a Part-Work Structure ($\mathbf{P}$, $\mathbf{G}$) we define: *(i)* The parents and childs of a part (work): $\mathbf{p}$, $\mathbf{q} \in \mathbf{P}$. $\mathbf{p}$ is *parent part* of $\mathbf{q}$ and $\mathbf{q}$ is *child part* of $\mathbf{p}$ iff there is $\mathbf{g} \in \mathbf{G}$ such that $\mathbf{q} \in \iota(\mathbf{g})$ and $\mathbf{p} \in \omega(\mathbf{g})$. *(ii) Parent/Child Generations* (generations 1, 2, ...). *(iii) Ancestry* (of a part): union of all parent generations (parts in which the part can be used). *(iv) Usage Trace*: a sequence of work templates connecting a part to a parent. *(v) Usage Bundle*: the union of all usage traces. *(vi) Posterity* (of a part) : union of all child generations (parts used to create the part). *(vii) Creation Trace*: a sequence of work templates connecting a child to a part. *(viii) Creation Bundle*: the union of all creation traces.
5. Given a Part-Work Structure ($\mathbf{P}$, $\mathbf{G}$) then we can construct a *place/transition tree* for each *product* ($\mathbf{p}$) in $\mathbf{P}$: take a *place* for each part in the *posterity* of $\mathbf{p}$, and a *transition* for each work in it. A transition fires (and marks its parent) iff all its childs are marked.

4.2 Cell-Order Structure

Given a part-work structure, then the important activities in phenomena flow are *delivery, transformation and absorption of parts* (instances of part templates) and *execution of works* (instances of work templates). A cell is a system capable of sustaining the delivery. transformation and absorption of parts and/or the execution of works (transforming information or material in response to orders). Cells can be classified according to the works they can sustain. These then determine the parts (the inputs of the works) that the cell has to acquire and the parts that the cell can deliver or distribute (output parts). For the latter parts – and in a system of autonomous cells – the cell can receive orders to which it responds by sending orders for child parts – to other cells –, and after delivery of the child parts, by executing the appropriate works.

Terminology
1. A *cell* is a system capable of sustaining: *(i)* the execution of a number of activities. *(ii)* the receipt of the input parts for its activities; *(iii)* the sending of orders for input parts which it needs for its activities; *(iv)* the receipt of orders for the parts it can produce; and *(v)* the delivery to the order sender of the parts which it creates (in response to orders).
2. An *order* is a document that is sent by one cell (the sender) to another cell (the receiver). An order is sent to ask the receiver to deliver the ordered part to the sender of the order.

Remarks
1. The definition of cell and order requires reference both to the mathematical system (a cell's capabilities relate to the division of labour and the subsequent coordination of results) and to phenomena flow. In a sense we can say that an order thanks its existence to the division of labour.
2. Atsumi (1993) gives an elaborate introduction of the *order cell* concept and its link to the division of labour.
3. Cells can be classified according to the works they sustain. These then determine the parts (the inputs of the works) that the cell acquires and the parts that the cell delivers or distributes (output parts). For the latter parts – and in a system of autonomous cells – the cell can receive orders to which it responds by sending orders for child parts – to other cells –, and after delivery of the childs, by executing the appropriate works.

Notations

1. Σ is the universe of cell templates; Ω is the universe of order templates; $\mathbf{C}, \ldots \subseteq \Sigma$; $\mathbf{O}, \ldots \subseteq \Omega$; $\mathbf{c}, \ldots \in \Sigma$; $\mathbf{o}, \ldots \in \Omega$; $c_1, c_2, \ldots \in \mathbf{c}$; $o_1, o_2, \ldots \in \mathbf{o}$.
2. Some functions (relations) on Σ, Π and Γ:
   (i) capability : $cap : \Sigma \to \Gamma^S : \mathbf{c} \mapsto cap(\mathbf{c})$
   (ii) inputs_to_cell_works : $\iota^c : \Sigma \to \Pi^S : \mathbf{c} \mapsto \iota^c(\mathbf{c}) \stackrel{\text{def}}{=} \bigcup_{\mathbf{g} \in cap(\mathbf{c})} \iota(\mathbf{g})$
   (iii) affected_by_cell_works : $\beta^c : \Sigma \to \Pi^S : \mathbf{c} \mapsto \beta^c(\mathbf{c}) \stackrel{\text{def}}{=} \bigcup_{\mathbf{g} \in cap(\mathbf{c})} \beta(\mathbf{g})$
   (iv) outputs_of_cell_works : $\omega^c : \Sigma \to \Pi^S : \mathbf{c} \mapsto \omega^c(\mathbf{c}) \stackrel{\text{def}}{=} \bigcup_{\mathbf{g} \in cap(\mathbf{c})} \omega(\mathbf{g})$
   (v) materiality: $mat : \Sigma \to \Pi^S : \mathbf{c} \mapsto mat(\mathbf{c}) \stackrel{\text{def}}{=} \omega^c(\mathbf{c}) \cup \iota^c(\mathbf{c}) \cup \beta^c(\mathbf{c})$
   (vi) part-work structure of the cell **c**: $pw : \Sigma \to \Pi^S \times \Gamma^S : \mathbf{c} \mapsto pw(\mathbf{c}) \stackrel{\text{def}}{=} (mat(\mathbf{c}), cap(\mathbf{c}))$
3. Some functions and relations link Π, Σ and Ω:
   (i) receiver: $\rho : \Omega \to \Sigma : \mathbf{o} \mapsto \rho(\mathbf{o})$
   (ii) sender: $\sigma : \Omega \to \Sigma : \mathbf{o} \mapsto \sigma(\mathbf{o})$
   (iii) part_ordered: $\pi : \Omega \to \Pi : \mathbf{o} \mapsto \pi(\mathbf{o})$
4. Some additional short hands:
   (iv) **a***cquired (or absorbed) parts of cell* **c** : $ap(\mathbf{c}) \stackrel{\text{def}}{=} \{\mathbf{p} | \exists \mathbf{o} : \pi(\mathbf{o}) = \mathbf{p} \wedge \sigma(\mathbf{o}) = \mathbf{c}\}$
   (v) **d***elivered parts of cell* **c** : $dp(\mathbf{c}) \stackrel{\text{def}}{=} \{\mathbf{p} | \exists \mathbf{o} : \pi(\mathbf{o}) = \mathbf{p} \wedge \rho(\mathbf{o}) = \mathbf{c}\}$
   (vi) **t***ransformed parts of cell* **c** : $tp \stackrel{\text{def}}{=} mat(\mathbf{c}) \setminus (ap(\mathbf{c}) \cup dp(\mathbf{c}))$

Definitions

1. Given a part-work structure (**P**, **G**). A *Cell Template* describes (all relevant) attributes of a cell. These include templates for the parts that are acquired, transformed, and delivered by the cell and templates for the works the cell can sustain. It is a tuple: $(\tau, cap : \{\mathbf{g}_1, \ldots, \mathbf{g}_s\}, dp : \{\mathbf{p}_1, \ldots, \mathbf{p}_k\}, ap : \{\mathbf{p}_{k+1}, \ldots, \mathbf{p}_{k+l}\}, tp : \{\mathbf{p}_{k+l+1}, \ldots, \mathbf{p}_{k+l+m}\})$
2. Given a set of parts **P** and a set of cells **C**. An *order template* describes (all relevant) attributes of an order. These are: a part (for which production activities must be planned), the sender (cell) of the order, and the receiver (cell) of the order. It is a tuple: $(\tau, \pi : \mathbf{p}, \sigma : \mathbf{c}, \rho : \mathbf{d})$.
3. A *Cell-Order Structure* is a tuple (**C**, **O**) such that: *(i)* $\mathbf{C} \subseteq \Sigma$; *(ii)* $\mathbf{O} \subseteq \Omega$; *(iii)* for all $\mathbf{o} \in \mathbf{O}$: $\{\sigma(\mathbf{o}), \rho(\mathbf{o})\} \subset \mathbf{C}$; *(iv)* for all $\mathbf{c} \in \mathbf{C}$: there exists $\mathbf{o} \in \mathbf{O}$ such that $\mathbf{c} = \sigma(\mathbf{o}) \wedge \mathbf{c} = \rho(\mathbf{o})$. In a cell-order structure all order templates concern cell templates of the set of cell templates, and all cell templates are involved in one or more order templates.

Remarks

1. Interaction among cells is by means of orders which request for the delivery of a part described by a part template, element of **P**. We restrict our treatment to orders as they are issued and handled in the computational flow of the cells, disregarding their effects in material flow.
2. It is only when considering the interflow and dynamic behaviour of a plant system that one must account for the full meaning of an order, i.e., that it is a request to create a (result) part *res* such that $\mathbf{p} \models res$ and that it can only be fulfilled when a sufficient number of all required input parts (as described by $\iota(\mathbf{g})$) are present) . Order fulfilment requires – in computational flow – the execution of the *creation bundle for the part* (as contained in the Part-Work structure (order explosion)). In phenomena flow it requires time, space, material, capability.

4.3 Plant Structure

To execute the work and produce/absorb the parts described in a part-work structure, one needs a set of cells such that all works in the part-work structure are sustained. These and other requirements are stated (statically) in a *plant structure* in which a part-work structure and a cell-order structure are joined.

Definition

1. A *Plant Structure* is a tuple $(\mathbf{C}, \mathbf{O}, \mathbf{P}, \mathbf{G})$ such that:
   (i) $(\mathbf{C},\mathbf{O})$ is a cell-order structure ;
   (ii) $(\mathbf{P},\mathbf{G})$ is a part-work structure.
   (iii) The cells in $\mathbf{C}$ can sustain all works in $\mathbf{G}$: $\bigcup_{\mathbf{c}\in\mathbf{C}} cap(\mathbf{c}) = \mathbf{G}$
   (iv) The cells in $\mathbf{C}$ can deliver and absorb, or affect all parts in $\mathbf{P}$: $\bigcup_{\mathbf{c}\in\mathbf{C}} mat(\mathbf{c}) = \mathbf{P}$
   (v) $\mathbf{c}$ ($\in \mathbf{C}$) can receive $\mathbf{o}$ ($\in \mathbf{O}$) if and only if it can distribute $\pi(\mathbf{o})$, $\pi(\mathbf{o}) \in dp(\mathbf{c})$:
   $$\rho(\mathbf{o}) = \mathbf{c}\ iff\ \pi(\mathbf{o}) \in dp(\mathbf{c})$$
   (vi) $\mathbf{c}$ ($\in \mathbf{C}$) can send $\mathbf{o}$ ($\in \mathbf{O}$) if and only if it can absorb $\pi(\mathbf{o})$: $\sigma(\mathbf{o}) = \mathbf{c}\ iff\ \pi(\mathbf{o}) \in ap(\mathbf{c})$

4.4 Plant Systems over a Plant Structure

A large number of conditions to be satisfied by the dynamic behaviour of manufacturing systems are statically – with reference to template structures only – expressed in a plant structure. Plant systems, or computational flows over a plant structure, are defined irrespective of whether the system is interfaced to phenomena flow or not. The dynamic behaviour of a plant system – capable of simulating shop floor operations – is defined by the procedures (or programmes) which it – its cell instances – executes in response to order instances. Computational flows over a plant structure will instantiate, transform and delete instances of templates much in the same way as a computational algorithm works with dynamic variables of data types (e.g. integers, reals, booleans) and (invokes) functions and procedures transforming their values.

In programming, the notions of *encapsulation* and abstract data types allows one to separate internal and external structure and behaviour of computational objects. In analogy one can separate internal structure and behaviour of a plant system on the one hand and the behaviour of the plant system as embedded in a market system on the other hand.

In the computational flow over a given plant structure we propose a distinction between the *specification* of a plant system (in a market) and its implementation. The specification determines in response to which order instances the plant system will produce and absorb part instances, and to which cell instances orders should be addressed. The *implementation* of a plant system determines how cell *instances* will explode order instances and how they will schedule the work instances. Combining the concept of plant structure and the definition of an *abstract data type* in CLU data abstractions (Guttag & Liskov, 1986) one can define:

Definitions

1. A *plant system specification* is a tuple PSS:
   $$(\mathbf{PS}, \{\mathbf{c}^{sales}, \mathbf{c}^{purchase}\}, \mathbf{O}^{sales}, \mathbf{O}^{purchase}, \mathbf{P}^{sales}, \mathbf{P}^{purchase}, PS, O\_h\_p) \quad \text{with:}$$
   (i) $\mathbf{PS}$ is a plant structure $(\mathbf{C}, \mathbf{O}, \mathbf{P}, \mathbf{G})$ such that: $\mathbf{P}^{sales} \subseteq \mathbf{P}$ and $\mathbf{P}^{purchase} \subseteq \mathbf{P}$;
   (ii) $\mathbf{c}^{sales}$ and $\mathbf{c}^{purchase}$ are cell templates;
   (iii) $\mathbf{O}^{sales}$ is a set of order templates such that: *(i)* for each $\mathbf{o} \in \mathbf{O}^{sales}$: $\rho(\mathbf{o}) = \mathbf{c}^{sales}$; and (ii) for each $\mathbf{p} \in \mathbf{P}^{sales}$ there is $\mathbf{o} \in \mathbf{O}^{sales}$ with $\pi(\mathbf{o}) = \mathbf{p}$;
   (iv) $\mathbf{O}^{purchase}$ is a set of order templates such that: *(ix)* for each $\mathbf{o} \in \mathbf{O}^{purchase}$: $\sigma(\mathbf{o}) = \mathbf{c}^{purchase}$; and (ii) for each $\mathbf{p} \in \mathbf{P}^{purchase}$ there is $\mathbf{o} \in \mathbf{O}^{purchase}$ with $\pi(\mathbf{o}) = \mathbf{p}$;
   (v) PS is an unbounded mathematical set of *plant system states* over $\mathbf{PS}$.

(vi) O_h_p is the specification of an order handling programme; its implementation will – in response to an order instance – cause the $PS - representation$ to go through a number of states in order to make it deliver a part instance: $O\_h\_p : (\mathbf{O}^{sales})^{\epsilon} \times PS \leadsto PS \times (\mathbf{P}^{sales})^{\epsilon}$

2. A *Plant System Implementation* PSI for a PSS (as defined before) is a tuple: $(PS.rep, O\_h\_p.body)$ with:
   (i) $PS.rep$ is the representation of PS (including an initial marking of PS). It includes a set of cell-instances such that all elements in $\mathbf{P}^{sales}$ can be produced;
   (ii) $O\_h\_p.body$ is the body of the order handling programme; it will handle the instances of $\mathbf{O}^{sales}$; invoke the appropriate instances of $\mathbf{O}$ and of $\mathbf{O}^{purchase}$; and handle the instances of $\mathbf{O}$ such that part instances are "produced" by the cell instances and the plant system.

Furthermore there must be a market system MS and a *order dispatch programme* O_d_p which causes the market system to supply the part instances in response to the instances of $\mathbf{O}^{purchase}$:

$$O\_d\_p : (\mathbf{O}^{purchase})^{\epsilon} \times MS \leadsto MS \times (\mathbf{P}^{purchase})^{\epsilon}$$

Remark
"$\leadsto$" is to be read "leadsto", it expresses our concern with the transit states a plant system or a market system passes through in response to an order instance.

Response to a Single Order Instance
The body of the order handling programme $O\_h\_p.body$ transforms $PS.rep$ according to the order instance received: *(i)* If $(o_i\epsilon)$ $\mathbf{o} \in \mathbf{O}^{sales}$ then $O\_h\_p.body$ will find out which cell instances can produce the ordered part and issue an instance of the relevant $\mathbf{o}' \in \mathbf{O}$. *(ii)* if $\mathbf{o} \in \mathbf{O}$ then $O\_h\_p.body$ (or the cell's autonomous order handling programme) will build a place/transition tree (or P/T tree) according to the creation bundle of the part ordered. A P/T tree is *planted* for each order a cell receives. The tree is *grown* (with empty places) when orders are exploded (in accordance with a cell-order structure and cell capability). Because of auxiliary parts, as required for executing the works, or waste parts the P/T tree should be extended into a P/T net. Places are marked after parts have been created (in phenomena flow, or by a discrete event simulator). After firing a transition, a parent place is marked, and the transition and all its child places can be *pruned*). The P/T net is *uprooted* when the works are executed and the finished part (result) can be delivered. After growing a P/T net, it must be pruned in a Last-In-First-Out manner. *(iii)* if $\mathbf{o} \in \mathbf{O}^{purchase}$ then $O\_h\_p.body$ will invoke an order dispatch programme of the market system.

Terminology
(1) The term *(computational) pulse* (in response to an order received by a cell instance (the *cell pulse*) or a plant system (the *plant pulse*)) denotes the combination of the planting, growing (including auxiliary and waste expansions), pruning and uprooting of the corresponding P/T net). (2) The term *flow* denotes the integration (or combination) of the pulses in response to a number of orders (for various products). (3) Extending a P/T net (auxiliary and waste expanded from a P/T tree) in response to an order is called *pulse explosion.* The activity of reducing the P/T net, typically after accomplishment of work, is called the *pulse implosion.*

Remark
Autonomous cells. For simplicity of presentation all computational response for an order is executed by a single procedure ($O\_h\_p.body$). As the P/T net construction in a cell is completely local, one could have each cell scheduling its own activities.

The Integration of Responses to Order Instances
After identifying the response by a plant system and a cell instance to a single order instance (a pulse), one must investigate the combination (or *integration*) of responses to a number of sequential or concurrent order instances. A *computational flow* is an integration of computational pulses. Planning strategies enrich/refine the order response procedure *O_h_p.body* such that a cell instance, or the plant system as a whole can work for more than one order instance at a time. The work instances in response to different order instances must be interleaved.

A wide variety of planning and scheduling strategies are available according to context and application domain; e.g., the instrument maker: one person does all the work for one order in sequence (marking the places in the P/T net one after one), he may decide to accept a new order only after the result for the previous order has been delivered. If the instrument maker works with apprentices (who can only do some of the activities) he may start working for different order instances at the same time, and try to reduce idle time. Order handling strategies that interleave the work for several order instances are defined on top of sets (*forests*) of P/T nets, as additional constraints, while respecting the precedence constraints expressed in the creation bundle of each part.

4.5 Plant Interflow System

In this step one must account for the interflow of computational flow with the shop floor activities ($\boxed{\aleph \models}$). The requirement of interfacing the computational flow to the operations of a factory implies that the plant system implementation should allow one to distinguish all relevant – according to operations control and monitoring – states of the factory (which is statically modelled by some plant structure). The discrete event simulator which would mark, prune and uproot P/T nets in cell instances is replaced by phenomena flow which changes the net in step with shop-floor operations. Phenomena flow changes include the execution of work during designated time slots, the receipt of parts resulting from earlier work, and the forwarding of parts to be utilized in further work. In defining the interface between a plant system and a shop floor we must ensure that the work, part and order flows are ordered in a time-space-material consistent manner, that the work required can be performed by the cells, that the task descriptions include sufficient details, etc. All these issues must be reflected in the τ attribute of the templates and at the shop floor.

Terminology
(1) The term *inter-pulse* denotes the interactive and synchronized action – in response to an order instance – of a plant system and the shop floor. (2) The term *inter-flow* denotes the combined action – in response to a set of orders – of a plant system and the shop floor.

5 CONCLUSIONS AND FUTURE WORK

A conceptual framework for the definition of information and command services for manufacturing enterprises and industry has been explained. The conceptual and practical validity of the proposed approach must be tested further. Work in the near future should focus on: (a) the derivation of particular models from generic ones; (b) the elaboration of generic models for markets, teams and industries; (c) the development of model execution services; (d) software tool support for the derivation, interfacing, validation, execution and evaluation of particular models; and (e) software tool support for the specification and implementation of interflow models capable of interflowing with enterprise and industry dynamics.

6 REFERENCES

Alting, L. and Jorgensen, J. (1993) The Life cycle concept as a basis for sustainable industrial production. *Annals of the CIRP*, 42/1, 1993.

ATLAS (1993) *ATLAS; Architecture, methodology and Tools for computer integrated LArge Scale engineering, Public Project Overview*, February 1993, ESPRIT Project 7280, CEC, Brussels.

Atsumi, R. (1983) A social human activity model as an information infrastructure system. In Yoshikawa, H. and Goossenaerts, J., editors, *Information Infrastructure Systems for Manufacturing, IFIP Transactions B-14*, Amsterdam, 1993. Elsevier Science B.V.

Browne, J., Sackett, P. and Wortmann, H. (1995) Industry requirements and associated research issues in the Extended Enterprise. In this volume.

CEN/TC310/WG1 (1994a) CIM systems architecture – enterprise model execution and integration services – statement of requirements. CR1832, CEN/CENELEC, Brussels, Belgium.

CEN/TC310/WG1(1994b) CIM systems architecture – enterprise model execution and integration services – evaluation report. CR1831, CEN/CENELEC, Brussels, Belgium.

CIMOSA (1993) ESPRIT Consortium AMICE, editor. *CIMOSA: Open System Architecture for CIM.* Springer Verlag, Berlin, 2nd, rev. and ext. edition, 1993.

ENV 40 003 (1990) ENV 40 003: Computer integrated manufacturing – systems architecture – framework for enterprise modelling. European prestandard, CEN/CENELEC, Brussels.

Goossenaerts, J. (1993) Enterprise Formulae and Information Infrastructures for Manufacturing. In: Yoshikawa, H. and Goossenaerts, J. (eds.): *Information Infrastructure Systems for Manufacturing*, IFIP Transactions B-14, Elsevier Science B.V. (North Holland), Amsterdam.

Goossenaerts, J. and Bjørner, D. (1994) Generic models for manufacturing industry. Technical report no. 32, UNU/IIST, Macau, December 1994.

Hammer, M. and Champy, J. (1993) *Reengineering the Corporation.* Harper Collins Publishers, Inc., New York, 1993.

Hertz, H.R. (1894) *Die Prinzipien der Mechanik,in neuem Zusammenhange.* Johann Ambrosius Barth, Leipzig, 1894.

Liskov, B. and Guttag, J. (1986) *Abstraction and Specification in Program Development.* The MIT Press, Cambridge, Massachusetts, 1986.

Matthiesen, M. (1994) Synopses of ESPRIT III Project 6706: MS^2O - Multi-Supplier/Multi-Site Operations, CEC, Brussels.

The RAISE Language Group (1992). *The RAISE Specification Language.* Prentice Hall, New York 1992.

Scheer, A.-W. (1989) *Enterprise-Wide Data Modelling – Information Systems in Industry.* Springer-Verlag, Berlin - Heidelberg.

Spur, G., Mertins, K. and Jochem, R. (1994). *Integrated Enterprise Modelling.* Beuth Verlag, Berlin.

Williams, T.J. (1993) The Purdue Enterprise Reference Architecture. In: Yoshikawa, H. and Goossenaerts, J.(eds.) *Information Infrastructure Systems for Manufacturing*, IFIP Transactions B-14, Elsevier Science B.V. (North Holland), Amsterdam.

Yoshikawa, H. (1981) General design theory and a CAD system. In T. Sata and E. Warman, editors, *Man-Machine Communication in CAD/CAM*, Amsterdam, 1981. Elsevier Science B.V. (North Holland).

12

A control-oriented dynamic model of discrete manufacturing systems

E. Canuto
Dipartimento di Automatica e Informatica, Politecnico di Torino,
Corso Duca degli Abruzzi 24, I 10129 Torino, Italy
e-mail: canuto@polito.it - fax: +39 11 654 7099

Abstract

The paper presents a mathematical model of discrete-manufacturing production systems in terms of discrete-time state equations; the model allows to analyse their dynamics under autonomous evolution and real-time feedback control.

Keywords

Discrete manufacturing systems, factory dynamics, production systems.

1 INTRODUCTION

In factories where discrete manufacturing takes place, objects of various types circulate and some of them are progressively transformed and assembled into commercial finished-products: raw materials, components, semifinished and finished products, fixtures, equipments, tools. The ensemble of such objects can be mathematically described (Canuto et alii, 1993) by a countable and finite set of *(manufacturing) objects*. Based on this concept, a *(manufacturing) algebra* has been introduced, capable of describing and manipulating, in a formal way and independently of any technological sector, the set of manufacturing operations to be carried out on such objects. Its key elements are:

- A generic definition of *manufacturing operations*, capable of encompassing any type of operation typical of discrete manufacturing processes, like assembling, disassembling, transporting, machining, setting-up, fixturing.
- Composition rules to aggregate elementary operations and to describe in a formal and compact way any complex network of operations designed to be fed by raw materials and to yield a set of finished-products.

The key elements of the algebra, i.e. objects and manufacturing operations, have been subsequently used to lay down the foundations of a dynamic model of the factory production processes. The basic concept, very often forgotten but of paramount importance in control theory, is the distinction between a factory plant and its production control system. It is the latter that makes the factory alive, launches and controls production processes, determines the inflow and the inflow-places of raw materials, dispatches transport operations, commands workstations and their manufacturing operations, decides drawing of finished products. The models developed in the present paper mainly concern the factory plant. Production control systems are barely mentioned, being introduced as subsystems capable of modifying the factory dynamics.

2 DEFINITIONS

Let us start with some definitions.

Manufacturing process. It is the output of the engineering work, defining the objects and the manufacturing operations needed to manufacture the desired finished-products. Manufacturing processes are not bounded to the factory where they will be performed. They are described with an *algebra,* capable of operating on the object set through algebraic operations transforming a set of objects into another set.

Factory. A factory is defined as a set of:

- manufacturing units: machine tools, presses, robots, or simply working stations where workers can perform specific operations;
- transport units;
- storage units: every factory space where objects can be placed, also temporarily.

Production system. It is defined as a pair: a factory with a *production plan* together with a *production control system* capable of ensuring the plan achievement. The *factory dynamics* describes the production systems, i.e. factory equipped with production plan and control.

Let us remark that *production controls* are often achieved though automatic devices and consequently the are embedded in the factory plant itself. It is however conceptually important to distinguish between the plant to be controlled and the control itself. To give a simple example, think to storage units that are managed either through a FIFO rule, first-in-first-out, either through a LIFO rule, last-in-first-out, or with random supply and drawing. The first two cases are storage units equipped with a control rule and possibly an automatic control device performing the rule; the rule and the relevant device must be retained separated from the storage unit.

3 THE ALGEBRA OF MANUFACTURING PROCESSES

3.1 The manufacturing objects

Types of objects

It is assumed that all the different object types circulating in the factory and entering the production process be elements of a countable and finite set $\boldsymbol{O}$, having cardinality n_k. In this way the set $\boldsymbol{O}$ can be ordered as a list and each object type is univocally defined by the index k. A basic assumption is that such a list, although very large (also thousands of hundreds of objects) be available.

Quantities of objects

Production processes in discrete-type manufacturing deal with quantities of objects: the stock level of a store, the input lot size to a manufacturing operation, the yield of a manufacturing operation, quantities of objects transported from one place to another. Quantities of objects are denoted by a vector $\boldsymbol{q}$ of size n_k, whose generic entry $q(k)$ is an integer number (positive or negative) denoting the available or missing quantity of the kth object type. The set of the quantity vectors is denoted by $\boldsymbol{Q}$ and it can be treated as an integer vector space.

3.2 The manufacturing operations

Definition

A generic manufacturing operation A has been defined to be the pair $A=(\boldsymbol{u},\boldsymbol{y})$ of an input quantity vector $\boldsymbol{u}$ (components) and an output quantity vector $\boldsymbol{y}$ (products). A set of manufacturing operations is denoted by $\boldsymbol{A}$. To each operation A specific functionals can be associated like the manufacturing time $\tau(A)$.

Product map

The concepts of *object quantity* and *manufacturing operation*, together with the *algebra* introduced in (Canuto et alii, 1993), allow to define for each list of finished-products a triple $M=\{\boldsymbol{O},\boldsymbol{Q},\boldsymbol{A}\}$, called *product map,* where $\boldsymbol{O}$ is the set of the objects needed for their production, $\boldsymbol{Q}$ is the quantity space of the objects and $\boldsymbol{A}$ is a finite set of the elementary manufacturing operations, needed to produce the list of finished-products. A generic operation A of $\boldsymbol{A}$ will be denoted also with A_s, s being the index of the operation in $\boldsymbol{A}$, $s=1,\dots,n_s$.

Using algebraic compositions (parallel and series) it is possible to create aggregate operations over the set $\boldsymbol{A}$. Given a subset $\boldsymbol{S}\{A_1,\dots,A_h,\dots,A_n\}\subset\boldsymbol{A}$, an aggregate operation over $\boldsymbol{S}$ is defined as an arbitrary composition of the such operations and it will be denoted by $W=c(A_1,\dots,A_h,\dots,A_n)=(\boldsymbol{u}(W),\boldsymbol{y}(W))$, where $\boldsymbol{u}(W)$ and $\boldsymbol{y}(W)$ are the input and output vectors of the aggregate operation. To each aggregate operation a quantity vector $\boldsymbol{m}(W)$ is also associated, whose generic element $m(h)$ equals the repetition number of the operation A_h in the aggregate operation W.

Using graphical symbols, any product map can be given a graphical representation, showing operations and object interconnections as well as object quantities. The manufacturing operations will be represented by rectangles; the object types by circles, the input and output vectors of a manufacturing operation by arrows connecting circles to rectangles and vice versa; each arrow is drawn with the object quantity aside.

A simple example

A simple example of product map with two finished-products is sketched in Figure 1; the following manufacturing times are assumed: $\tau(A_1)=4$, $\tau(A_2)=5$, $\tau(A_3)=4$, $\tau(A_4)=6$, $\tau(A_5)=6$, $\tau(A_6)=8$, $\tau(A_7)=5$, $\tau(A_8)=7$.

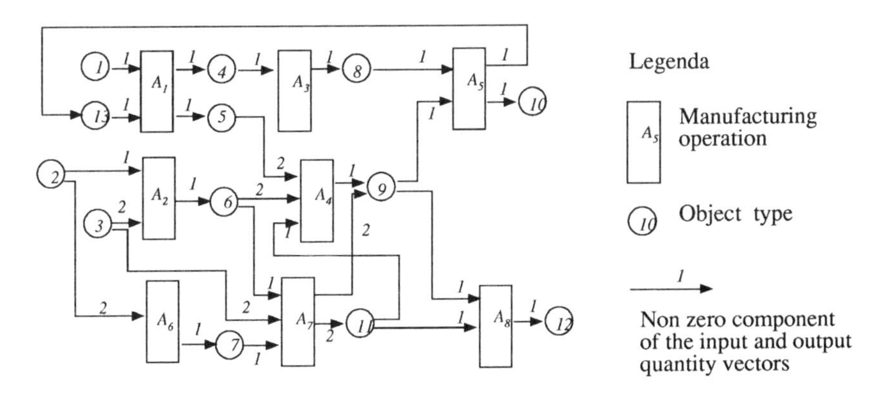

Figure 1 A simple product map.

Object classification

Given a product map, it is possible to classify objects into four classes according to definitions given in (Canuto et alii, 1994).

- Raw materials: no elementary operation of the map includes them as input objects; in the above example they are indexed by $k=1,2,3$.
- Semifinished objects: they are both employed or better consumed (input objects) and produced (output objects) by elementary operations; in the above example they are indexed by $k=4,5,6,7,8,9,11$.
- Reusable materials: they are objects which are employed without being consumed by elementary operations and consequently they may re-circulate; in the above example they are indexed by $k=13$.
- Finished-products: they are objects which are only produced by elementary operations and do not appear as input objects; in the above example they are indexed by $k=10,12$.

Manufacturing plan.

A specific composition $W(\mathbf{A})$ made from the whole set of elementary operations of $\mathbf{A}$ can be defined as a manufacturing plan: it yields as product the output quantity vector $\mathbf{y}(W)$, it has a Bill of Materials equal to the input quantity vector $\mathbf{u}(W)$ and a manufacturing time $\tau(W)$. In case of more than one finished-products the vector quantity $\mathbf{y}(W)$ defines the product mix. Moreover the repetition vector $\mathbf{m}(W)$ defines how many times each operation has to be repeated. Thus a manufacturing plan can be defined as a quintuple $\{W(\mathbf{A}),\mathbf{u}(W),\mathbf{y}(W),\mathbf{n}(W),\tau(W)\}$.

Specifically, given an integer product mix $\mathbf{p}$, the aggregate operation $W(\mathbf{A})$ having as output vector the product mix, i.e. such that $\mathbf{y}(W)=\mathbf{p}$, is in general not unique. If $W(\mathbf{A})$ exists, it can range from compositions reducing the working time (when operations are mostly composed in parallel) to compositions reducing norms of the input quantity vector $\mathbf{u}(W)$ (when operations are mostly composed in series).

Using the previous example, given the mix: $p(10)=4x$ and $p(12)=2y$, x and y being integers ≥ 0, the aggregate operation:

$$W=(\alpha A_5+\eta A_8)(\delta A_4)(\alpha A_3+\gamma A_7)(\alpha A_1+\beta A_2+\gamma A_6) \quad (1)$$

can be shown to possess the minimum-norm Bill of Materials, i.e. $\boldsymbol{u}(W)$ includes only raw objects and fixtures, i.e. $u(1)=4x$, $u(2)=2y+6x$, $u(3)=4y+12x$, $u(13)=4x$, $u(i)=0$, $4\leq k<13$. Moreover, within the the whole set of aggregate operations having a minimum-norm Bill of Materials, W can be shown to possess the least manufacturing time $\tau(w)$:

$$\tau(W)=max\{\tau(A_5),\tau(A_8)\}+\tau(A_4)+max\{\tau(A_3),\tau(A_7)\}+max\{\tau(A_1),\tau(A_2),\tau(A_6)\}=26 \quad (2)$$

Finally, the repetition vector $\boldsymbol{m}(W)$ yielding the product mix $\boldsymbol{p}$ holds:

$$m(A_1)=m(A_3)=m(A_5)=\alpha=4x,\ m(A_2)=\beta=5x+y,\quad m(A_6)=m(A_7)=\gamma=x+y,\ m(A_4)=\delta=2x,\ m(A_8)=\eta=2y. \quad (3)$$

4 THE ELEMENTS OF THE FACTORY DYNAMICS

In (Canuto et alii, 1993) it has been argued that a factory can be modelled as a network made of three different mathematical elements:

- storage units,
- transport units,
- manufacturing units.

Storage units correspond to space volumes in the factory building, where objects can be temporarily laid up for any convenience.

Transport units connect storage units and, whenever required by the production control system, transfer objects from one place to another. Transport operations are specified by the type and quantity of the objects to be moved from the origin to the destination places.

Manufacturing units draw objects to be manufactured, from a specific storage unit (the input store), and lay up their manufactured product in the output store. Each unit can carry out a set of admissible operations: the operations to be performed are selected and commanded by the production control system.

The factory model is assumed to evolve only at discrete times; in other words the evolution of the production process is sampled only at time instants separated by a finite time interval T, called sampling step. Therefore the factory mathematical model will be of the following type:

$$\boldsymbol{x}(t+1) = \boldsymbol{F}[\boldsymbol{x}(t), \boldsymbol{u}(t)], \quad \boldsymbol{x}(t_0)=\boldsymbol{x}_0 \quad (4)$$

being $\boldsymbol{x}$ the state vector, $\boldsymbol{u}$ the input vector, $\boldsymbol{x}_0$ the initial state. Only the state and the input vectors are assumed to vary in time; the factory layout is assumed to be time-invariant. A generic time interval $[\tau,\tau+T)$ during which the t-th sample of the factory state is obtained will be indicated as the t-th time step.

Factory modelling, still being always an approximate description, can be made at various levels of detail depending upon the scope. Here, since the main goal is a dynamic analysis, the frequency band of interest has to be defined. In a production process, different transient phenomena can be observed: fast transients related to start-stop of manufacturing and transport units, slow transients due to fluctuations of object quantities in storage units and in the whole factory.

A common treatment of both fast and slow dynamics would make the model too involved, without a specific advantage. When focusing on fast dynamics, usually it is not necessary to look at the whole factory. Instead it is preferable to analyze the single manufacturing units, modelling transitions between different tooling equipments or setups or, for a specific setup, transitions between operative and wait states. When the interest is on slow dynamics, i.e. on drifts and trends of materials and semifinished products in storage units, one should model the whole factory, but could also greatly simplify the model of manufacturing and transport units, by neglecting their fast transients.

In both cases, it is of paramount importance to choose the right time unit or sampling step T. A rule of thumb is to make it sufficiently shorter than the time constants of the transient phenomena to be described, but longer than the transients to be neglected. Hence, to investigate slow factory dynamics, the sampling step has to be longer than manufacturing and transport times and such that, during any single step, the units can carry out several operations without accounting for starts and stops of the single operations.

4.1 The storage units

The basic assumption is that in the factory a finite list $\boldsymbol{S}$ of places where to store objects be available. Each place will be a storage unit, indicated by the index i, $i \leq n_i$, where n_i is the cardinality of $\boldsymbol{S}$. Storage units are dynamic elements modelled by adders. The list of the object quantities, or stock level, available during the t-th time step, is the state vector of the unit and is indicated by the n_k-sized vector $\boldsymbol{x}(i,t)$. The current value of $\boldsymbol{x}(i,t)$ is the results of a sequence of storage and drawing operations occurred in the past time steps $\tau < t$. To each storage unit a very small subset of objects is usually assigned: such an assignment problem must be considered as a planning problem which is here retained solved.

To model the whole factory, the state vectors of the single storage units are composed into a bi-dimensional vector $\boldsymbol{x}(t)$, sized $n_i \times n_k$, whose generic component will be $x(i,k;t)$, i being the storage unit index and k the object type. In a similar way, any list of object quantities, which is referred to the whole set of the storage units - like drawn or stored quantities - will be denoted by bi-dimensional vectors $\mathbf{y}$.

The state equations describing the evolution at discrete times t of the whole factory stock level will be:

$$\boldsymbol{x}(t+1) = \boldsymbol{x}(t) + \boldsymbol{y}_S(t) - \boldsymbol{y}_D(t), \quad \boldsymbol{x}(t_0) = \boldsymbol{x}_0 \tag{5}$$

having denoted with $\boldsymbol{y}_S(t)$ and $\boldsymbol{y}_D(t)$ the amount of objects respectively supplied to and drawn from all the storage units during the t-th time step.

The fact that the state vectors of the storage units are highly sparse should not be a worry; reduction of sparsity becomes an issue only when software is designed and implemented; thus it shall be solved with appropriate design methods.

4.2 The manufacturing units

The basic assumption is that the set of manufacturing units $\boldsymbol{W}$ be finite; the generic unit will be denoted with the index r and their total number with n_r. Each manufacturing unit can perform a very small subset of the elementary operations belonging to $\boldsymbol{A}$; the problem of assigning such subset of operations to each unit is still a planning problem which is retained here solved. We remark that the operation set $\boldsymbol{A}$ is finite, the operation index is denoted with s and the total number of operations with n_s.

Manufacturing units operate upon commands of the production control system; in other words, as soon as the command of performing the operation type $A_s=(\boldsymbol{u}_s,\boldsymbol{y}_s)$ has been received, the unit will draw at predefined time instants the input objects included in $\boldsymbol{u}_s$ and it will supply, always at predefined time instants, the output objects included in $\boldsymbol{y}_s$. Drawing and supply will be made from/to predefined storage units.

If fast dynamics has to be described, a sampling step shall be selected which is less than the time intervals between the command reception and the drawing and supply times; in this case a manufacturing unit behaves like a set of delay lines. Instead, if only slow dynamics has to be described, drawing and supply times are assumed to be indistinguishable and happening during the same sampling step; their net effect on the input and output stores will become apparent only at the next step. Under such assumption, manufacturing units are no more dynamic elements and the factory state reduces to stock level, i.e. the object quantities in the storage units during a sampling step t.

In case of slow dynamics, the command vector $\boldsymbol{u}(t)$ is a bi-dimensional vector, whose generic component $u(r,s;t)$ equals the number of operations of type s have been commanded to unit r at step t.

The net effect of the command vector on the stock level $\boldsymbol{x}(t)$ is given by the matrix product $\boldsymbol{M}\boldsymbol{u}(t)=(M_S-M_D)\boldsymbol{u}(t)$, where:

- M_S is a multidimensional matrix whose generic element $w_D(i,k;r,s)$ equals the quantity of the objects of type k supplied to storage unit i during the performance of the manufacturing operation s on the unit r.
- M_D is similar to M_S and its elements equal the object quantities drawn from the different storage units during the performance of the different manufacturing operations.

The matrix M and the vector $\boldsymbol{u}$ are called, respectively, *manufacturing matrix* and *manufacturing control vector*. It should be clear that the previous description assumes that each manufacturing unit possesses a production plan and a control system capable of fulfilling commands during the sampling step T; any deviation from such an assumption has to be retained a modelling error.

4.3 The transport units

The construction of a generic model, valid for all the transport units, is a very complex task since transport techniques can vary widely from factory to factory. Different transport systems require different models, but material handling and transport are usually regarded as quite independent problems, to be solved by designing specific *transport subsystems* capable of ensuring that part handling will be free of queues which might deteriorate production plans. But if any transport system is considered to possess its own control system and hence is able to perform the commanded operations in well defined time intervals, the same considerations made for manufacturing units apply. The only difference is that the transport operations are a particular type of the manufacturing operations defined by the algebra; they are identity operations since $\boldsymbol{u}_l=\boldsymbol{y}_l$, i.e. the input and output quantity vectors of a transport operation A_l are equal.

The set of transport units is assumed to be finite and denoted with $\boldsymbol{H}$; a generic unit will be indicated by h, $h=1,...,n_h$. Each transport unit is able to perform a limited number of transport operations from store to store, being commanded by the factory production control system. Also in this case fast and slow dynamic models are possible; in case of slow dynamics transport units are static elements, since drawing and supply of single operation are assumed to be concentrated during a single sampling step.

The command vector $\boldsymbol{v}(t)$ is a bi-dimensional vector having $v(h,l;t)$ as a generic element. The effect of a set of commanded transport operations during a step T on the factory stock level is modelled by the matrix product $H\boldsymbol{v}(t)$. The matrix H and the vector $\boldsymbol{v}$ are called, respectively, *transport matrix* and *transport control vector*.

4.4 A simple example of factory network

In Figure 2 a simple factory network for implementing the product map of Figure 1 is illustrated. Each manufacturing unit is represented by a numbered square, $r=1,...,10$, with the assigned operations on one side. Each storage unit is indicated by a numbered circle, $i=1,2,3,4$. Each transport unit by a double square, $h=1,2,3$, having the assigned operation on one side.

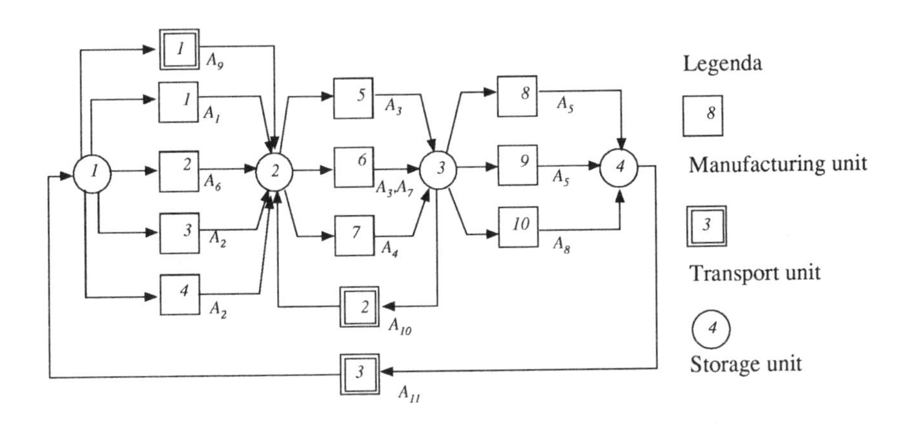

Figure 2 A simple factory network

Let us note that storage units have to be considered without any supplying or drawing rule, i.e. random access; thus any assigned object can be supplied or drawn at any time step in the needed quantity. The elements of the factory and the relevant objects and operations are reported in Table 1.

Table 1 The factory elements.

Storage units		*Manufacturing units*		*Transport units*		
Index i	*Assigned objects k*	*Index r*	*Admissible operations s*	*Index h*	*Admissible operations l*	*Object k*
1	1,2,3,13	1	1	1	9	3
2	3,4,5,6,7,11	2	6	2	10	11
3	8,9,11	3	2	3	11	13
4	10,12,13	4	2			
		5	3			
		6	3, 7			
		7	4			
		8	5			
		9	5			
		10	8			

5 THE EQUATIONS OF THE SLOW DYNAMICS

5.1 The state equations

The equations

Since only the slow phenomena of factory dynamics will be accounted for, a sampling step T sufficiently long with respect to manufacturing and transport times is assumed. At time step t, the state $\boldsymbol{x}(t)$ measures the stock level of the whole factory. It is assumed that the transport and manufacturing control vectors $\boldsymbol{u}(t)$ and $\boldsymbol{v}(t)$, to be actuated during the time step, have been computed by the control system taking into account the current factory plant capacity, hence ensuring that all the operations will be completed during the same step. The state of the storage units at the next time step $t+1$ holds:

$$\boldsymbol{x}(t+1)=I\boldsymbol{x}(t)+H\boldsymbol{v}(t)+M\boldsymbol{u}(t)+\boldsymbol{r}(t)-\boldsymbol{p}(t), \quad \boldsymbol{x}(0)=\boldsymbol{x}_0 \tag{6}$$

where the n_k-sized vectors $\boldsymbol{r}$ and $\boldsymbol{p}$ denote respectively the input flow of raw objects and the output flow of finished-products and I the identity matrix. Each state component $x(i,k;t)$ is integer (positive or negative) and each input component is a non negative integer.

Example

Next the manufacturing matrix M and the transport matrix H are detailed. On the left side the bi-dimensional row indices (i,k) are listed, i being the storage unit index and k the object type. On the upper side, the bi-dimensional column indices (r,s) and (h,l) are listed, r being the manufacturing unit index, s the manufacturing operation index, h the transport unit index and l the transport operation index.

	(r,s)											(h,l)		
(i,k)	1,1	3,2	4,2	2,6	5,3	6,3	6,7	7,4	8,5	9,5	10,8	1,9	2,10	3,11
1,1	-1	0	0	0	0	0	0	0	0	0	0	0	0	0
1,2	0	-1	-1	-1	0	0	0	0	0	0	0	0	0	0
2,3	0	0	0	0	0	0	-2	0	0	0	0	2	0	0
1,3	0	-2	-2	0	0	0	0	0	0	0	0	-2	0	0
2,4	1	0	0	0	-1	-1	0	0	0	0	0	0	0	0
2,5	1	0	0	0	0	0	0	-2	0	0	0	0	0	0
2,6	0	1	1	0	0	0	-1	-2	0	0	0	0	0	0
2,7	0	0	0	1	0	0	-1	0	0	0	0	0	0	0
3,8	0	0	0	0	1	1	0	0	-1	-1	0	0	0	0
3,9	0	0	0	0	0	0	2	1	-1	-1	-1	0	0	0
4,10	0	0	0	0	0	0	0	0	1	1	0	0	0	0
3,11	0	0	0	0	0	0	2	0	0	0	-1	0	-1	0
2,11	0	0	0	0	0	0	0	-1	0	0	0	0	1	0
4,12	0	0	0	0	0	0	0	0	0	0	1	0	0	0
1,13	-1	0	0	0	0	0	0	0	0	0	0	0	0	1
4,13	0	0	0	0	0	0	0	0	1	1	0	0	0	-1
					M								H	

(7)

State controllability

Since the state matrix is the identity matrix, the state controllability only depends on the input matrices. To this end a compact control matrix $B=[M\ H]$ is introduced, together with a single control vector $\boldsymbol{u}$ including manufacturing and transport commands. Then the state equation is rewritten as:

$$\boldsymbol{x}(t+1)=I\boldsymbol{x}(t)+B\boldsymbol{u}(t)+\boldsymbol{r}(t)-\boldsymbol{p}(t), \quad \boldsymbol{x}(0)=\boldsymbol{x}_0 \tag{8}$$

Definition. A state component $x(i,k;t)$ is completely controllable if the row (i,k) of B contains at least a pair of components $b(i,k;r_-,s_-)$ and $b(i,k;r_+,s_+)$, with the property: $b(i,k;r_-,s_-)<0$, $b(i,k;r_+,s_+)>0$.

According to this definition, there must exist at least two command components, one supplying and the other drawing objects, assuring that the state can reach any admissible value. The vector of the completely controllable components will be denoted by $\boldsymbol{x}_C$.

Definition. A state component $x(i,k;t)$ is partially (input/output) controllable if the (i,k)th row of B contains only negative or positive components.

According to this definition, some objects can only be supplied or drawn upon factory commands; this is the case of finished-products (output factory objects) and raw materials (input factory objects). State components which are either completely or partially controllable are simply referred to as controllable. The vectors of partially controllable components will be denoted by $\boldsymbol{x}_I$ (input) and $\boldsymbol{x}_O$ (output); such components become completely controllable if the input vectors $\boldsymbol{r}$ and $\boldsymbol{p}$ are under control.

Definition. A state component $x(i,k;t)$ is uncontrollable if the (i,k)th row of B is null. The vectors of uncontrollable components will be denoted by $\boldsymbol{x}_N$.

It is now possible to introduce the following state decomposition. The state equation (8) can be decomposed in the following form:

$$\begin{vmatrix} \boldsymbol{x}_C(t+1) \\ \boldsymbol{x}_I(t+1) \\ \boldsymbol{x}_O(t+1) \\ \boldsymbol{x}_N(t+1) \end{vmatrix} = \begin{vmatrix} I & 0 & 0 & 0 \\ 0 & I & 0 & 0 \\ 0 & 0 & I & 0 \\ 0 & 0 & 0 & I \end{vmatrix} \begin{vmatrix} \boldsymbol{x}_C(t) \\ \boldsymbol{x}_I(t) \\ \boldsymbol{x}_O(t) \\ \boldsymbol{x}_N(t) \end{vmatrix} + \begin{vmatrix} B_C \\ B_I \\ B_O \\ 0 \end{vmatrix} \boldsymbol{u}(t)+\boldsymbol{r}(t)-\boldsymbol{p}(t) \tag{9}$$

where $B_I \geq 0$ and $B_O \leq 0$.

Definition. The state equation (8) is reduced to minimal form if and only if all the state components are either completely or partially controllable.

For instance, the matrices M and H in equation (7) correspond to a state vector in minimal form. The previous controllability definitions can be applied separately to transport and manufacturing controls.

5.2 The steady state solution

The steady state equation

Under steady conditions, the input vectors (i.e. the commands dispatched by the control system) and the state vector of the storage units are constant. That means:

$$\boldsymbol{u}(t)=\underline{\boldsymbol{u}},\ \boldsymbol{v}(t)=\underline{\boldsymbol{v}},\ \boldsymbol{x}(t)=\underline{\boldsymbol{x}},\ \ \boldsymbol{r}(t)=\underline{\boldsymbol{r}},\ \boldsymbol{p}(t)=\underline{\boldsymbol{p}} \tag{10}$$

Applying such equalities to state equation (8), the following steady state condition is obtained:

$$H\underline{\boldsymbol{v}} + M\underline{\boldsymbol{u}} + \underline{\boldsymbol{r}} - \underline{\boldsymbol{p}} = \boldsymbol{0} \tag{11}$$

setting up a constraint between transport and manufacturing commands, raw objects supply and production yield during a generic time step.

In Section 3 the concept of manufacturing plan of a product map has been introduced. Manufacturing plans are independent of factory network, being only related to objects and manufacturing operations. In practice they assume a factory network to have only one random-access store and at least one manufacturing unit for each operation. Due to single-storage assumption a manufacturing plan does not need transport units. An assignment problem translates the manufacturing plan into a production plan by assigning objects to actual storage units and manufacturing operations to actual manufacturing units; at this stage provision is made also of transport operations and units. If the assignment is not unique as in the above example, i.e. the same operation can be assigned to more manufacturing units, the problem includes some degrees of freedom to be solved. The following computation steps are necessary to produce a steady state production plan $\{\underline{r}, \underline{p}, \underline{u}, \underline{v}\}$, given the quintuple of the relevant manufacturing plan $\{W(A), \boldsymbol{u}(W), \boldsymbol{y}(W), \boldsymbol{n}(W), \tau(wW\}$,

- Given the product mix $\boldsymbol{y}(W)$ and the Bill of Materials, the object assignment to storage units determines the vectors $\underline{r}$ and $\underline{p}$.
- Given the repetition vector $\boldsymbol{m}(W)$, the operation assignment to manufacturing units determines the vector $\underline{u}$ and the matrix W, after solution of the degrees of freedom.
- If the vector $W\underline{u} + \underline{r} - \underline{p} = \underline{h}$ is null, no transport operation and consequently no transport unit is needed. Otherwise they must be defined, resulting in the matrix H. Finally the transport vector $\underline{v}$ is solved; $\underline{v}$ is called the steady-state transport plan.

Example

Coming back to the previous example, assume the mix variables introduced in Section 3 to have the following values: $x=1$ and $y=1$. The corresponding steady-state solution is reported in the following equation, by listing the values of the steady-state vectors $\underline{r}$, $\underline{p}$ and $\underline{h}$.. In the same equation the manufacturing control vector $\underline{u}$ resulting from the assignment of the plan repetition $\boldsymbol{m}(W)$ is reported. The assignment criteria are not mentioned here, but they mainly correspond to force balance of manufacturing times. To this end the manufacturing time vector $\underline{\tau}$ required by each pair (unit, operation) and computed from the manufacturing times defined in Section 3 is reported; the total manufacturing time of each pair (unit, operation) has been computed by assuming series composition on each unit.

(i,k)	$\underline{r}$	$\underline{p}$	$\underline{h}$
1,1	4	0	0
1,2	8	0	0
2,3	16	0	-4
1,3	0	0	4
2,4	0	0	0
2,5	0	0	0
2,6	0	0	0
2,7	0	0	0
3,8	0	0	0
3,9	0	0	0
4,10	0	4	0
3,11	0	0	-2
2,11	0	0	2
4,12	0	2	0
1,13	4	0	-4
4,13	0	0	4

(r,s)	$\underline{u}$	$\underline{\tau}$
1,1	4	16
3,2	3	15
4,2	3	15
2,6	2	16
5,3	3	12
6,3	1	4
6,7	2	10
7,4	2	12
8,5	2	12
9,5	2	12
10,8	2	14

(12)

Different considerations apply to the above solution.

- A transport plan $\underline{v}$ is needed since $\underline{\boldsymbol{h}}\neq\boldsymbol{0}$. Indeed three different objects, namely $k=3,11,13$, need to be transported between storage units with a quantity listed in the same vector $\underline{\boldsymbol{h}}$. The transport units together with the transport matrix H satisfy the requirements of $\underline{\boldsymbol{h}}$. The transport plan is obtained after having defined the input and output vectors of each transport operation: in this case the following transport plan holds: $\underline{v}=[2\ 1\ 2]^T$.
- The manufacturing times of the vector $\underline{\tau}$ can be used to select the sampling step of the model; the manufacturing time of the production plan is obtained by assuming that operations on different manufacturing units are composed in parallel and operations on the same one are composed in series. Applying algebra notations to the pair (r,s)=(unit,operation) the following aggregate operation, called production cycle, $\underline{W}$ result:

$$\underline{W} = (1,1)^4 + (3,2)^3 + (4,2)^3 + (2,6)^2 + (5,3)^3 + (6,3)(6,7)^2 + (7,4)^2 + (8,5)^2 + (9,5)^2 + (10,8)^2 \qquad (13)$$

It is easy to verify that the manufacturing time holds $\tau(\underline{W})=16$. If also transport operations can be performed in parallel, such a time interval can be used to fix the sampling step T, i.e. $T\geq\tau(\underline{W})$.

5.3 Dynamic properties

Open-loop control.

Recall the factory state equation (8) and assume it has been reduced to minimal form:

$$\boldsymbol{x}(t+1) = I\boldsymbol{x}(t) + B\boldsymbol{u}(t) + \boldsymbol{r}(t) - \boldsymbol{p}(t), \quad \boldsymbol{x}(t_0) = \boldsymbol{x}_0 \qquad (14)$$

First, assume that the input vectors $\boldsymbol{u}(t)$, $\boldsymbol{r}(t)$ and $\boldsymbol{p}(t)$ are planned independently of the current state; e.g. they are the result of an optimized off-line plan. As long as there is no conflict with the physical constraints on the state vector $\boldsymbol{x}(t)$, $0\leq\boldsymbol{x}(t)\leq\bar{\boldsymbol{x}}$ or, more generally, as long as the commands are kept within their feasibility range, the above state equations remain linear and all their eigenvalues are unitary or equal to the complex value $(1,0)$. In this case, the state equation (14) correspond to a bank of N adders, one for each state component; they are interconnected through the input vectors, but not interacting at the state level. Hence the following simple result can be stated.

Result. If the input vectors $\boldsymbol{u}(t)$, $\boldsymbol{r}(t)$ and $\boldsymbol{p}(t)$ in equation (14) are state-independent (open-loop control), the same equation is bounded-input unstable, i.e. at least one bounded input exists producing an unbounded state vector. This means that the state variables, i.e. the stock levels, are subject to uncontrolled drifts.

Note that such a situation, corresponding to an open-loop production control, is unrealistic, since any real factory neither operates nor could operate in this way.

Perfect closed-loop control

If instead the command vectors $\boldsymbol{u}(t)$, $\boldsymbol{r}(t)$ and $\boldsymbol{p}(t)$ are state-dependent, i.e. the production control is closed-loop, then the eigenvalues of the state equations are moved away from the open-loop values and the state dynamics is modified. In this case, the state equation (14), modified as follows, represent an autonomous dynamic system:

$$\boldsymbol{x}(t+1) = \boldsymbol{x}(t)+\boldsymbol{B}\boldsymbol{u}[\boldsymbol{x}(t)]+\boldsymbol{r}[\boldsymbol{x}(t)]-\boldsymbol{p}[\boldsymbol{x}(t)], \ \boldsymbol{x}(t_0)=\boldsymbol{x}_0 \qquad (15)$$

The resulting dynamics highly depends on the type of feedback control strategy, whose complete study is outside the scope of the present paper. Only very preliminary results will be presented without formal proof.

Definition. A feedback control strategy $\boldsymbol{u}[\boldsymbol{x}(t)]$, $\boldsymbol{r}[\boldsymbol{x}(t)]$, $\boldsymbol{p}[\boldsymbol{x}(t)]$ is perfect if it shifts the eigenvalues of equation (15) to the origin $(0,0)$.

In general perfect control strategies are not unique. Under perfect control, no matter which is the initial state value, are capable of reaching their steady-state condition in a finite number N of sampling steps. It can also be proved that the transient interval N is the same interval necessary to produce the earliest finished-product, by starting production with zero initial state (zero stock level).

Push and FIFO control strategy

The following simple closed-loop control strategies are now assumed:

- The transport command vector $\boldsymbol{v}(t)$ is computed according to a *push strategy,* i.e. the semifinished products, as soon as they are available, are moved and queued into the input storage unit of the manufacturing units, where they have to undergo the subsequent manufacturing operations detailed in the product map.
- The manufacturing command vector $\boldsymbol{u}(t)$ is computed according to a *FIFO strategy,* i.e. the next operation of any manufacturing unit is decided by the first objects of the input queues, which are managed with a *first-in-first-out* rule.

Fixture-free product map.

Under such strategies, if the product map does not envisage re-circulation of objects, that is to say, if *fixtures* or objects to be re-utilized do not exist, the state equations take a *tree-like* structure, free of interactions capable of generating complex dynamic effects. The output stores of any manufacturing unit are immediately emptied as soon as the manufacturing units have fed them. They are therefore characterized by a zero eigenvalue. The input stores instead, if their current state value is positive, are emptied at a constant rate, which is independent of the state value or, in other words, independent of the queue of the waiting operations. Hence, their eigenvalues remain blocked at the open-loop value $(1,0)$, with a clear tendency toward process instability.

This is a well-known fact. If one keeps on feeding the production process at the rate of the first production stage, that is at the rate of the operating units supplied without limitations by the raw object quantity $\boldsymbol{r}$, the production will end up into instability (intermediate stores exploding) in all cases in which the bottleneck will not exactly coincide with the first stage manufacturing units.

In order to avoid instability, the supply of raw materials should be made dependent upon the state of the input stores of the bottleneck workstations. In this case the design of the closed-loop control shall coincide with a delay-line control, but the design will be not at all trivial if a single, well defined bottleneck does not exist.

Re-circulating product map.

Whenever *fixtures* exist, such to have a closed-loop re-circulation, all stores in the loop are automatically stabilized. The loop of stores having re-circulating objects cannot however be modelled by linear state equations, because the feedback control strategy is inherently non linear. Some preliminary results can already be stated:

- The dynamic properties of such a production process depend on the number of the available fixtures.
- The upper stock level of the stores distributed along the production process can not go beyond the total number of available fixtures.
- The production rate decreases as far as the number of the available fixtures is reduced. When the total number of fixtures becomes large enough, the production rate reaches the bottleneck rate.

6 CONCLUSIONS

The model presented in this paper appears a simple and at the same time powerful tool for the analysis of the dynamic properties of manufacturing plants and of their real-time production control systems.

7 REFERENCES

Canuto, E., Donati F. and Vallauri M. (1993) Factory modelling and production control. *International Journal of Modelling and Simulation*, **13**, 162-166.

Canuto E., Donati F. and Vallauri M. (1994) A new approach to modelling manufacturing systems, in *Proc. 27th ISATA Conf. on Lean Agile Manufacturing in the Automotive Industries*, Aachen (Germany), 317-324.

8 BIOGRAPHY

The author is currently associate professor at Politecnico di Torino, where he teaches Automatic Control and Industrial Automation.

His basic research activity is concerned with the automatic control of uncertain and complex systems. He also applies theory and methods of automatic control in different fields: discrete manufacturing, urban traffic, robots and measuring machines calibration, spacecraft attitude estimation and control, industrial process control, economics.

His present research activity in the discrete manufacturing field is partly done within the ESPRIT Basic Research Project HIMAC 8141, supported by the European Community and coordinated by the an Italian company, EICAS Automazione spa, Torino.

He is author of more than 70 publications.

PART SIX

Manufacturing System Analysis

A NEW TOOL FOR MODULAR MODELLING: THE GENERALISED AND SYNCHRONISED STOCHASTIC PETRI NETS

O.DANIEL, Z.SIMEU-ABAZI, B.DESCOTES-GENON
LAG-ENSIEG, BP 46, 38402 Saint Martin d'Hères CEDEX, FRANCE
Tél: +33 76 82 64 15, Fax: +33 76 82 63 88
e-mail:abazi@lag.grenet.fr

Abstract

The association between Stochastic Petri Nets and Markov chains constitute a powerful tool to perform a manufacturing system analysis. Unfortunately, the markovian's models obtained for complex manufacturing systems are so large (combinatory explosion) that their storage and analysis are very expensive and very long. To contain this explosion we are interested in modular modelling This paper deals with a new tool call Generalised and Synchronised Stochastic Petri Nets (GS^2PNs).

Keywords

Stochastic Petri Nets, Decomposition, Modular modelling, Model reduction, manufacturing systems.

1. INTRODUCTION

The models obtained for large and complex systems are often very heavy. Thus, they loose one of the most important advantage of Petri Nets which is the readability. Some tools like Coloured Petri Nets have been developed in Jensen (1981, 1982, 1983), and authorise to make a strong reduction of the number of places and transitions of the models. This sort of Petri Nets is very efficient to decrease the system's structural complexity but doesn't simplify their analysis. Another very important point for the markovian analysis that we want to realise is that we need to generate the marking graph of the Petri Net model which is often very large. The continuous Petri Nets developed by R.David and H.Alla in David (1987, 1988, 1989) bring an answer for combinatory explosion. In this sort of nets the number of tokens in a place can be a real and not only an integer. The interest of this tool is that we don't build a marking graph but an evolution graph which is smaller. The continuous Petri Nets offer an interesting answer to states combinatory explosion but, at the moment, they don't provide any way to perform an analysis. Our aim is to develop a tool based on SPNs and that authorises modular modelisation.

2. THE GS^{2}PNs - DEFINITIONS

The idea of decomposition as already been wildly explored. Many studies have yet been realised on that topic by Beounes (1984), Noyes (1987) or Jungnitz (1992).

The approach that we propose is different because it doesn't generate decomposable models but directly decomposed systems. Furthermore, our decomposition is not structural but functional.

Definition 2.1

Let R = $\{R_1, R_2, ..., R_t\}$ a set of GSPNs.

R is a GS^{2}PN if $\forall$ i $\in \{1,2,...,t\}$ then:

- $Ri = <P_i, T_i, Pré_i, Post_i, M0_i, temp_i, !S_i, ?S_i>$ where:
- $Pi = \{P1_i, P2_i, ..., P_{ni}\}$, finite set of places of R_i with $|P_i| = n_i$.
- $Ti = \{T1_i, T2_i, ..., P_{mi}\}$, finite set of transitions of R_i with $|T_i| = m_i$.
- $Préi = Pi \times Ti \rightarrow \{0,1\}$.
- $posti = Ti \times Pi \rightarrow \{0,1\}$.
- $M0_i = (M0(P1_i), M0(P2_i), ..., M0(Pn_i))$ the initial marking of R_i.
- $tempi = \{\lambda 1_i, \lambda 2_i, ... \lambda m_i\}$ set of temporisations associated to the transitions of R_i.
- $!S_i$ = set of synchronisation signals sent by R_i.
- $?S_i$ = set of synchronisation signals awaited by R_i.

with:

$(P1 \cap P2 \cap ... \cap Pt) = \emptyset$.

$(T1 \cap T2 ... \cap Tt) = \emptyset$.

$(P1 \cup P2 ... \cup Pt) = P$ the set of places of R.

$T1 \cup T2 \cup ... \cup Tt = T$ the set of transitions of R.

□

The signals are sent by places when they contain one token (or more) and are received by transitions. Thus, we can find four sorts of transitions which are shared into the following four sets:

- *Tin*={immediate and not synchronised transitions}.
- *Ttn*={timed and not synchronised transitions}.
- *Tis*={immediate and synchronised transitions.
- *Tts*={timed and synchronised transitions}.

We shall represent immediate transitions with an horizontal line and timed transition with an horizontal thick bar.

Definition 2.2

A synchronised transition is *authorised* if the synchronisation signal that it waits for is present.

□

Definition 2.3
A no synchronised transition is firable if and only if it's validated.

□

Definition 2.4
A synchronised transition is firable if it's both validated and authorised.

□

As mentioned above, the places send the signals which will be received by the transitions.

Definition 2.5
We shall call « *sending place* », any place which occupation by at least one token induces the emission of a signal.

□

Example 2.1
The place Pi of net of the picture 2.1 is a sending place.

Definition 2.6
We shall call « *receptive transition* » or « *synchronised transition* », any transition which firing is linked to the presence of a synchronisation signal.

□

Example 2.2
The transition Tj of the net of the picture 2.1 is receptive.

Graphically, the emission (reception) of a signal by a place (transition) will be represented with a broken arrow leaving (arriving) the place (on the transition

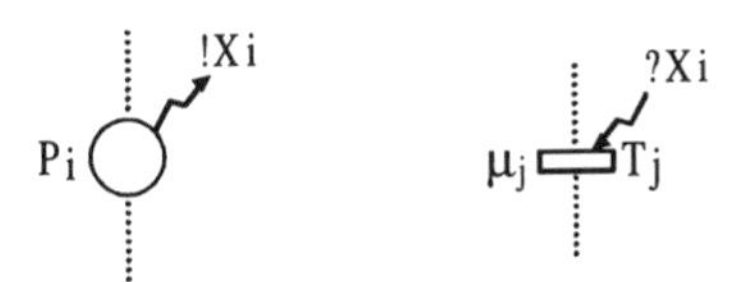

Picture 2.1 Emission of Xi by Pi, reception of Xi by Tj

Thus, let consider a GS^2PN R constituted by two GSPNs R1 and R2. Some transitions of R2 are synchronised with signals sent by places of R1 and vice versa. We can remark that the sending places of R1 (R2) are the ones which marking influence the evolution of R2 (R1).

3. EVOLUTION RULES AND ALGORITHM

The problems linked to the modelisation with GS^2PNs are principally encountered when trying to interpret the conflicts. For classical GSPNs, the only problematic case is the one for which many immediate transitions are simultaneously validated. The GS^2PNs are quite similar. If in the conflict are implied only transitions $\in$ *Tin*, then they are required to belong to the same net (we'll say « *internal conflict* ») and, the conflict can be solved by the classical way which consists in using priorities or switching distributions (see Marsan (1984) or Marsan (1987)). If in the conflict appear some immediate and synchronised transitions, two ways are possible:

- If the conflict concerns transitions of the same net, there is an internal conflict (see above).
- If the conflict concerns transitions from many different nets we'll speak then of « *extended conflict* ».

The extended conflicts are more difficult to solve. The picture 3.1 shows an example of an extended conflict.

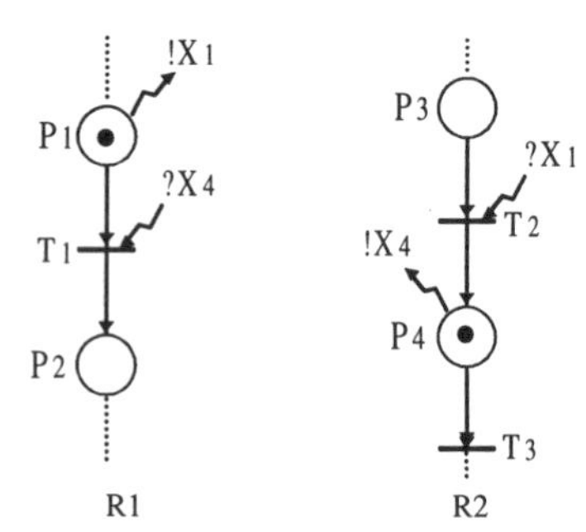

Picture 3.1 Extended conflict

This system represents two GSPNs (R1 and R2) synchronised by a send receive. R2 waits for a message from R1 (X1) to go on with its evolution. R1 after the emission of X1 waits for the answer from R2 (X4).

Let suppose that a token is deposed in P3. Transition T2 is then validated but not authorised because signal X1 is not present. Therefore, if a token is deposed in P1, signal X1 is sent and T2 is then authorised. We fire T2 and we reach the marking P1P4. There is here an extended conflict between T1 and T3. In fact, if we fire T3 first then the emission of signal X4 will be stopped and transition T1 will be unauthorised. On the other hand, if we fire T1 first then transition T3 is still firable. To solve this sort of conflict we will use the VFP interpretation which has been proposed in Jacot (1992). This interpretation consists in making the balance-sheet between the tokens produced and consumed by the firing of the transitions in conflict. If this balance-sheet is positive or null then we can fire all the transitions simultaneously. Thus if two transitions are in conflict and if one of them is a reading transition, then the VFP interpretation authorises to fire them simultaneously (see picture 3.2).

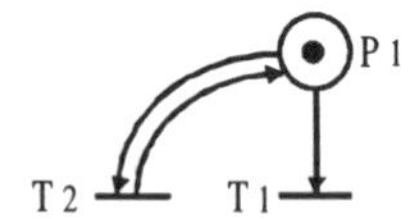

Picture 3.2 Conflict which needs the VFP interpretation

In the picture 3.2, transitions T1 and T2 are in conflict. T2 takes a token in P1 and puts it back at once (balance-sheet = 0). T1 consumes one token (balance-sheet = -1). The global balance-sheet is: 0-1=-1 and there is just one token in P1. This way, we can fire T1 and T2 simultaneously.

Now, let observe the effect of this interpretation on the GS^{2}PNs. The conflict of picture 3.1 may be seen as following.

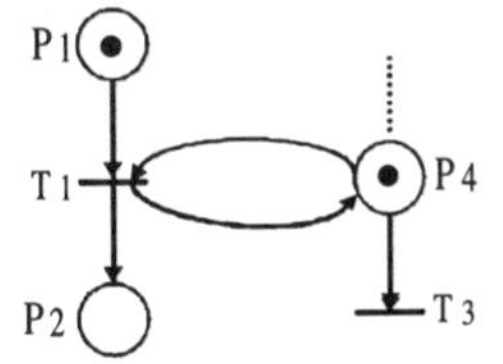

Picture 3.3 conflict of picture 3.1 with classical PN's

In fact, the synchronisation signal X4 corresponds to a reading of the contents of place P4 by transition T1. The VFP interpretation authorises us to fire T1 and T3 simultaneously.

Definition 3.1
In the case of an external conflict, and according to the VFP interpretation, we fire simultaneously all the transitions.

□

Picture 3.4 shows another case of conflict that can be solve with the VFP interpretation.

Picture 3.4 A conflict a: with GS^{2}PNs, b: with PN's

X1 represents the reading of the contents of P1 by T2 and X2 is the reading of the contents of P2 by T1. T1 consumes one token (balance-sheet = -1) and T2 consumes one token too (balance-sheet = -1). The global balance-sheet is: -1 - 1 = -2 and we dispose of exactly two tokens. Thus, we can fire simultaneously T1 and T2.
Picture 3.5 shows an other case of conflict that the VFP interpretation can't solve.

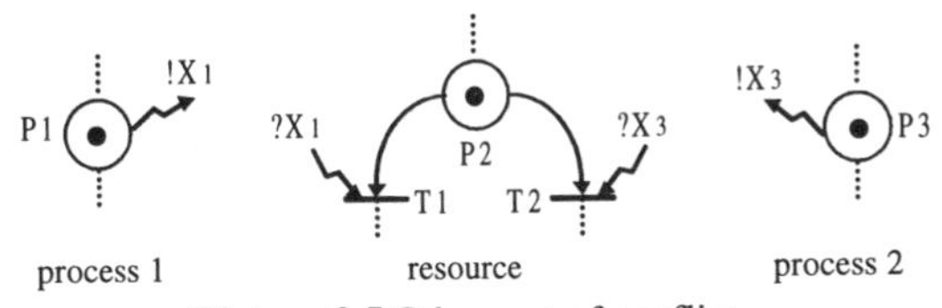

Picture 3.5 Other sort of conflict

This case is the resource share. Two processes require the same resource and create a conflict between. The global balance sheet of this configuration is negative and we can't fire T1 and T2 simultaneously. Fortunately, T1 and T2 belong to the same net and the conflict is an internal conflict that we can solve by using switching distributions.

Now, let see the GS^2PNs evolution rules.

A GS^2PN possesses two sorts of markings which are « *tangible markings* » and « *wanishing markings* ». According to the sort of marking is the system in, we shall have two sorts of emitted signals. If the system is in a vanishing marking the emitted signal will be a *pulse* (Dirac $\delta(t)$). If the system is in a tangible state, the emitted signal is *a rectangular signal* (Heaviside $H(t)$) that endures till there are no more tokens in the concerned places.

Definition 3.2

The reception of a « pulse signal » by a transition authorises the passage of only one token. □

The evolution of a GS^2PN is subject to the three following rules.

R1: If a marking M validate only timed transitions we use the same evolutions rules than for classical SPNs.

R2: If a marking M validate only immediate transitions, we fire them simultaneously after solving the external and internal conflicts.

R3: If a marking M validate both timed and immediate transitions, we consider only the immediate ones and then apply the rule R2.

Remark 3.1

A signal received by a transition can be a logical combination of signals emitted by others nets.

The picture 3.6 shows three examples of signals logical combinations

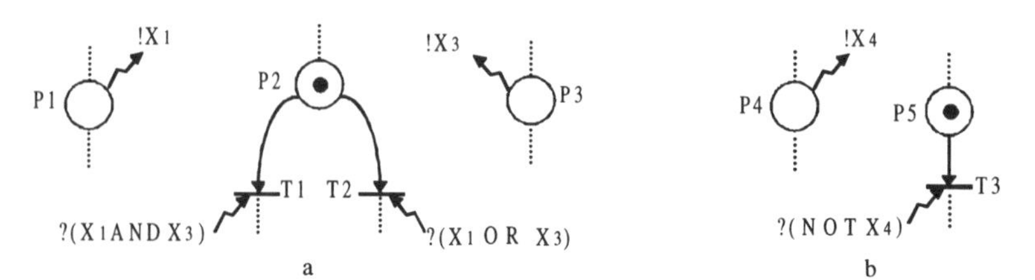

Picture 3.6 Logical combinations of synchronisations signals

For the picture 3.6a, T1 will be fired if the signals X1 and X3 are simultaneously present and T2 will be fired if X1 or X3 is present. Notice that if X1 and X3 are simultaneously present, there will be an internal conflict that we solve by using a switching distribution. For example, a switching distribution should be:

$$Pr\ \{fire\ T1 = 1/3\} \qquad Pr\ \{fire\ T2 = 2/3\}$$

The picture 3.7 give some equivalencies between GS^2PNs and classical PNs.

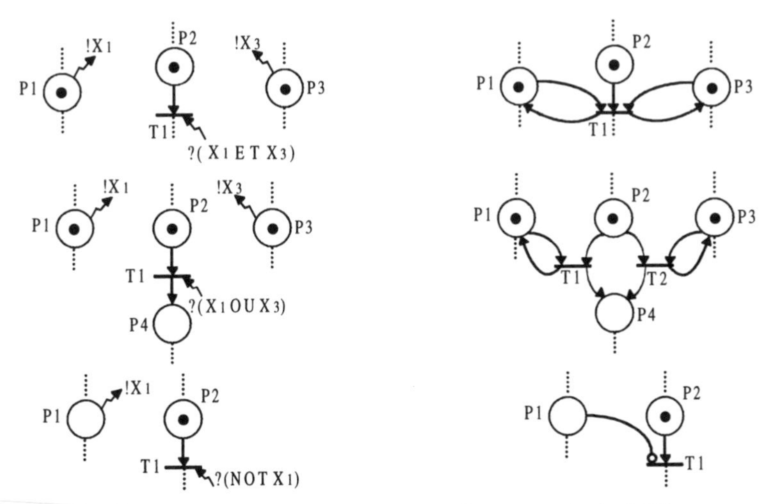

Picture 3.7 Equivalencies between GS²PNs and classical PNs

The following algorithm is the interpretation algorithm for the GS²PNs.

ALGORITHM I

step1: Build list L1 of transitions ∈ {Tin ∪ Ttn} and which are validated for M.

step2: Build the list L2 of transitions ∈ {Tis ∪ ts} and which are validated for M.

step3: Build list L3 of transitions ∈ {Tis ∪ Tts} and which are authorised for M.

step4: Build L4 = L2 ∩ L3 the list of transitions ∈ {Tis ∪ Tts} and which are both validated and authorised for M.

step5: Build L5 = L1 ∪ L4 the list of all the transitions which are firable for M.

step6: If L5 contains both immediate and timed transitions then consider only the immediate ones. Apply rule R3 and obtain:

$$L6 = L5 - \{Ti / Ti \in \{\{Ttn \cup Tts\} \cap L5\}\}$$

step7: Solve the conflicts between transitions in the list L6. We obtain L7 the list of firable transitions.

step8: Firing of all the transitions with the rules mentioned above.

step9: Return to **step1**.

4. FUNDAMENTAL EQUATION

The classical PNs have got their fundamental equation that realise the iring. But, what about GS²PNs We shall work on the following GS²PN.

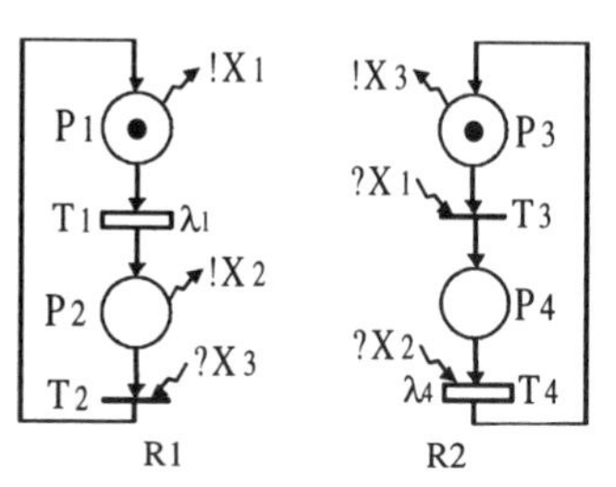

Picture 4.1 Example of GS2PN

Let R a GS²PN composed by two GSPNs R1 and R2. The incidence matrix W and the initial marking M0 are given:

$$W = \begin{bmatrix} -1 & 1 & 0 & 0 \\ 1 & -1 & 0 & 0 \\ 0 & 0 & -1 & 1 \\ 0 & 0 & 1 & -1 \end{bmatrix} \qquad M0 = \begin{bmatrix} 1 \\ 0 \\ 1 \\ 0 \end{bmatrix}$$

Let apply the algorithm I:

step1: L1 = {T1}
step2: L2 = {T3}
step3: L3 = {T3}
step4: L4 = L2 ∩ L3 = {T3}
step5: L5 = L1 ∪ L4 = {T1, T3}
step6: L6 = L5 - {T1} = {T3}
step7: No conflicts. L7 = L6 ={T3}
step8: The new marking is calculated by firing T3 and is P1P4.

The incidence matrix is diagonal by blocks. Each block represents one of the two nets constituting R. M0 is also constituted by two blocks which each one represents the initial marking of one of the two nets constituting R. We notice that the ith block of the incidence matrix and the ith block of the marking corresponds both to the same net Ri. Let apply the classical PN's fundamental equation to W and M0 with a firing sequence $S^T = [0\ 0\ 1\ 0]$ (firing of T3). We notice that the firing sequence S^T is also constituted by blocks.

$$M1 = M0 + W.S = \begin{bmatrix} 1 \\ 0 \\ 1 \\ 0 \end{bmatrix} + \begin{bmatrix} -1 & 1 & 0 & 0 \\ 1 & -1 & 0 & 0 \\ 0 & 0 & -1 & 1 \\ 0 & 0 & 1 & -1 \end{bmatrix} \cdot \begin{bmatrix} 0 \\ 0 \\ 1 \\ 0 \end{bmatrix} = \begin{bmatrix} 1 \\ 0 \\ 1 \\ 0 \end{bmatrix} + \begin{bmatrix} 0 \\ 0 \\ -1 \\ 1 \end{bmatrix} = \begin{bmatrix} 1 \\ 0 \\ 0 \\ 1 \end{bmatrix}$$

We find, without any surprise, exactly the previous result. The classical PNs' fundamental equation can always be used and we can generalise this result to a GS²PN constituted by *t* GSPNs.

This result is interesting but, isn't it possible to take advantage of the modular nature of the

GS^2PNs to find an adapted and more rapid fundamental equation?
Let realise the same calculus but on R1 and R2 separately. Let evaluate M11 and M12 the markings reached by R1 and R2 after the firing of T3. The incidence matrix and the initial markings of R1 and R2 are given:

$$W1 = \begin{bmatrix} -1 & 1 \\ 1 & -1 \end{bmatrix} \qquad W2 = \begin{bmatrix} -1 & 1 \\ 1 & -1 \end{bmatrix}$$

$$M01 = \begin{bmatrix} 1 \\ 0 \end{bmatrix} \qquad M02 = \begin{bmatrix} 1 \\ 0 \end{bmatrix}$$

The new marking can be evaluate separately because R1 and R2 are two classical PNs. As T3 belongs to R2, the two firing sequences S1 and S2 that we must apply to R1 and R2 are respectively $s_1^T = [0,0]$ and $s_2^T = [1,0]$. Thus, we obtain:

$$M11 = M01 + W1 \cdot S1 = \begin{bmatrix} 1 \\ 0 \end{bmatrix} + \begin{bmatrix} -1 & 1 \\ 1 & -1 \end{bmatrix} \cdot \begin{bmatrix} 0 \\ 0 \end{bmatrix} = \begin{bmatrix} 1 \\ 0 \end{bmatrix} = M01$$

$$M11 = M01 + W1 \cdot S1 = \begin{bmatrix} 1 \\ 0 \end{bmatrix} + \begin{bmatrix} -1 & 1 \\ 1 & -1 \end{bmatrix} \cdot \begin{bmatrix} 0 \\ 0 \end{bmatrix} = \begin{bmatrix} 1 \\ 0 \end{bmatrix} = M01$$

This new marking is the one that we expected. The calculation of the successive markings reached by the GS^2PN R can be realised by using the classical fundamental equation on each net separately. This result is more interesting than the previous one. In fact, it is more efficient and simple to apply the classical fundamental equation on r ($r \le t$) little nets $\left(n_i \times m_i\right)$ than on a big one $\left(\sum_{i=1}^{t} n_i \times \sum_{i=1}^{t} m_i\right)$. Furthermore, and practically, the fundamental equation is applied less than r times. In fact, the markings of the nets for which no transitions are fired are not modified. We shall not consider them.

Definition 4.1

If a GS^2PN is constituted by t nets then we shall say that its order is t and we shall talk about $t\text{-}GS^2PN$.

□

Property 4.1

Let R be a $t\text{-}GS^2PN$.

$$\left.\begin{cases} n_i \text{ the number of places of Ri} \\ m_i \text{ the number of transitions of Ri} \end{cases}\right\} i \in (1,2,..,t)$$

The **step8** of the algorithm1 can be performed as following:

- Build the firing sequence Si for each net Ri as:

$$\left.\begin{cases} Si_j = 1 & \text{if } Tj \in L7 \\ Si_j = 0 & \text{else} \end{cases}\right\} \text{with } Tj \in Ri$$

- Apply the fundamental equation to each net Ri separately, if at least one element of the firing sequence Si is positive.

$$M_i' = M_i + W_i \cdot S_i$$

□

5. INVARIANTS

To perform thestructural validation we must prove that each place of the GS^2PN is *bounded* and that each transition of the GS^2PN is *alive*. This structural validation is usually realised by the research of the *p-semi-flots* and *t-semi-flots*.

A p-semi-flot is a vector F (ni × 1) as: $F^T \cdot W = 0$.

A t-semi-flot is a vector S (mi × 1) as: $W \cdot S = 0$.

We can extract, from the p-semi-flots the list of the bounded places and, from the t-semi-flots the list of the alive transitions. If all the places are bounded we shall say that the net is bounded and if all the transitions are alive we shall say that the net is alive.

for example and for the net of the picture 4.1, if F^T=[f1 f2 f3 f4] and S^T=[s1 s2 s3 s4] then we obtain:

$$F_1^T = \begin{bmatrix} 1\,1\,0\,0 \end{bmatrix} \text{ and } F_2^T = \begin{bmatrix} 0\,0\,1\,1 \end{bmatrix}$$

$$S_1^T = \begin{bmatrix} 1\,1\,0\,0 \end{bmatrix} \text{ and } S_2^T = \begin{bmatrix} 0\,0\,1\,1 \end{bmatrix}$$

We can then extract from F1 and F2 the two following minimal marking invariants:

$$\begin{Bmatrix} M(P1) + M(P2) = M0(P1) + M0(P2) = 1 \\ M(P3) + M(P4) = M0(P3) + M0(P4) = 1 \end{Bmatrix}$$

We can also extract from S1 and S2 the two following minimal firing invariants:

$$\left\{ T1T2 \text{ and } T3T4 \right\}$$

In the same way that for the fundamental equation, couldn't we use the modular nature of the GS^2PNs to deduce its invariants from the invariants of its sub-nets?

5.1. MARKING INVARIANTS

This calculus is evident for the marking invariants. In fact, as there are no possible transfer of tokens from one sub-net to an other, it is sufficient to prove that each sub-net is bounded.

Let Fj be a p-semi-flot of the sub-net Rj ($j \in (1,2,..,n)$). It is by definition a column vector (nj×1). Let extend this vector to the GS^2PN R by addicting αpj zeros above and βpj zeros below as:

$$\begin{Bmatrix} \alpha p_j = \sum_{i=1}^{j-1} ni \quad \text{if} \quad j > 1, \quad \text{else } 0 \\ \beta p_j = \sum_{i=j+1}^{n} ni \quad \text{if} \quad j < n, \quad \text{else } 0 \end{Bmatrix}$$

We then obtain the vector $F'^T j = \begin{bmatrix} 0 \cdots 0 \; Fj \; 0 \cdots 0 \end{bmatrix}$ which is a p-semi-flot of the net R. The marking invariants of each sub-net Rj are also marking invariants of the GS^2PN R and this without any modification.

Example 5.1

For the GS²PN of the picture 4.1 the p-semi-flots of R1 and R2 are:

$$F1 = \begin{bmatrix} 1 \\ 1 \end{bmatrix} \qquad F2 = \begin{bmatrix} 1 \\ 1 \end{bmatrix}$$

We extend these two p-semi-flots as explain above and we obtain:

$$\alpha p_1 = 0 \qquad \text{and} \qquad \beta p_1 = \sum_{i=2}^{2} ni = n2 = 2$$

We must add no zero above and two zeros below and obtain $F1'^T = [1\ 1\ 0\ 0]$ *which is a p-semi-flot of R.*

$$\alpha p_2 = \sum_{i=1}^{1} ni = n1 = 2 \qquad \text{and} \qquad \beta p_2 = 0$$

We must add 2 zeros above and no zero below and obtain $F2'^T = [0\ 0\ 1\ 1]$ *which is a p-semi-flot of R.*

5.2. FIRING INVARIANTS

The search of firing invariants can be perform by applying a similar method than for marking invariants. This method consists in searching the t-semi-flots for each sub-net separately. To each t-semi-flot corresponds a firing invariant of the sub-net. The firing invariants of the GS²PN are, without any modification, those of its constituting sub-nets.

On an other hand, the GS²PN's t-semi-flots are deduced from the sub-nets' t-semi-flots by :

Let Sj be a t-semi-flot of the sub-net Rj ($j \in (1,2,..,n)$). It is by definition a column vector ($mj \times 1$). Let extend this vector to the GS²PN R by addicting αtj zeros above and βtj zeros below as:

$$\left\{ \begin{array}{l} \alpha t_j = \sum_{i=1}^{j-1} mi \quad \text{if} \quad j > 1, \quad \text{else } 0 \\ \beta t_j = \sum_{i=j+1}^{m} mi \quad \text{if} \quad j < m, \quad \text{else } 0 \end{array} \right\}$$

We then obtain the vector $S'^T j = [0 \cdots 0\ Sj\ 0 \cdots 0]$ which is a t-semi-flot of the net R.

Knowing the firing invariants of a GS²PN is not enough to prove its vivacity. At the moment, it doesn't exist a mathematical method which prove the vivacity for GS²PNs. So, we propose two necessary conditions which, if they are not verified, prove that a GS²PN is not alive.

Necessary condition n°1

To be alive, a GS²PN needs necessary that all its constituting sub-nets are alive. The proof of the vivacity of each sub-net can be perform separately by searching its t-semi-flots.

The fact that each sub-net is alive is necessary but not sufficient to prove the vivacity of the GS²PN. In fact, it is easy to find some structures that induce problems. The picture 5.1 is one example of this sort of structure that we have named « *cross synchronisation* ».

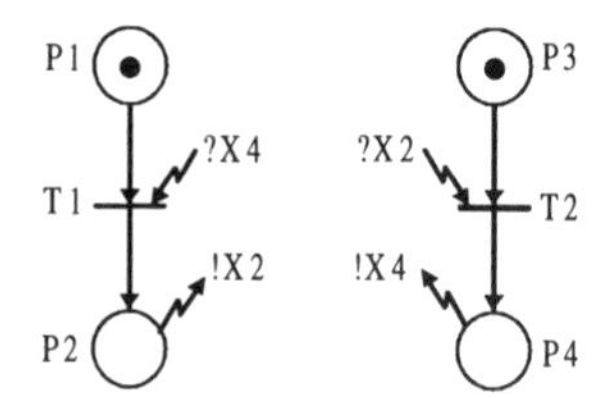

Picture 5.1 Cross synchronisation

The notion of cross synchronisation needs some restrictions. In fact, the fact that the GS^2PNs of the picture 5.2a and picture 5.2b contain cross synchronisations is not sufficient to affirm that they are dead.

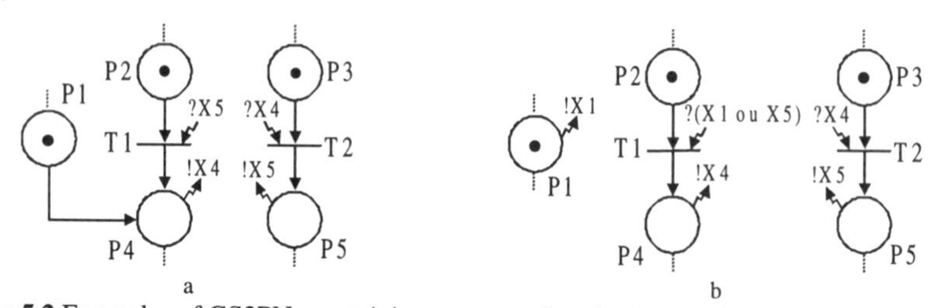

Picture 5.2 Examples of GS2PNs containing cross synchronisations and not necessary deads

Then we introduce,by the following definition, the notion of « *strong cross synchronisation* ».

Definition 5.1

A cross synchronisation is called *strong* if the two following conditions are verified :

1. The places which send of cross synchronisation signals are only filled by the transitions that they lock.
2. The cross synchronisation signals are wether simple or builded with the logical AND of other signals.

□

We can deduce from this a necessary condition that a GS^2PN must possess to be alive:

Necessary condition n°2

To be alive, a GS^2PN must not contain any strong cross synchronisation.

If a GS^2PN doesn't contain any strong cross synchronisations we can use the same method than for

Example 6.2

For the GS^2PN of the picture 4.1 the t-semi-flots of R1 and R2 are:

$$S1 = \begin{bmatrix} 1 \\ 1 \end{bmatrix} \qquad S2 = \begin{bmatrix} 1 \\ 1 \end{bmatrix}$$

We extend these two t-semi-flots as explain above and we obtain:

$$\alpha t_1 = 0 \qquad \text{and} \qquad \beta t_1 = \sum_{i=2}^{2} mi = m2 = 2$$

We must add no zero above and two zeros below and obtain $S1'^{T} = [1\ 1\ 0\ 0]$ *which is a t-semi-flot of R.*

$$\alpha t_2 = \sum_{i=1}^{1} mi = m1 = 2 \qquad \text{and} \qquad \beta t_2 = 0$$

We must add 2 zeros above and no zero below and obtain $S2'^{T} = [0\ 0\ 1\ 1]$ *which is a t-semi-flot of R.*

The two necessary conditions are verified and it is not possible to affirm (absolutely) that the GS2PN of the picture 4.1 is alive. However, the simplicity of this net authorises us to affirm (empiricaly) that it is alive.

6. BLACK BOX REPRESENTATION

A GS2PN may be seen as constituted by a set of black boxes. Each of those black box contain a GSPN which evolution depends on signals sent by the other black boxes, and sending also synchronisation signals to the other black boxes. The manufacturing systems are well adapted to this sort of representation. In fact, a manufacturing system can be consider as constituted by a set of machines separated by stocks. We have defined generic models for machines and stocks and a manufacturing system can be easily represented with a sort of « construction game ». Generic models for machines an stocks are given by picture 6.1a and picture 6.1b.

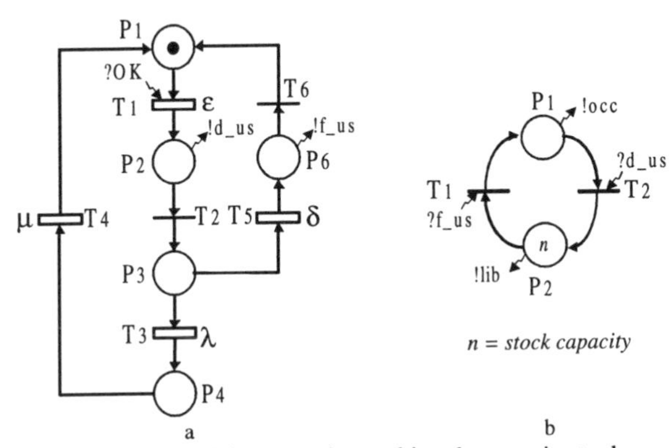

Picture 6.1 a: generic machine, b: generic stock

The picture 6.2a and the picture 6.2b represent the generic models for machines and stocks as black boxes that can be used to describe any manufacturing system

Picture 6.2 generic black boxes for, a: machines, b:stocks

This kind of representation authorises to add other generic models for other function like maintenance or rounting scenari.

7. CONCLUSION

In this paper is presented a new tool called GS^{2}PN. This tool is very flexible and authorises a modular description of a system. We can describe each functionality separately and add finally the synchronisations signals.
The evolution rules are simple and more rapid because we work only on little nets and, on an other hand, because only a few sub-nets are concerned for each firing.
The principal reproach with this tool is that we sometimes must add some sets « sending place and immediate transition » to create the emission of a signal. But, to minimise this problem, we must say that the markings thus generate are vanishing markings and don't appear in the marking graph.
We have developed an algorithm which authorises to generate automatically the markovian process of a system describe with GS^{2}PNs. The markovian model obtained is also modular and we have developed an iterative algorithm which extracts the RAM's parameters from this representation.

8. BIBLIOGRAPHY

Jensen 1981, K.Jensen, « Coloured Petri Nets and the invariant-method » Theorical Computer Science 14, pp 317-336 (1981).

Jensen 1982, K.Jensen, « High level Petri Nets »,third European workshop on application and theory of Petri Nets, Varenna, septembre 1982.

Jensen 1983,K.Jensen, « High level Petri nets », Informatik Fachberichte, vol.66, pp.166-180, 1983

David 1987, R.David, H.Alla, « Continuous Petri nets », 8th European Workshop on application and theory of Petri nets, Saragossa (E), juin 1987.

David 1988], R.David, H.Alla « modelling of productions systems by continuous Petri nets », 3rd Int.conf. CAD/CAM, CARS & FOF88, Detroit, USA 1988.

David 1989, R.David, H.Alla: « Du grafcet aux réseaux de Petri ». Traité des nouvelles technologies série Automatique. Hermès 1989.

Beounes 1984, C.Beounes, « Analyse de la sûreté de fonctionnement de systèmes informatiques complexes par réseaux de Petri », rapport de recherche du LAAS n°83053 octobre 1983, révisé le 22 novembre 1984.

Noyes 1987, D.Noyes, « Approches méthodologiques pour l'aide à la conception et à la conduite des systèmes de production », Thèse de l'Institut National Polytechnique de toulouse, novembre 1987.

Jungnitz 1992, H.J.Jungnitz, « Approximation methods for stochastic Petri nets », Rensselaer Polytechnique Institute, Electrical, Computer, and Systems Engineering Department Troy, New York 12180-3590, may 1992.

Marsan 1984, M.A.Marsan, G.Conte, and G.Balbo, « A class of generalized stochastic Petri nets for the performance évaluation of multiprocessor systems ». ACM Transactions on computer systems, volume 2, n°1, pp.93-122, may 1984.

Marsan 1987, M.A.Marsan, G.Balbo, G.Chiola, and G.Conte, « Generalized stochastic Petri nets revisited: random switches and priorities », in proc.int, workshop on Petri nets and performance models, 1987, pp.44-53.

Jacot 1992, L.Jacot, « Les règles de fonctionnement des Réseaux de Petri revisitées et généralisées », LAG, note interne n° 92-231, mai 1992 révisée en juillet 1992.

14

Petri net modelling for dynamic process planning

Dimitris Kiritsis
Swiss Federal Institute of Technology - Lausanne
DGM-IMECO-LCAO, CH-1015 Lausanne, Switzerland
Dimitris.Kiritsis@imeco.dgm.epfl.ch

Abstract

In this paper a generic dynamic model for process planning is presented, based on Petri net technology. The proposed Petri net model 1) represents manufacturing knowledge of the type of precedence relations constraints 2) represents dynamically the process planning procedure itself, and 3) encapsulates all possible process planning solutions (process plans). Furthermore, process planning simulation can be performed directly on the Petri net, and finally, reachability analysis will give the complete set of solutions (process plans).

Keywords

Process planning, Petri nets, simulation, reachability analysis, CAPP, manufacturing constraints

1. INTRODUCTION

Knowledge based approach predominated the research work on CAPP during the last years. A recent review showed that knowledge representation and dynamic modelling techniques are key topics of intelligent and integrated process planning systems (Kiritsis, 1993). Dynamic process planning is indispensable for its integration with production planning, the immediate higher level of the global manufacturing planning hierarchy (Iwata and Fukuda, 1992).

Petri nets have been extensively used for modelling Discrete Event systems and FMSs. For a review of Petri net applications in manufacturing and a complete list of related references see the paper of Cecil, et.al (1992). Tönshoff et.al. (1987), Srihari and Emerson (1990), and Kruth and Detand (1992) presented Petri net models integrating job-shop scheduling and process planning. In these works, process planning is seen as part of a more global production planning system and more attention is paid to job-shop scheduling and production planning modelling by using Petri nets

In the present paper our attention is concentrated into the process planning problem and a new structure of a Petri net model for dynamic process planning is proposed. This model is generic in the sense that its construction is based on a set of standard generic rules and its graphic representation is similar for any part to be processed. With the proposed method two tools are used for model analysis and solutions finding: 1) the simulation tool can show visually, on the net graph, non-desirable conflict situations. Simulation on Petri nets is performed by firing it from its initial marking and observing tokens traveling through the net. 2) The reachability analysis tool gives all possible solutions (process plans) dynamically included in the Petri net.

2. THE PROCESS PLANNING PROBLEM

In part machining, process planning is the act of preparing detailed machining operation instructions to transform an engineering design to a final functional workpiece. The detailed plan contains the route, actions, machining parameters, machines and tools required for production.

The main input to the process planner could be an engineering drawing or a CAD model providing geometrical and technological information about the designed mechanical part. In an engineering drawing design entities are geometrical entities with technological attributes. In feature-based design systems, design entities are the so called manufacturing features. In both systems only finished part information is provided. The definition of all intermediate phases and states of any design entity is a task of the process planner.

The result of one machining operations is an intermediate or the final state of a design entity (geometrical surface or manufacturing feature) while the result of the process plan is the final part.

We accept that for each design entity (surface or manufacturing feature) one or more candidate operations may exist.

Whatever the design approach, the process planner's work is to choose or determine the best machining operations or processes and their sequence, able to realize the designed part, respecting the desired quality at the best cost.

A more or less complete list of process planning functions includes:

- selection of machining operations
- selection of tools
- selection of machine tools
- grouping of operations
- selection of fixturing systems
- sequence of machining operations
- determination of machining data
- generation of tool paths and NC programs
- calculation of machining times and costs
- document generation (process plan sheets)

In this paper, the process planning problem refers to that of ordering the execution of a number of machining operations given 1) the machining data and associated actions for each machining operation and 2) a set of constraints.
Machining operations that can be executed in a machining center could be classified as following:

1. preparation
2. roughing
3. semi-finishing
4. finishing
5. post-finishing

From the global set of machining data and assuming that our machine tool reference is a modern machining center, the most influencing the planning aspects of the problem are:

1. cutting tool data and tool change actions
2. part positioning data and machining face change actions (table rotations)
3. the cutting operation itself.

The tool changes and table rotations are time consuming actions and influence the quality of the machined part.
The given constraints are:

1. precedence relations (PR) among design entities (DE), consequently, among machining operations (MO)
2. each machining operation must be processed only once.

The above information could come from many different sources: human experts (interactive input), data bases (CAPP systems of variant type), expert or knowledge based systems, and it can be summarized in machining tables like Table 1 where lines refer to design entities and contain candidate machining operations for each machining category (columns):

Table 1 Table associating design entities to machining operations

	PREPA-RATION	**ROUGHG**	**1/2-FINISH**	**FINISH**	**POST-FINISH**
DE_1		MO_1 MO_2	MO_1	MO_1 MO_2	
DE_2	MO_1	MO_1 MO_2		MO_1	MO_1
....					
DE_n				MO_1 MO_2	

A worthwhile note at this point is that in such a table some operations are already planned partially: within a row, left-column operations must be executed before right-column operations.
Generally, precedence constraints can be summarized in precedence graphs like this:

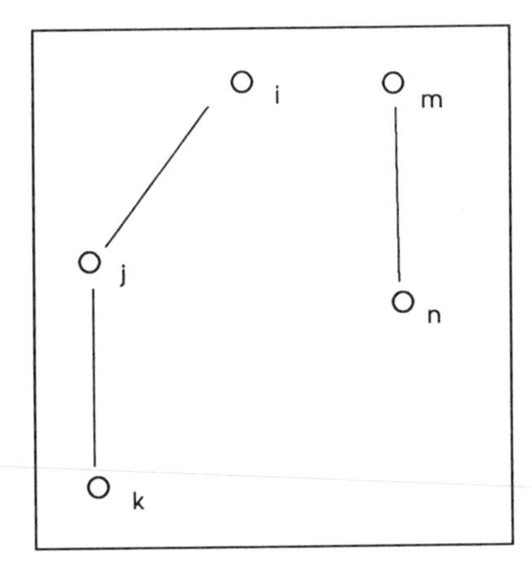

Figure 1 Precedence graph.

3. PETRI NET MODELLING

A Petri net consists of places and transitions which are linked to each other by directed arcs, with some arcs directed from places to transitions (input arcs), and some arcs directed from transitions to places (output arcs). A Petri net can be described as a bipartite directed graph whose nodes are a set of places and a set of transitions.
Places represent passive system components which store "tokens" and take particular states. Graphically, places are represented by circles.
Transitions represent the active system components which may produce, transport or change "tokens". For each transition there is a set of input places and a set of output places Graphically, transitions are represented by rectangles.
Arcs connect places with transitions and represent the relations between them. The arc's direction indicates the flow of information (token flow) through the net.
A transition is enabled if there is at least one token in each of its input places. When enabled, a transition removes a token from each input place and adds a token to each output place.
Tokens represent and carry information.
Marking of a Petri net is the position of tokens in the net at any instant in time. A given marking of a Petri net defines which transitions are fireable. The firing of a transition moves the net to a new marking.

A marking μ' is said to be reachable from a marking μ if there is a sequence of intermediate markings (and transitions) leading from μ to μ'. The set of all reachable markings from μ is called reachability set and can be represented by a reachability graph.
In this paper, we assume that:

1. only one transition can be "fired" at a time, and
2. the number of tokens in every place does not exceed one (safe Petri net).

To construct the Petri net model of a process planning problem we first analyze it and all necessary machining operations and associated resources are determined. All possible precedence relationships among machining operations are recognized and established. This is very important because it confines the number of possible solutions. For the same reason groups of operations to be executed under the same conditions are established if possible.
The following rules are applied for the proposed Petri net modelling method for process planning:

1. Each machining operation (O_i) or a well established group of them is represented by a transition, (T_i).
2. Alternative (candidate) machining operations (O_{ij}) for a design entity, if any, are represented by transitions (T_{ij}) using the same input and output places.
3. There is a common input-output (dynamic) place (ControlPlace) with an initial token (initial marking) for all transitions representing a)the evolution of the state of the processed part (its state after every fired transition/machining operation), and b) the evolution of the state of the machine tool (tool and table position).
4. For each transition T_i create one output place with no successor transition. This end-place (EP_i), after receiving one token from the corresponding transition after its firing, indicates that this transition has already been fired and cannot be fired again.
5. For each successor transition T_k of a transition T_i create one output place (CP_{ik}) of T_i which is an input place for the corresponding successor transition T_k. This kind of place represents the type of knowledge given by a precedence relation constraint.
6. All arcs are weighted by 1.

A Petri net model constructed according to the above rules:

- represents accurately and dynamically the process planning procedure for a given mechanical part
- provides a graphic tool for knowledge representation of the type of precedence relations constraints, represented by the relation: transition-output place-successor transition
- provides a powerful simulation tool for process planning, simulating both machining operation sequence and state of part and machine tool
- gives all possible solutions/process plans by simulation tracing or reachability analysis

Figure 2 shows graphically the constructing elements and the basic generic structure of the proposed Petri net model for process planning.

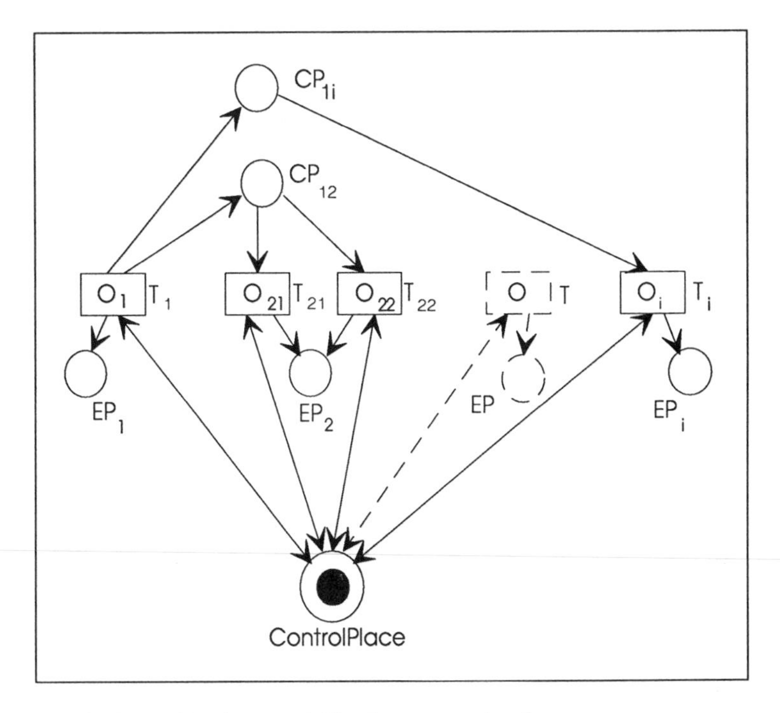

Figure 2 Principles of Petri net modelling for process planning.

4. AN EXAMPLE

The following example shows the principles of the proposed Petri net modelling method for process planning.
Consider the part of Figure 3. It is an "academic" prismatic workpiece consisting of five Design Entities (surfaces) to be machined in the same set-up, on the same machine tool (machining center).

4.1. Part Analysis

For each surface of this part, defined here as design entities (DE), the machining operations summarized in the machining table Table 2 are required. The DE 2 and 4 should be associated, i.e. machined with the same tool at the same time.

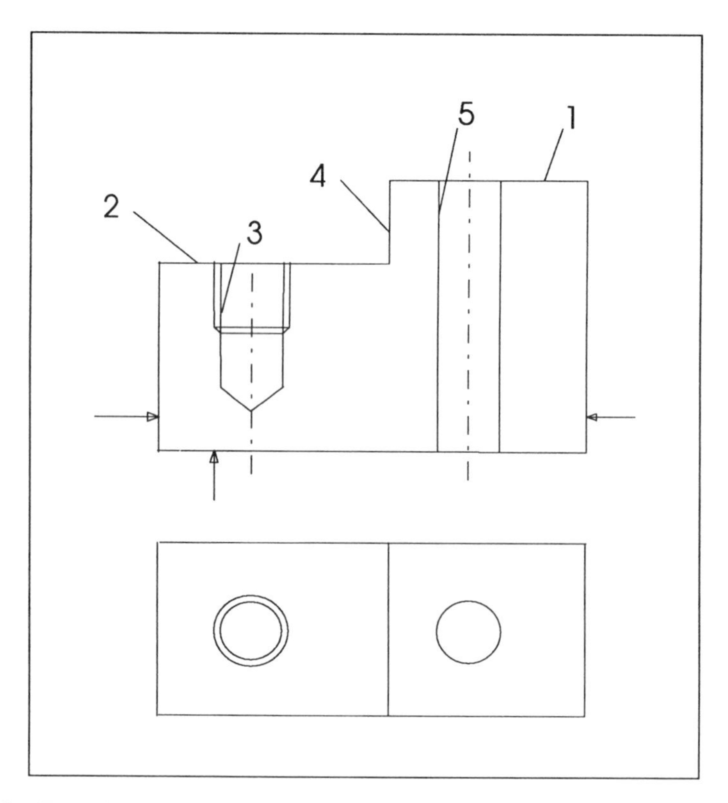

Figure 3 Example part.

Table 2 Machining table of the example of Figure 3

	PREPA-RATION	ROUGH	1/2-FINISH	FINISH	POST-FINISH
DE_1		1R1 1R2		1F1 1F2	
DE_2 **DE_4**		24R		24F	
DE_3	3C	3D		3T	
DE_5	5C	5D			

where C stands for Centering, R for Roughing (milling), F for Finishing (milling), D for Drilling, and T for Tapping.

Candidate machining operations are defined according to the recognized possibilities to realize a DE in terms of resources (tools, machine-tools, etc.): for this example we assume 1R1 and 1R2 as candidate operations for roughing DE_1, and, 1F1 and 1F2 as candidate operations for finishing DE_1.

The precedence relation constraints that should be respected are:

- operation ***1F*** must be executed ***after*** operation ***1R and after*** operation ***5D***
- operation ***24R*** must be executed ***after*** operation ***1R***
- operation ***24F*** must be executed ***after*** operation ***24R and after*** operation ***3T***
- operation ***3C*** must be executed ***after*** operation ***24R***
- operation ***3D*** must be executed ***after*** operation ***3C***
- operation ***3T*** must be executed ***after*** operation ***3D***
- operation ***5C*** must be executed ***after*** operation ***1R***
- operation ***5D*** must be executed ***after*** operation ***5C***

which are summarized into the following precedence graph:

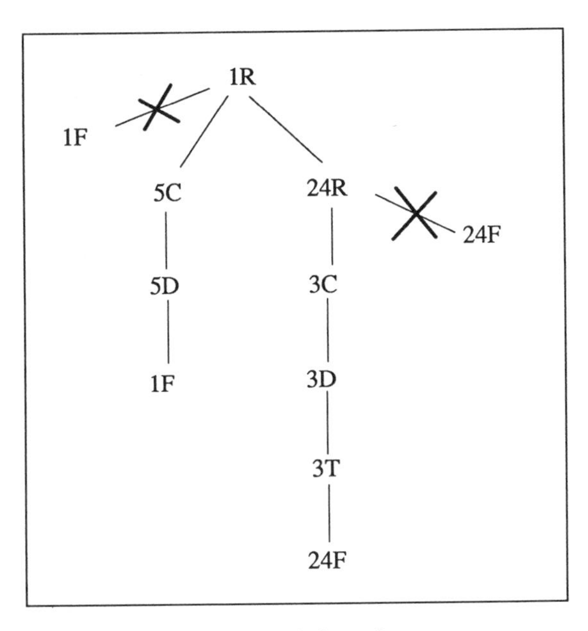

Figure 4 Precedence graph of the example of Figure 3.

By inspection of the precedence graph, we eliminate superfluous precedence constraints if any. In the above graph, for example, we can eliminate the direct precedences 1R-1F and 24R-24F because they are covered indirectly by the precedences 1R-5C-5D-1F AND 24R-3C-3D-3T-24F respectively. The elimination of superfluous precedence constraints is useful since it simplifies the corresponding Petri net without changing the final result.

4.2. Petri Net Modelling

The corresponding Petri net model is shown in Figure 5. The initial marking of this Petri net consists of one token in the common place "ControlPlace". We can simulate process planning by firing this Petri net. A process plan can be obtained after a complete simulation cycle by tracing the information carried by the token of the ControlPlace during its traveling through all transitions. Of course simulation is not an efficient method to find all possible process plans. Instead, it is a very powerful tool to detect and correct "solvable" conflict situations, if any, by adding extra constraints when this seems to be possible.

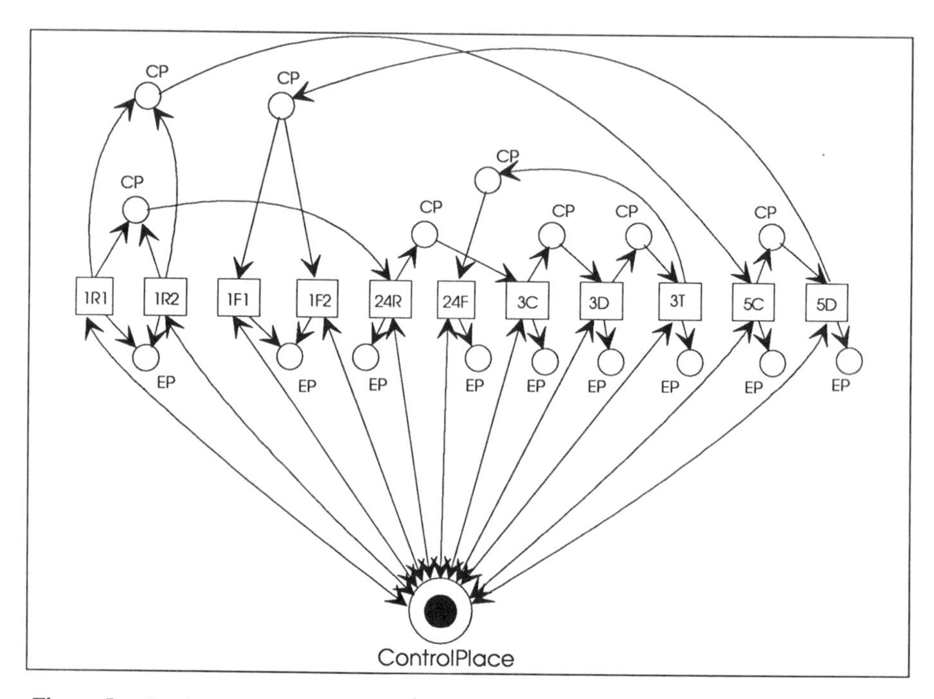

Figure 5 Petri net model of the example of Figure 3.

4.3. Reachability Analysis

After defining a Petri net model we can perform reachability analysis (Peterson, 1981) and calculate all alternative solutions. It is not possible to do reachability analysis manually, especially if the Petri net is big. We need to use software tools that perform reachability analysis automatically. Such a tool is the PROD system from the Helsinki University of Technology (Grönberg et.al., 1993). But since, with PROD, it is not possible to restrict the number of tokens of a place to one, we have to construct the equivalent of the Petri net of Figure 5 which respects PROD assumptions. This new Petri net is shown in Figure 6 and it is totally equivalent with the original Petri net of Figure 5 from the modelling and simulation point of view. Its difference with the original net is that the common place "ControlPlace" is replaced by more input places, one for each main transition. Transition corresponding to alternative candidate operations have one common input place. Input places are connected with their corresponding transitions by uni-directed arcs from the place to transition. By this method we guarantee that each transition will be fired only once. In the net of Figure 5 where the "ControlPlace" is connected with all transitions by bi-directed arcs, this is guaranteed by the assumption that the number of tokens at any place is restricted to one. The corresponding reachability graph of the Petri net of Figure 6 is shown in Figure 7. All data for creating this reachability graph were calculated automatically by using PROD.

The reachability graph represents also all possible transition sequences (arcs of the reachability graph), 84 totally, in other words all possible process plans. A list of a part of them is given in Figure 8. This list is automatically created by applying a depth-first search algorithm (Baase, 1978) to the reachability graph of Figure 7.

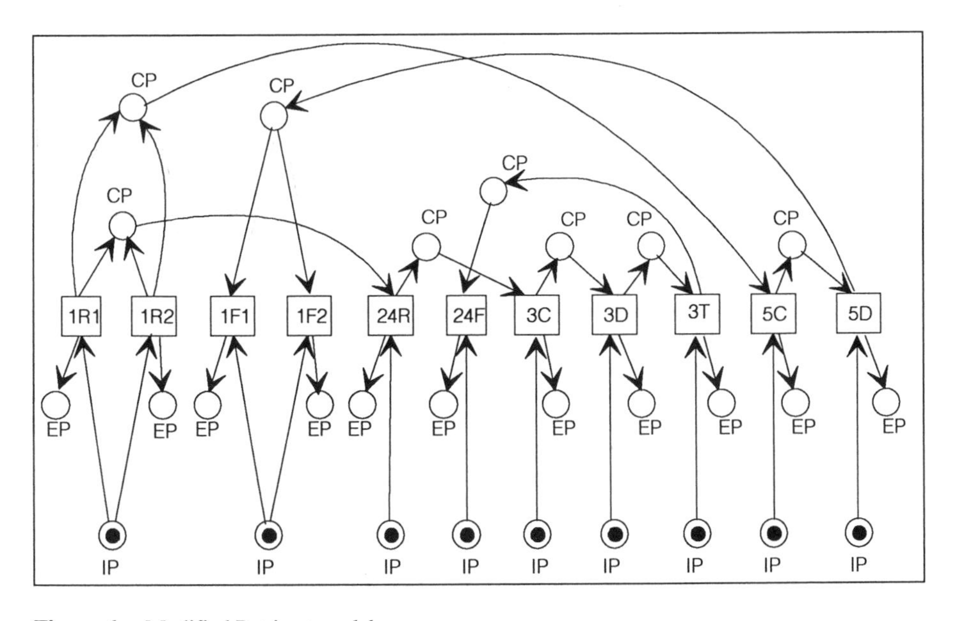

Figure 6 Modified Petri net model.

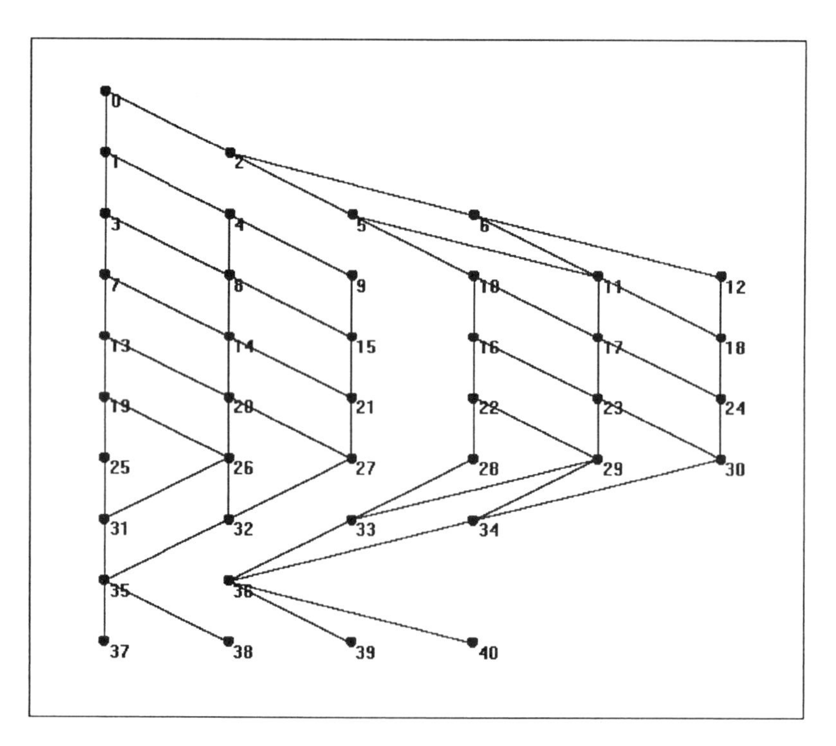

Figure 7 Reachability graph of the Petri net model of Figure 6.

1) T1R1 ->T24R ->T3C ->T3D ->T3T ->T24F ->T5C ->T5D ->T1F1
2) T1R1 ->T24R ->T3C ->T3D ->T3T ->T24F ->T5C ->T5D ->T1F2
3) T1R1 ->T24R ->T3C ->T3D ->T3T ->T5C ->T24F ->T5D ->T1F1
4) T1R1 ->T24R ->T3C ->T3D ->T3T ->T5C ->T24F ->T5D ->T1F2
5) T1R1 ->T24R ->T3C ->T3D ->T3T ->T5C ->T5D ->T24F ->T1F1
6) T1R1 ->T24R ->T3C ->T3D ->T3T ->T5C ->T5D ->T24F ->T1F2
7) T1R1 ->T24R ->T3C ->T3D ->T5C ->T3T ->T24F ->T5D ->T1F1
8) T1R1 ->T24R ->T3C ->T3D ->T5C ->T3T ->T24F ->T5D ->T1F2
9) T1R1 ->T24R ->T3C ->T3D ->T5C ->T3T ->T5D ->T24F ->T1F1
10) T1R1 ->T24R ->T3C ->T3D ->T5C ->T3T ->T5D ->T24F ->T1F2
11) T1R1 ->T24R ->T3C ->T3D ->T5C ->T5D ->T3T ->T24F ->T1F1
12) T1R1 ->T24R ->T3C ->T3D ->T5C ->T5D ->T3T ->T24F ->T1F2
13) T1R1 ->T24R ->T3C ->T5C ->T3D ->T3T ->T24F ->T5D ->T1F1
14) T1R1 ->T24R ->T3C ->T5C ->T3D ->T3T ->T24F ->T5D ->T1F2
15) T1R1 ->T24R ->T3C ->T5C ->T3D ->T3T ->T5D ->T24F ->T1F1

Figure 8 Process plans as transition sequences (15 out of 84).

4.4. Finding An Optimum Solution

It is possible that not all of the given by a reachability graph solutions are interesting. Especially in the case we have a lot (thousands) of possible solutions the finding of an optimum solution becomes combinatorial and there is a need of an optimization method using one or more heuristic criteria based on rules concerning machining actions executed in the corresponding transitions. This issue will be discussed in another paper.

5. CONCLUSIONS

The basic ideas of a graphic approach to process planning by using Petri nets have been discussed in this paper. A generic Petri net dynamic model of process planning has been presented and applied to an example part. The use of the proposed model is twofold. Firstly, it is a simulation tool used to detect possible problems during process planning. Secondly, reachability analysis of the Petri net gives all possible process plans for a given part under a set of given constraints. Consequently, an optimum process plan can be found by applying an optimization method to the result of the reachability analysis based on one or more heuristic criteria.

Future work will include the application of search algorithms directly to the proposed model and the introduction of time to its elements (Lee and DiCesare, 1992, Shen, et.al., 1992) The combination of Petri nets with multi-agent techniques seems to be also a promising domain of research in this direction.

Another interesting topic should be the investigation of the use of the proposed Petri net model as an element of an integrated higher level Petri net model of a production planning system (Tönshoff, et.al., 1987, 1989). This should permit the search of process planning solutions in real time according to the availability of resources at a given time (Just In Time process planning).

6. ACKNOWLEDGMENTS

The author would like to thank Abdelkader Belhi for his help in the example, Norbert Ebel for proposing the depth-first search algorithm, and Kimmo Varpaaniemi for his help to understand and work with PROD.

7. REFERENCES

Baase, S. (1978) *Computer Algorithms: Introduction to Design and Analysis,* Addison-Wesley Publishing Company Inc.

Cecil, J.A., Srihari, K. and Emerson, C.R. (1992) A review of Petri Net Applications in Process Planning, *The International Journal of Advanced Manufacturing Technology*, **7**, 168-177.

Grönberg, P., Tiusanen, M. and Varpaaniemi, K. (1993) PROD - A Pr/T-Net Reachability Analysis Tool, *Technical Report*, Series B, No. **11**, Helsinki University of Technology, Digital Systems Laboratory.

Iwata, K. and Fukuda, Y. (1989) A New Proposal of Dynamic Process Planning in Machine Shop, *CIRP International Workshop on Computer Aided Process Planning*, 73-83, IFW, Hannover University.

Kiritsis, D. (1995) A Review of Knowledge-Based Expert Systems for Process Planning, *The International Journal of Advanced Manufacturing Technology,* accepted for publication.

Kruth, J.P. and Detand, J. (1992) A CAPP System for Nonlinear Process Plans, *Annals of the CIRP*, **41/1**, 489-492.

Lee, D.Y. and DiCesare, F. (1992) FMS Scheduling Using Petri Nets and Heuristic Search, *Proceedings of the 1992 IEEE International Conference on Robotics and Automation,* 1057-1062.

Peterson, J.L. (1981) *Petri Net Theory and the Modeling of Systems*, Prentice-Hall Inc, Englewood Cliffs, NJ.

Shen, L., Chen, Q., Luh, J.Y.S., Chen, C. and Zhang, Z. (1992) Truncation of Petri Net Models of Scheduling Problems for Optimum Solutions, *Proceedings of the JAPAN/USA Symposium on Flexible Automation,* ASME, **2**, 1681-1688.

Srihari, K. and Emerson, C.R. (1990) Petri Nets in Dynamic Process Planning, *Computers Industrial Engineering*, **19**, 447-451.

Tönshoff, H.K, Beckendorff, U. and Anders, N. (1989) FLEXPLAN-A Concept for Intelligent Process Planning and Scheduling, *CIRP International Workshop on Computer Aided Process Planning*, 87-106, IFW, Hannover University.

Tönshoff, H.K, Beckendorff, U. and Schaelle, M. (1987) Some Approaches to Represent the Interpedence of Process Planning and Process Control, *Proceedings of the CIRP Seminars, 19th CIRP Seminar on Manufacturing Systems, Computer Aided Process Planning*, Pennsylvania State University, also in *Manufacturing Systems*, **18/2**, 93-113.

8. BIOGRAPHY

Dimitris Kiritsis was born in Greece in 1957. He received a diploma (1980) and a PhD (1987) in mechanical engineering from the University of Patras, Greece. His primary research interests are in computer-aided manufacturing, computer-aided process planning and high precision interpolation methods for CNC. He is a senior researcher at the CAD/CAM Laboratory of the Swiss Federal Institute of Technology at Lausanne, Switzerland.

PART SEVEN

Manufacturing System Coordination and Integration

15

Coordination Approaches for CIM

Moira C. Norrie, Martin Wunderli, Robert Montau, Uwe Leonhardt,
Werner Schaad, Hans-Jörg Schek
Swiss Federal Institute of Technology
ETH Zentrum, CH-8092 Zurich, Switzerland,
{norrie,wunderli,schaad,schek}@inf.ethz.ch,
{montau,leonhard}@vmeth.ethz.ch

Abstract

We propose a general architecture for Computer Integrated Manufacturing (CIM) based on the coordination, rather than integration, of component systems. The coordination process is achieved through inter-system dependencies controlled by a central, global coordinator. Coordination, like integration, may be either data- or application-oriented. In the case of data-oriented coordination, multidatabase technologies may be exploited to maintain global data consistency. For application-oriented coordination, the global coordinator uses operational dependencies as a basis for the invocation of methods in remote systems. We examine each of these orientations in detail and then provide a comparison of approaches. Specifically, we describe two prototype systems developed in the context of the CIM/Z project.

Keywords

integration methods, coordination model, repository technology

1 INTRODUCTION

Traditional approaches to Computer Integrated Manufacturing (CIM) have tended to focus on either a total or a partial integration of component systems. The integration approach may be either data- or application-oriented – or some combination of both. In the case of data-oriented integration, all or some of the component systems' data are stored in some form of logical or physical central database. A logical central database is one in which data may be distributed but there is a single, global schema through which the data is accessed. With application-oriented integration, systems are tightly-coupled either by means of direct calls from one application system to another, or, by encapsulating the component systems in a single, global application system.

Generally, the main problems of the integration approach are loss of component system autonomy and a lack of support for system evolution. There are many aspects of system autonomy some of which may be compromised by a particular integration approach. For example, total data integration based on a single global schema is not only expensive in terms of the integration effort but also forces a single enterprise model and this might be considered as a sacrifice of data autonomy. Controlling access to shared data objects by storing them in a central database means that the availability of the central database is critical to the continued working of the component systems; this can be described as a loss of operational autonomy.

System evolution can occur through the addition of new component systems or the replacement of existing component systems. To support evolution, it is essential that the general architecture is flexible and that component system dependencies are minimised and isolated. Some forms of integration are too inflexible in that inter-system working is hard-coded and not amenable to change.

In an effort to minimise loss of autonomy and maximise flexibility, we advocate a looser coupling of component systems based on coordination rather than integration. A central, global coordinator responds to actions in one component system by delegating actions to one or more other component systems. For example, the activities of a Computer-Aided Design (CAD) system and a Production Planning System (PPS) system may be coordinated in such a way that the deletion of a CAD assembly results in the coordinator requesting the deletion (or invalidation) of related part lists in the PPS system.

As far as possible, component systems continue to operate as before with coordination performed "behind the scenes". Any changes to the user view of component system functionality arise not from the coordination process itself but rather only in cases where the functionality may be enhanced either in terms of local system extensions or through the introduction of global applications.

The coordination process is based on inter-system dependencies which are stored in the global coordinator's database using a global representation scheme. These dependencies may be viewed, interrogated and updated with relative ease. An important factor is that the global coordinator is not critical to individual system operation. It is the assumption that the coordination activity occurs when significant changes are made to local component system data, e.g. checking-in a design object, or, when global queries are evaluated. Further, we assume that the coordination process is not time-critical.

Just as integration approaches may centre either on data or operations (or some combination of both), coordination approaches may be based on either data dependencies or operational dependencies. In the former case, coordination is achieved through the maintenance of global data consistency. In the latter case, methods in one component system are mapped to methods in one or more other component systems through a global object method server.

In section 2, we discuss the general requirements of CIM in terms of supporting cooperative working and we summarise various approaches to the realisation of CIM systems. Section 3 presents the coordination approach in further detail by introducing a general coordination architecture and an example of the coordination process. Sections 4 and 5 detail two coordination approaches which have been investigated in the context of the CIM/Z project (Integration of Databases and CIM Component Systems) at ETH Zurich (Schaad, Montau, Wunderli, Leonhardt & Lüthi 1993). The first of these is an approach based on data dependencies and the exploitation of multidatabase technologies and is described in section 4. The second approach, based on operational dependencies, is described in section 5. Section 6 provides some general comments on the two approaches. Concluding remarks are given in section 7.

2 CIM REQUIREMENTS AND APPROACHES

The manufacturing process spans many departments and activities of an enterprise. The department structure reflects the decomposition and distribution of activities and, correspondingly, separate application systems have been developed to support the activities of these departments. Enterprise structure is not the only reason for the evolution of independent application systems. Manufacturing systems cover a wide range of activities, both technical and management, and these may have very different requirements in terms of the amount and type of data, the functionality, the run-time environment and performance. Consider the two main tracks of information flow shown in figure 1.

The economic and administrative track comprises data which is usually organised within complex but monolithic applications purpose-built for the tracking of the complete economic life cycle of a product. Because of the integrated nature of these so called Production Planning Systems (PPS) (or Management and Resource Planning Systems (MRP)), *data exchange* and *data consistency* between the different production phases has not been a problem. Even current efforts to decompose and modularise these applications does not effect the data consistency and the capability to exchange data between the modules. So we can regard data exchange and data consistency as guaranteed on this track of information flow.

On the technical track, we find a quite different and much more complex situation. Here, the applications are highly specialised. The primary purpose of a component system is not to exchange or synchronise data but rather to fulfil, in an efficient way, a specific application task, such as the design activity (CAD) or the programming of numerically controlled machines for a CAM process. Often data exchange consists of physically transferring files on disks between departments or data may even have to be reentered because of the heterogeneity of the computing environment.

A slightly different view of the information flows in the production process is taken by Scheer (Scheer 1990). Scheer divides the production process of a product into two phases, the *planning phase* and the *manufacturing phase* which is reflected in both the economic and the technical track. Figure 2 shows in more detail the required tasks (and departments) of the manufacturing process and the supporting component systems.

CAE
CAD
CAP
CAM
CAQ
economic and administrative information flow
technical information flow
PPS
General Planning
Detail Planning
Con-troling
Order-checking

Figure 1: Data sources in manufacturing

It has been found that the decisions taken in the planning phase, although directly causing only about 10% of the total production costs, have a high indirect effect on the total cost of a product because of their influence on the manufacturing phase. At the start of the manufacturing phase, all decisions for the ergonomics, functionality and production costs of a product have already been taken, either directly or indirectly. So the main purpose of this phase is to fulfil the directives defined by the planning phase.

Increasing national and international competition has led to the shortening of product life cycles and an increase in both product complexity and the importance of cost efficiency and flexibility in production. As a result, the necessity of standardised product data exchange and automatic data consistency control has become a prerequisite of a successful enterprise.

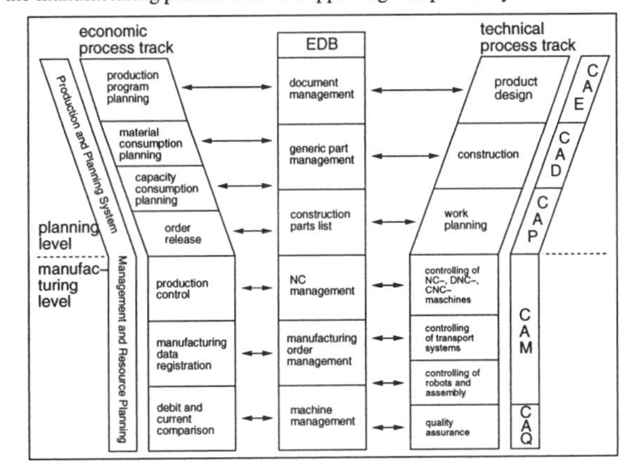

Figure 2: Data sources in manufacturing

The problem of *product data exchange* between CAx component systems has been tackled by different national and international standards. As examples, we mention the 'International Graphics Exchange Standard' (IGE 1988), the 'Standard d'Échange et de Transfert' (SET 1984, SET 1985) and country specific subsets like the 'Verband

Deutscher Automobilindustrie — IGES Subset' (VDA 1987*a*) and the 'VDA Flächenschnittstelle' (VDA 1987*b*); these are currently used for the exchange of data but are more or less restricted to 2D/3D geometric data. However, the upcoming ISO Standard 10303 STEP/EXPRESS (STE 1992), which covers the exchange of product model data in total, is intended to replace all the currently used product data exchange standards and so to solve the problem of different, partly overlapping, partly disjoint formats. Note that these standards are *not meant* to ensure data consistency across CAx systems' data. Nevertheless, they can serve as the foundation for tools built for that purpose.

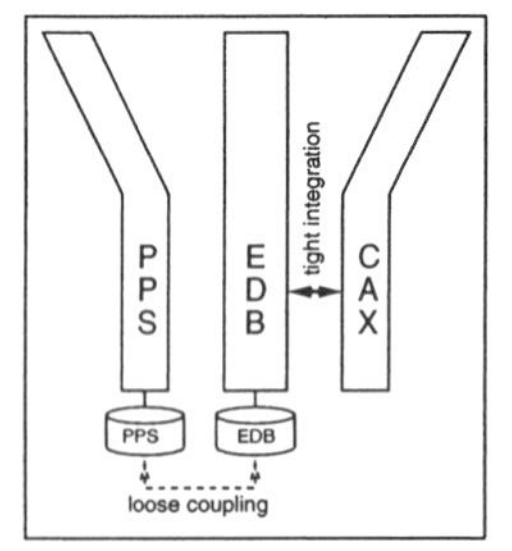

Figure 3: Current EDBs

In the classic CIM strategy, data could be integrated into one central database. However, in practice, such a total integration is not achieved mainly because of variation in requirements, complexity and the self-evident need for the use of existing systems. Additionally, PPSs are very large and well-established and therefore must be retained. For these reasons, the economic and administrative data is kept separate from the technical data. This is commonly referred to as the Engineering Database (EDB) approach and is illustrated in figure 3.

The problems with this approach are twofold. First, the EDB is critical to the operation of all technical component systems. Any failure occurring in the EDB means that work on component CAx systems cannot proceed. Second, usually there is little or no coordination between the PPS and the EDB and therefore global consistency of the CIM system is left to management control. Further, as discussed before, the integration of all the technical data into one system may result in a certain degree of inflexibility in terms of replacing component systems with alternative products.

The recommendation is to leave component systems with their local data and only integrate some of it into the EDB as illustrated in figure 4. To avoid an unnecessary replication of data, the EDB may consist primarily of references to local data objects and any data required to support extensions to the functionality of component systems. An example of such a functionality extension would be to allow users of a CAD system to query part information from other component systems to locate parts with designs which can be modified to give new designs. Due to the fact that only 25% of a design is new in terms of functionality, reusability can be applied in at least 75% of the design task.

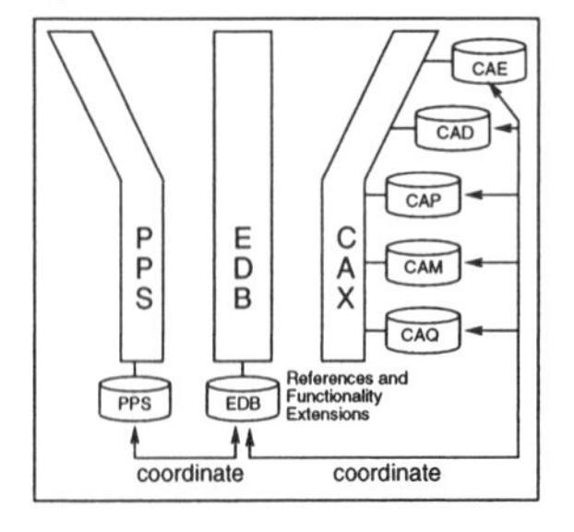

Figure 4: Recommended data handling for EDBs

For global consistency in a CIM system based on this approach, coordination is required between the technical CAx systems and the EDB and also between the EDB and the PPS. To achieve this, we propose a coordination architecture which minimises the degree of integration thereby increasing the autonomy of component systems and flexibility in terms of the introduction and replacement of component systems. This approach is based on inter-system dependencies which are stored in a central coordinator in a global representation language. Dependencies among systems can easily be queried and updated. We describe the coordination approach in further detail in the following section. In the context of the CIM/Z project, we have investigated two basic coordination approaches: one is data-oriented in that it is based on data dependencies and the other is method-oriented in that it is based on operational dependencies. Prototype systems have been developed for each of these two approaches and these are described in sections 4 and 5 respectively.

Other related projects include the following. The concept of using an integration database was used in a project at IBM Germany, at the Heidelberg Scientific Centre (Brosda 1992, Brosda & Herbst 1992). The use of enhanced database support (e.g. by using complex objects) for CIM was investigated at IBM Almaden Research Centre, San Jose (Lorie & Bever 1987). The integration of CAD and Production Planning Systems through the use of update dependencies was studied in a project at the University of Maryland (Mark, Roussopoulos & Cochrane 1994). Achieving coordination through the use of a central, integrated document and order management system was investigated in the DOCMAN project at the Technical University of Aachen (Eversheim & Grosse-Wienker 1991).

Our data-oriented approach relates to ActMan (Jablonski, Ruf & Wedekind 1988, Jablonski, Ruf & Wedekind 1990), a project at the University of Erlangen-Nürnberg, Germany. This project investigated a form of bilateral coordination between different CIM component systems. An important goal of our project is that we want to increase the local autonomy for the component systems, whereas, in ActMan, a tight coupling between the component systems was chosen. For example, ActMan uses a central, global database where globally important data is replicated. We avoid this approach and keep, whenever possible, all the data in the data management system of the component system. We then increase the local autonomy by using an advanced transaction model rather than the two-phase commit protocol as used, for example, in ActMan. Further differences stem from the fact that both projects use different concepts for global constraint management.

3 COORDINATION ARCHITECTURE

In this section, we outline the general architecture of a coordination system and the functionality of the two main kinds of component through which coordination is achieved; these are the CIM Agents and the Global Coordinator. Figure 5 shows the general coordination architecture; for simplification we show only two component systems.

As an example of coordination, we consider a CAD system and a PPS. The CAD system supports the design activity and stores information about the various CAD drawings consisting of CAD assemblies and their components. One of the PPS' tasks is to manage information about the structure of parts. There is a dependency between the two systems; part data referred to in the CAD drawings must exist in the PPS. If a part is discontinued, then we must ensure that designers cannot reference this part in future designs – and we must somehow inform designers that existing designs using this part are no longer valid. This is an example of an inter-system dependency which is the basis of the coordination process and will be stored in the Coordination Repository.

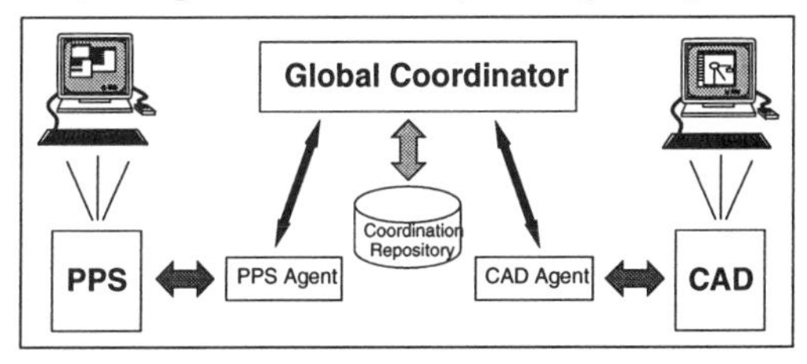

Figure 5: General Coordination Architecture

The main task of the Global Coordinator is to ensure, with the help of the CIM Agents, the consistent state of the CIM system. The Global Coordinator does this by coordinating activities of the various component systems according to various inter-system dependencies. Each component system has a CIM Agent which monitors local activity and notifies the Global Coordinator of any actions that are pertinent to the inter-system dependencies. Based on these dependencies, the Global Coordinator will delegate necessary actions to one or more component systems via their CIM Agents.

A CIM Agent provides the coordination interface for a component system. As illustrated in figure 6, a CIM Agent must be able to communicate with its component system and also with the Global Coordinator. One of the major functions of a CIM Agent is to map between local representations of operations and data to the global representation scheme. Local system dependency is therefore isolated in part of the local CIM Agent and the rest of the coordination components are general and not specific to particular component systems.

A CIM Agent may have its own repository to record information about which local operations and data are relevant to coordination, logs of local activities and any other information required to perform its part of the coordination task.

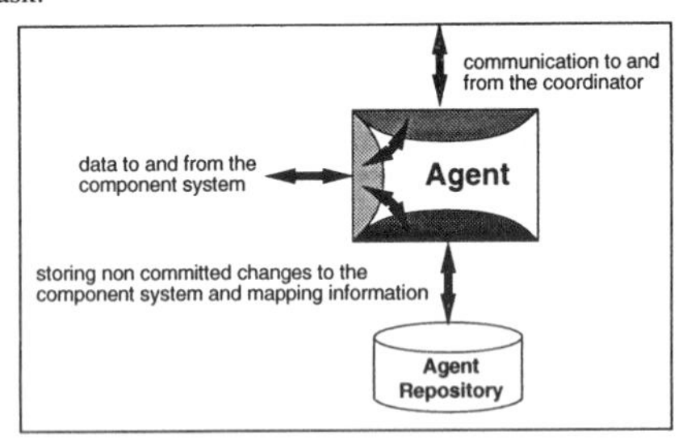

Figure 6: General Architecture of Agent

Global transaction management is a prerequisite to ensure that all information about changes and the corresponding coordination activities reach the relevant parties and are acted upon. Since a component system may not be a database application system and therefore may be without transaction and recovery support, it may be necessary that its CIM Agent takes over the role of providing some form of logging and atomicity control. The issue of global transaction management is an important part of the CIM/Z project and has been investigated in the context of the data-oriented coordination approach which is discussed in section 4.

The coordination architecture presented in this section is very general and does not specify either the form of inter-system dependencies or the methods of communication between the various component systems, their CIM Agents and the Global Coordinator. We now go on to discuss two specific architectures – one based on data dependencies and one based on operational dependencies.

4 COORDINATION THROUGH DATA DEPENDENCIES

In the data-oriented coordination approach, the activities of the CIM component systems are coordinated through the maintenance of system-wide data consistency. Here we describe a particular data-oriented coordination prototype, CIM/DB, that has been implemented as part of the CIM/Z project (Norrie, Schaad, Schek & Wunderli 1994*a*, Norrie, Schaad, Schek & Wunderli 1994*b*). CIM/DB adopts a "database approach" to the CIM problem and exploits multidatabase technologies. To illustrate the approach, we return to the example of the CAD and the PPS systems discussed in the previous section. There is a data dependency between the two systems; parts referred to in the design assemblies of the CAD system must exist in the PPS. Such a dependency is expressed as a global data constraint and the Global Coordinator must be informed of any updates within a component system that may result in a violation of such a constraint.

The Global Coordinator stores in a coordination repository information about the various component systems, their schemas, the global constraints and reactions on constraint violations. In addition, it is necessary for the coordinator to store information about the individual relationships between data objects of the component systems. For example, it must record the relationship between a particular part list of the PPS and a particular design of the CAD system. When the Global Coordinator is informed of local component changes, it delegates actions to be taken by the appropriate CIM Agents in order that consistency be maintained.

The Coordination View is a conceptual model of the Global Coordinator's data. It is expressed in terms of the semantic data modelling language NIAM/RIDL (Wintraecken 1989). Figure 7 shows the part of the Coordination View that describes object types and the constraints in which they are involved. The solid circles represent entities while boxes represent relationships between those entities. The dashed circles represent attributes. For example, each ISA constraint is represented by an *ISA_Constraint* entity which has an attribute *ISA_Constraint_Type* that indicates the category of ISA constraint (disjoint, total etc.). Bars above relationship boxes are used to specify relationship cardinalities. For example, the placement of a single bar above the relationship between *ISA_Constraint* and *ISA_Constraint_Type* indicates that the relationship is one-to-many, i.e. an ISA constraint has a type, but there may be many ISA constraints of that type. The 'V' on the *ISA_Constraint* edge indicates that every ISA constraint must have such an attribute.

Each global constraint is represented by a constraint entity and is classified as either a relationship constraint (*Relationship_Constraint*) or a subtyping constraint (*ISA_Constraint*).

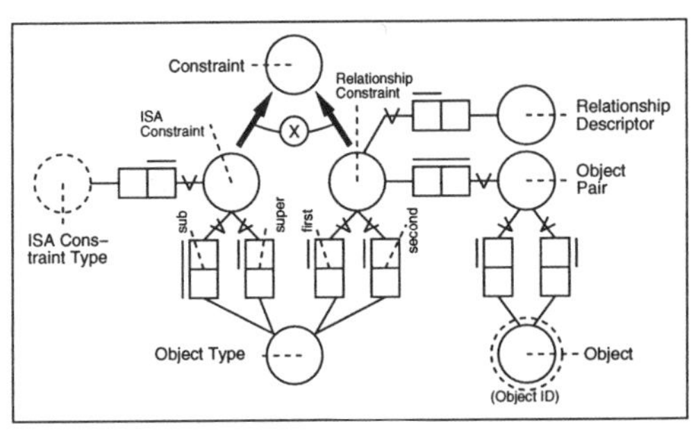

Figure 7: Part of the Coordination View

An example of a relationship constraint is shown in figure 8: A CAD assembly must have at least one, and possibly many, associated part lists while a part list is associated with exactly one CAD assembly.

In addition to the representation of global constraints, the coordinator repository must record the actions to be taken on the violations of such constraints. For example, it might associate with the constraint of figure 8 that a violation of this constraint resulting from the deletion of a particular CAD assembly should invoke actions to delete all of the associated part lists. Then the coordinator would send messages to the appropriate CIM Agents requesting them to initiate local operations to delete the relevant part list objects. This would all be performed within a global transaction to ensure that either all of the local deletions are performed, or, the original deletion of the CAD assembly would be undone and the user initiating the CAD delete operation would be informed of the failure of the delete and some compensating local action would be performed under the auspices of the local CIM Agent.

The inter-system data dependencies can be thought of as having both static and dynamic parts. The static part is expressed as global constraints that should be satisfied by the system. The dynamic part specifies actions to be taken by the Global Coordinator to restore global consistency in the event that an action in one component system causes one or more constraints to be violated. Further details of the data dependencies and the language used to define them is given in (Norrie & Wunderli 1994).

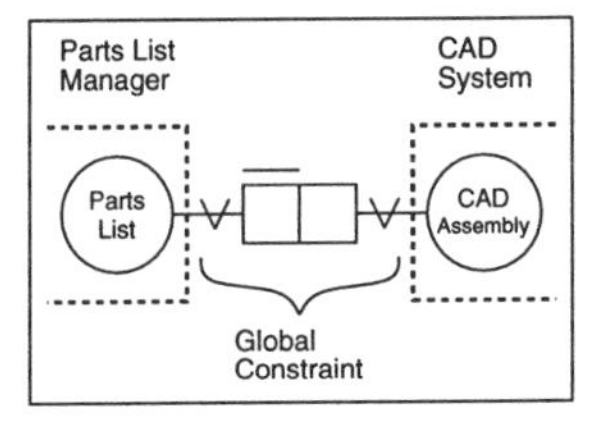

Figure 8: Relationship Constraint

The actions known to the Global Coordinator are high-level logical operations which are mapped by the CIM Agents to one or more local operations. For example, the Global Coordinator may request the deletion of an object in a component system. It is possible that this is mapped directly to a database delete operation on the local object. However, it is also possible that, in a specific component system, it is mapped to a sequence of local operations that involve manual intervention such as requesting a local system administrator to perform the delete and notifying local users.

The actions to be taken by the Global Coordinator in the event of a global constraint violation are specified in the Global Coordinator's repository. It would be very inconvenient if every reaction description for every combination of constraint and violating action had to be specified individually. For this reason, default reactions for the various forms of constraint and violating actions are predefined so that they ensure a consistent database state in terms of the global data model.

It is the responsibility of a CIM Agent to notify the Global Coordinator of updates on objects relevant to the global data constraints. This communication requires an exchange model and language, such as STEP/EXPRESS (STE 1992), for communication between the coordinator and agents — but we shall not detail this here. The CIM

Agent has a local NIAM/RIDL schema which describes those parts of the local data set relevant to system-wide consistency; this defines the Coordination Interface of the component system in terms of globally important objects.

The interface between a CIM Agent and its component system is less flexible than those between a CIM Agent and its repository and between the agent and the coordinator since it involves interfacing to existing systems. To understand how such an interface can be realised, we outline one of the component systems of CIM/DB — the CAD system Pro/Engineer (Par 1993). The objects managed by this CAD system are, for example, assemblies, parts and drawings. Initially, a user either calls a search/retrieve operation to load an existing object or they create a new object. An object in memory can be modified by calling a sequence of predefined update operations. There are also operations to query an object and to create reports about its components. On completion of their task, the user can call a store operation to save the changes to the CAD object.

To perform the coordination task, a CIM Agent has to take over the role of a "spy". It has to monitor the operations that are executed and, based on its model of globally important objects, must recognise operations of global interest. For example, for the part lists objects stored in the PPS, it is important to know when a part or subassembly is added to or deleted from an assembly in the CAD system. Then the CIM Agent must be aware of calls to operations to create, assemble, disassemble or delete parts and subassemblies.

In the case of Pro/Engineer, the CIM/DB component system is enhanced with procedures that are triggered whenever the associated operation is executed. Such a notification procedure can be installed so that it is called either before or after the execution of the associated operation; it sends a notification message to the CAD Agent with the name and the parameters of the operation. The agent receives the message, records it in its repository and evaluates the message. It specifies the operation and its parameters in terms of the global data model ready for communication to the Global Coordinator.

The CIM Agent also has to respond to actions delegated by the global coordinator. The operations and objects of the coordinator's request must be mapped into the language of the component system. For example, consider the case where the coordinator sends a message to the agent of the CAD system to remove a part from an assembly. The agent searches for the local names of that part and assembly in its repository. It calls a read operation for this assembly to load it into the memory and then a disassemble function is called to remove the part from the assembly. Finally, the modified assembly is stored. Information is saved in the repository in order that the user can be informed of the changes when the assembly is next accessed.

The issue of transaction management is how to ensure the atomicity and durability of global transactions, specifically in the case where one or more component systems do not support transaction management. A global transaction arises when a CIM Agent informs the coordinator of an operation on a globally important object. Some part of the coordination activity may not complete successfully either because of failure or due to the coordinator or some component system denying a requested action. Atomic transaction execution is the only guarantee that a change performed in one component system is propagated to the other component systems. The durability of global transactions ensures that once such a coordination activity has been completed successfully, its effects will not be lost.

If a component system is using a DBMS, the same DBMS may be used to store the agent's repository. A modification performed on both objects in the component system and objects in the repository can be performed inside a transaction thereby ensuring local atomicity. Additionally, the use of the component system's DBMS for the agent's repository avoids the replication of data. In the case where all component systems use a DBMS, the problem of building transaction management on top of existing DBMSs corresponds to the multidatabase problem (Breitbart, Garcia-Molina & Silberschatz 1992).

Component systems without any DBMS normally also do not support transactions. Some of these systems write all operations and their parameters to trail files. If the system crashes, these files are read after the restart and all operations are re-executed. This helps to avoid the loss of data locally but does not guarantee consistency if the component system has been involved in coordination activities. We therefore consider a slightly extended multidatabase problem where not all component systems support transactions.

If a component system has no transaction support, its CIM Agent monitors the system and logs local operations in its repository which may be considered stable as it is managed by a DBMS. In the case of a local abort, the agent checks the log after restart. If there is an entry about the start of an operation, but no entry about the completion of this operation, then the abort must have happened during the execution of the operation. It is possible that a global transaction had already been started in which case a *begin-of-transaction* would have been written to the log. The agent, together with the coordinator, can determine the state of the global transaction and, using a global commit protocol, terminate the global transaction. If the global transaction was completed successfully, the agent has to ensure that the local operation also completes. This is done by checking the version of the relevant globally important objects and, if necessary, re-executing some parts of the local operation. If the global transaction was aborted, the agent has to undo any local changes by reinstating the old version or, as an alternative, restarting the global transaction and completing the operation. We have built a CIM Agent for Pro/Engineer based on this approach.

The second aspect of transaction management is to isolate transactions executing in parallel. Rather than using strict two-phase locking (2PL) together with two-phase commit (2PC), which would severely restrict the local autonomy, we use open-nested transactions (Weikum & Schek 1992). Open nested-transactions have the advantage that subtransactions can be committed early, thereby allowing other transactions to access the objects in the database. Because, in some cases, committed subtransactions have to be compensated if the global transaction aborts, the system has to ensure that compensation will be possible. This is done by using multi-level transaction management (Weikum 1991, Beeri, Bernstein & Goodman 1989). By exploiting the semantics of the transactions, the global transaction manager decides what subtransactions are compatible and can be executed together. We have extended this idea so that we can have a mixed execution of global and local transactions (Schaad & Schek 1993, Schek, Weikum & Schaad 1991). This is necessary if we want to use this concept for existing component systems.

5 COORDINATION THROUGH OPERATIONAL DEPENDENCIES

The use of data dependencies, as described in the previous section, is only one way to obtain consistency in a CIM environment. Alternatively, one can focus on operational dependencies to ensure consistency by coordinating at the level of application methods. For this, it is necessary to know how data is accessed by the operations in a component system.

In order to discuss this operational approach of coordinating CIM component systems, the idea of global objects and methods must be reaffirmed. An object is of global interest if it is important for more than one component system. Therefore, access to global objects requires system-overlapping actions. For example, the insertion of a global object in the CAD system may require another insert in the PPS.

Any access to objects requires a method and these methods are usually kept in the method bench of an application. A method is defined as a self-contained function which performs an operation belonging to an object. These object methods include every operation to create, delete or manipulate an object (for example creating a point or a line) (Shooman 1983). In the case of relational database systems, an object is a tuple of a relation and an associated method could be a simple insert procedure. For complex objects which affect more than one relation, the corresponding method will also be more complicated involving access to several relations.

To achieve the various aims of connecting component systems in the CIM environment, a number of requirements have to be fulfilled. Primarily, a component system must be able to distinguish local and global objects and their methods. In the case of accessing a global object method, an appropriate action in the coordinator has to be initiated by sending the relevant information on the object, the required object method, their parameters and the initiator to the Global Coordinator. The coordinator then has to determine and initiate the requisite local methods in all of the involved component systems. During the execution of such a global object method, the task of the Global

Coordinator is to ensure the ACID-properties (atomicity, concurrency, isolation, duration) of transactions in order to perform global commits or aborts.

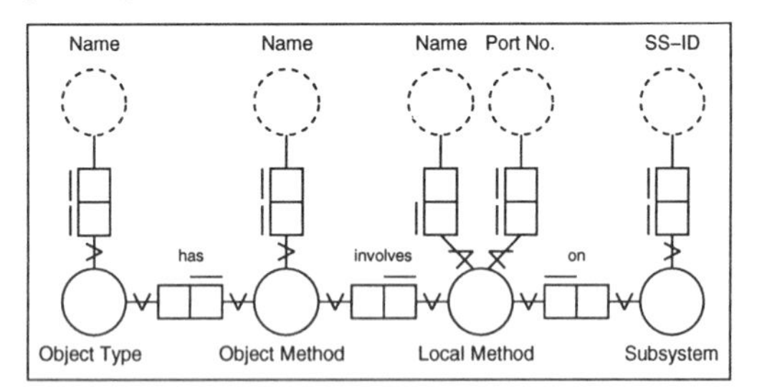

Figure 9: Metamodel for method server

To enable calls of methods between different systems, the Global Coordinator requires knowledge about the coherence between the systems, the relationship between global methods and local methods and the functionality of managing the control flow between the systems. In addition, the Global Coordinator must be able to find out where the required local methods reside in order to access them. Therefore, the Global Coordinator must store information about the global object methods associated with a global object and the involved local methods with their location, which is mapped to a port number, and the identifier of the component system to which they belong. The attachment of actions to component systems should be done in such a way that it is easy to view and change them and hence support system evolution. For that reason, the operational dependencies are described through the relationship of global and local methods as shown in the simplified NIAM-model for the Global Coordinator of figure 9.

Through this model, maximum flexibility of connection can be attained. If the Global Coordinator receives a request for a global object method, it only has to look for the required local object methods and initiate them. Further, the addition or deletion of a component system only requires a simple update to the Global Coordinator's database once the required local method has been implemented. For example, assuming a relational database for the Global Coordinator, only a single tuple has to be changed. In the case of migrating a component system from one computer system to another, only the location addresses (Port No.) of the local methods have to be updated. However, we note that, while it is very simple to realise a Global Coordinator in this way, it disregards possible transaction and communication problems.

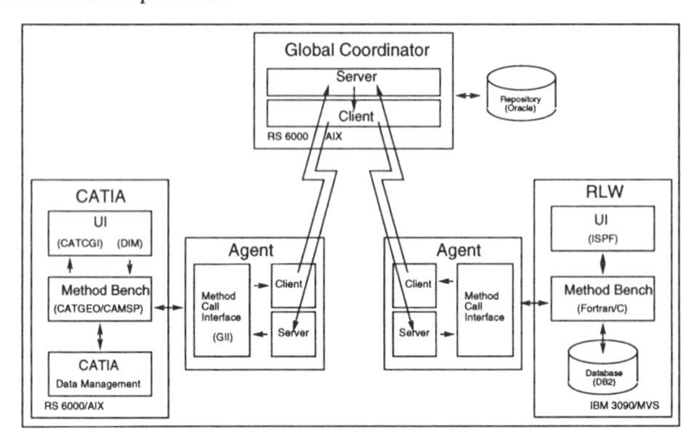

Figure 10: General architecture for a method server

To enable a system-overlapping access of methods, a reliable mechanism for the heterogeneous environment of current CIM systems is required. The standardised Remote Procedure Calls (RPC) (Bloomer 1992) can be used; these allow a method call in a remote computer system with different hardware and operating system from the source system. The RPCs are based on the TCP/IP network protocol, a common standard for data transfer in a heterogeneous computer environment. Almost every system provides an interface for network communication based on TCP/IP and the IP protocol is still the only way to obtain a really dynamic routing in a network (Gorys & Brauch 1993) which is a prerequisite for this coordination approach.

RPC use the client-server principle and the general architecture of coordination given in section 3 is elaborated in figure 10. The architecture indicates how RPC can be embedded in the coordination concept for an example of two component systems which will be discussed later.

A component system must know the network address of the Global Coordinator which must have an independent server process waiting for any calls to global object methods. The Global Coordinator must be able to call all the associated local methods, even if they are located on different hosts. Every CIM Agent must have a server process running which is prepared for any call of the local methods. These server processes can be accessed from the Global Coordinator by their internet addresses (SS-ID) and their port numbers.

In addition to the remote call of methods, the data part, which is the method parameters, has to be transferred. In a heterogeneous environment, there may be different ways of representing the standard data types like integer, character, etc. For this, the XDR standard from SUN-Microsystems is useful; it defines routines to translate between the standardised external representation and C-Variable-types which may be stored in different system-dependent codes such as ASCII or EBCDIC.

Another problem is the difference in the parameters of the local methods from those of the global method. A global valid format for the parameters of a global method is required. This global format for parameters consists of the particular parameters, their data type and their position in the transferred data stream. To start a global method, the client of the initiating agent has to map the format of the parameters from the local format to the global one. This global format can be described as the global language which the coordinator and every involved component system of a global method must understand. In the event of a global request arriving at a CIM Agent, the server part has to map the incoming data stream to the format required by the associated local method.

To illustrate how this coordination approach based on operational dependencies can be used, we now consider two examples. The first one deals with the maintenance of global integrity constraints and the second one gives an example of how the functionality of a component system can be extended. The component systems are the CAD system CATIA running on an RS/6000 environment and an EDB-like prototype, RLW, developed at the CADETH-centre of ETH Zurich. RLW supports the management of part data, including structural data for part lists as well as classification data and the management of documents.

For the first example, consider the insertion of a new part in a CAD model during a CATIA session and the following release of the technical drawing. In the case of adding a new part to the CAD model of a group drawing, the part data inserted must also be inserted in the part data management of the RLW-system. To achieve a mechanism for consistency over geometrical data in CAD and non-geometrical data stored in a relational database, the functionality of CATIA had to be extended. All of the existing geometrical elements can be grouped together persistently to the user-defined data type or feature "part" through an additional CATIA-function using the GII (Graphical Interactive Interface) application interface of CATIA (GII 1988, CAT 1992). For such a defined part, the relative data, such as name, part number, material and manufacturing restrictions, and the structural data for the part list can be inserted.

Since the relative method is a global one, the action has to be sent as a request for a global method (insert part) to the coordinator. The coordinator determines that for this global method, the local methods of RLW to insert part data and insert the associated structural data are both required and initiates them accordingly.

After this action, the user might want to finish his work on the CAD model and release it. Again this operation is of global interest and, after notification, the coordinator activates local methods of RLW in the modules for part and document management to update the status of all related data. Naturally, in the CAD system, the access possibilities of the model also have to be changed in order to ensure the consistency of the component systems cannot be violated by future changes.

In contrast, the second example focuses on the extension of component system functionality. Consider the case of searching for a similar part during a CAD session. The user has to describe the characteristic features of the required part. This he does using shape properties in the case of a single part or using functional properties in the case of a group part. This search request has to be sent through the Global Coordinator to RLW where the relative parts can be determined in the classification module. The result of this query is the set of part identifications and some characteristic features and these are returned to the CAD user. Note that RLW supports a part number mapping which maps the internal identification to an external part number system which conveys more information to the user (Montau 1993, Montau 1994). After accepting a part in CATIA, the global method to find out the identification of the CAD model related to that part occurs and this initiates a query in the document management of RLW. In the case of success, the model identification is given back to the CATIA-application and the geometry of the part can be shown. If the user is satisfied with the result, he obtains a copy of all the CAD data in a so-called detail workspace of his current model. In this way, the costs and the time to market can be reduced as there is no further need to design redundant parts; this reduces the cost of tools for manufacturing and tests, spare parts etc. In general, the coordination of CIM component systems, especially in the field of CAx- and EDB-systems, can result in a lot of rationalisation.

6 DISCUSSION

Any evaluation of the coordination concept in general, and of the data- and application-oriented approaches in particular, has to be based on the requirements for CIM systems. The basic requirement is the capability to connect systems in a heterogeneous hardware and operating system environment; it is important that product data can reside in the component system in which it is created. This stems from a need for the component systems to be able to exploit optimisations in accessing their specialised kind of data. The independence of local tasks and the buffering of global requests are additional desirable requirements. Last, but not least, the incorporation of CAD systems to ensure consistency of geometric and non-geometric data is one of the major problems to be dealt with and has to be examined in detail.

One of the general advantages of coordination approaches, in contrast to integration based approaches, is the resulting autonomy of component systems and the elimination of the necessity to build an integration database and the consequent reduction of the integration effort. Such an integration database would often result in an enterprise-wide model which may be too expensive to produce and too difficult to attain. The amount of data redundancy is minimised in the coordination approach and, further, the availability of component systems is not dependent on the availability of a central database.

As discussed in section 5, the application-oriented approach based on operational dependencies is simple to realise and, additionally, facilitates the extension of component system functionality and the inclusion of global applications. Furthermore, there is no need to create global object identifications for new objects of global importance. For a migration of component systems to other hosts, only minimal updates of the system identification in the coordinator's database is necessary and so maximum flexibility can be achieved. To standardise the global object method interface in order to transfer parameters, an exchange format such as STEP can be used. However the specific architecture presented does not support transaction management to ensure the consistency and durability of updates in the event of system or communication failures. To cater for this, the agents would require additional database functionality in line with those of the data-oriented approach presented in section 4.

There are certain forms of inter-system dependencies that can be represented in the data-oriented approach but not in the application-oriented approach. For example, many-to-many relationships between objects of different component systems can be represented by the data dependencies of the Coordination View as described in section 4; but there is no way to represent these through method mappings. Further, its use of default reactions for violations of global data consistency mean that the system is more amenable to ad-hoc forms of coordination. On the negative side, the data-oriented approach is less direct in its coordination of user tasks and is less well-suited to providing specific extensions of component system functionality. Therefore, it is suggested that the data-oriented approach may be most appropriate in the case of PPS–EDB coordination while the operational one is preferred for functionality extension of CAx and EDB systems (compare figure 4).

For both coordination approaches, the following check list could be used to judge the ease with which a component system could be incorporated into a CIM system. Ideally, a component system should have

- accessibility of the component system's data at the logical level on which this access can be performed, e.g. whether a CAD Agent 'sees' SQL statements or operations on CAD objects;
- the possibility of a control transfer in the case of operations on globally important objects;
- freedom from side effects for all accessible object methods;
- the possibility to use a commonly used communication protocol (e.g. TCP/IP and sockets/RPC);
- the ability to transform a component system's data into a form suitable for network transfer.

7 CONCLUSIONS

We have presented two coordination approaches for CIM systems which support cooperative working through the coordination of component system activities as opposed to the traditional approaches based on the integration of data and/or applications. The advantages of a coordination approach stem from the resulting looser coupling of component systems thereby yielding maximal local autonomy and flexibility. An important factor is to support the evolution of CIM systems in terms of the addition and/or replacement of component systems. This is achieved by basing coordination on inter-system dependencies which are stored in a central repository using a global representation language. These dependencies can then be queried and updated directly.

The two approaches described, with their respective prototypes, illustrate two extreme orientations arising from whether inter-system dependencies are based on data or operations. Of course, there is a broad spectrum between these two extremes in which aspects of the two orientations can be combined. We are continuing to investigate these two approaches and their relative merits and consider how best to apply them and combine them in future CIM systems.

ACKNOWLEDGEMENTS

The work described in this paper is part of the project "Integration of Databases and CIM subsystems" (CIM/Z) funded by KWF (Swiss Federal Commission for the Advancement of Scientific Research). We thank our partners on this project for their contributions to the project as a whole and the CIM/Z system in particular; they are ABB Informatik AG and Sulzer Informatik AG.

REFERENCES

Beeri, C., Bernstein, P. & Goodman, N. (1989), 'A Model for Concurrency in Nested Transaction Systems', *Journal of the ACM*.

Bloomer, J. (1992), *Power Programming with RPC*, O'Reilly & Associates, Inc., Sebastopol, CA.

Breitbart, Y., Garcia-Molina, H. & Silberschatz, A. (1992), 'Overview of Multidatabase Transaction Management', *VLDB Journal*.

Brosda, V. (1992), Data Integration of Heterogenous Applications - A Technique for CIM System Implementation, Technical report, IBM Germany, Heidelberg Scientific Center.

Brosda, V. & Herbst, A. (1992), Die Integrationsdatenbank - Ein Ansatz zur Datenintegration im CIM-Umfeld, Technical report, IBM Germany, Heidelberg Scientific Center, Tiergartenstr. 15, D-69121 Heidelberg.

CAT (1992), *CATIA Base – Geometry Interface Reference Manual, SH50-0091-04.*

Eversheim, W. & Grosse-Wienker, R. (1991), Document Management Architecture - A Concept for Integration of Distributed Application Systems in Manufacturing and Engineering, *in* 'Proc. of the 3rd International Symposium on Systems Research, Informatics and Cybernetics, Baden-Baden, Germany'.

GII (1988), *CATIA – Graphics Interactive Interface (GII) Reference Manual, SH50-0020-0.*

Gorys, L. T. & Brauch, A. (1993), *TCP/IP Arbeitsbuch: Kommunikationsprotokolle zur Datenübertragung in heterogenen Systemen*, 3rd edn, Hüthig Verlag, Heidelberg.

IGE (1988), *Initial Graphics Exchange Specification (IGES).* Version 4.0.

Jablonski, S., Ruf, T. & Wedekind, H. (1988), Implementation of a Distributed Data Management System for Manufacturing Applications, *in* 'Proc. of the IEEE Int. Conf. on Computer Integrated Manufacturing (CIM)', pp. 19–28.

Jablonski, S., Ruf, T. & Wedekind, H. (1990), Concepts and Methods for the Optimization of Distributed Data Processing, *in* 'Proc. of the IEEE Second Int. Symposium on Databases in Parallel and Distributed Systems, Dublin', pp. 171–180.

Lorie, R. & Bever, M. (1987), Database Support for Computer Integrated Manufacturing, Technical report, IBM Research Division, San Jose.

Mark, L., Roussopoulos, N. & Cochrane, R. (1994), Update Dependencies in the Relational Model, submitted for publication.

Montau, R. (1993), 'Integritätsgewinn für die Sachnummerung anhand semantischer Datenmodellierung', *Konstruktion* **45**(10), 321–328.

Montau, R. (1994), 'Sachnummernabbildung: Zugriffsmechanismen und Integritätskontrolle für das Engineering Daten Management', *CIM-Management* **10**(6), 40–44.

Norrie, M. C., Schaad, W., Schek, H.-J. & Wunderli, M. (1994*a*), CIM through Database Coordination, *in* 'Proc. of the Int. Conf. on Data and Knowledge Systems for Manufacturing and Engineering, Hongkong'.

Norrie, M. C., Schaad, W., Schek, H.-J. & Wunderli, M. (1994*b*), Exploiting Multidatabase Technology for CIM, Technical Report 219, ETH Zurich.

Norrie, M. C. & Wunderli, M. (1994), Coordination System Modelling, *in* P. Loucopoulos, ed., 'Proc. of the 13th Int. Conf. on the Entity Relationship Approach, Manchester, UK', Springer, pp. 474–490.

Par (1993), *Pro/ENGINEER User Manuals.*

Schaad, W., Montau, R., Wunderli, M., Leonhardt, U. & Lüthi, A. (1993), Integration von Datenbanken und CIM Subsystemen, Technical report, ETH Zürich. Zwischenbericht KWF Projekt 2308.2.

Schaad, W. & Schek, H.-J. (1993), Federated Transaction Management Using Open Nested Transactions, *in* 'Proc. of the DBTA-Workshop on Interoperability of Database Systems and Database Applications, Fribourg, Switzerland'.

Scheer, A.-W. (1990), *Computer Integrated Manufacturing – Der computergesteuerte Industriebetrieb*, 4th edn, Springer, Berlin.

Schek, H.-J., Weikum, G. & Schaad, W. (1991), A Multi-Level Transaction Approach to Federated DBS Transaction Management, *in* 'Proc. of the 1th Int. Workshop of Interoperability in Multidatbase Systems, Kyoto, Japan'.

SET (1984), *Standard d'Échange et de Transfert.* Revision 1.1.

SET (1985), *Automatisation Industrielle Représentation externe des donneés de definition de produits: Specification du Standard d'Échange et Transfert.* Version 85-08.Z68-300.

Shooman, M. L. (1983), *Software Engineering: design, reliability and management*, Computer Science, Mac Graw-Hill, New York.

STE (1992), *ISO DIS 10303, Product Data Representation and Exchange.*
VDA (1987*a*), *Festlegung einer Untermenge von IGES Version 3.0 (VDA-IS).* VDMA/VDA 66319.
VDA (1987*b*), *VDA Flächenschnittstelle (VDA-FS).* Version 2.0.
Weikum, G. (1991), 'Principles and Realization Strategies of Multilevel Transaction Management', *ACM TODS.*
Weikum, G. & Schek, H.-J. (1992), Concepts and Applications of Multilevel Transactions and Open Nested Transactions, *in* A. Elmagarmid, ed., 'Database Transaction Models for Advanced Applications', Morgan Kaufmann, San Mateo, CA, chapter 13, pp. 515–553.
Wintraecken, J. (1989), *The NIAM Information Analysis Method: Theory and Practice*, Kluwer Academic Publishers.

BIOGRAPHIES

Moira Norrie received a B.Sc. in Mathematics from the University of Dundee, an M.Sc. in Computer Science from Heriot-Watt University and a Ph.D. from the University of Glasgow. She is currently a Senior Research Associate in the Database Research Group at the Swiss Federal Institute of Technology (ETH) Zurich. Previously, she has held research and lecturing positions at a number of institutions including the Universities of Edinburgh, Glasgow and Stockholm. Her main research interests are in the area of object data models and semantic interoperability.

Martin Wunderli holds a Master's Degree in Computer Science from the Swiss Federal Institute of Technology (ETH) Zurich and is now a Ph.D. student at ETH. His main interests are foundations of data models, especially object data models. In the CIM/Z project, he has been working on modelling issues since June 1992.

Robert Montau studied mechanical engineering at the University of Karlsruhe. Since 1990, he has been a research assistant at the Institute for Construction and Design Methods of the Swiss Federal Institute of Technology (ETH) Zurich (head: Prof. Dr.-Ing. M. Flemming). His major areas of interest are product data management, CAx-technologies and computer integrated design models.

Uwe Leonhardt studied computer science at the University of Karlsruhe and completed his studies with a Diploma Work at the Swiss Federal Institute of Technology (ETH) Zurich. Since 1993, he has been working as research assistant at the Institute for Construction and Design Methods of ETH (head: Prof. Dr.-Ing. M. Flemming). His major areas of interest are product data management, CAx-technologies and computer communication.

Werner Schaad is a Ph.D. student at the Swiss Federal Institute of Technology (ETH) Zurich. His research interests include transaction theory in database systems and distributed and federated database systems. He has built a prototype system of a federated database system using the concept of multi-level transactions. In the CIM/Z project, he has been working on transaction issues since 1991.

Hans-Jörg Schek received an M.Sc. in Mathematics and a Ph.D. in Civil Engineering from the University of Stuttgart. He is currently a Professor of Computer Science at the Swiss Federal Institute of Technology (ETH) Zurich where he is the head of the Database Research Group. Between 1972 and 1983, he was with the IBM Heidelberg Scientific Centre. From 1983 to 1988, he was a Professor of Computer Science at the Technical University of Darmstadt. His main research interests are transaction management and advanced storage services.

16

Integration of industrial applications: The CCE-CNMA approach

P. Pleinevaux
Computer Engineering Dept.
Swiss Federal Institute of Technology, Lausanne
EPFL-DI-LIT
CH-1015 Lausanne, Switzerland
pleinevaux@di.epfl.ch

Abstract

Integration of industrial applications has been and still is a major problem for industrial enterprises. Development of new manufacturing applications and integration of existing ones is hampered by a number of problems, such as heterogeneity of hardware and operating systems, complexity and non harmonisation of application programming interfaces, diversity of communication protocols, user interfaces and databases involved in the applications.

The ESPRIT CCE-CNMA has specified, implemented and validated an open and portable platform for integration of industrial applications. This platform, called CCE (CIME Computing Environment), hides the diversity in communication protocols, databases and access methods to the user. It provides high level interfaces that allow the user to concentrate on his application domain and not on the way to program his application.

In this paper, we review the problems faced during integration, describe the CCE architecture and discuss the relationship between CCE and the CIMOSA integrating infrastructure.

Keywords

Application integration, CIMOSA, CNMA, MAP, CCE.

1 INTRODUCTION

The ESPRIT CCE-CNMA project introduced in 1993 an open platform for the integration of industrial applications. Called CCE (CIME Computing Environment), this platform is based on a standard communication infrastructure - CNMA (Communications Network for

Manufacturing Applications) - and offers the appropriate services, tools and administration to develop and execute applications in the manufacturing and process control environments.

The purpose of this paper is to introduce this platform, presenting its architecture, its properties, its relationship with the CIMOSA integrating infrastructure and indicating the current status of the project.

Proprietary platforms like Digital's BASEstar or IBM's PFS/DAE exist on the market. These platforms however are not open in the sense that their interfaces are proprietary and that they are not available from different suppliers. CCE on the contrary is an open platform, available from three different vendors on a number of machines and operating systems.

The paper is organized as follows: Section 2 describes the problems found by system integrators when developing CIME applications. Section 3 describes the CCE architecture while Section 4 presents its main properties. Section 5 discusses the relationship between CCE and the CIMOSA integrating infrastructure. We conclude with a presentation of the problems that must be dealt with in future versions of CCE (Section 6).

2 PROBLEMS AND USER REQUIREMENTS

CIMOSA distinguishes three levels of integration provided by CIM (Computer Integrated Manufacturing): business integration, application integration and physical integration (AMICE, 1992). For six years, in the period 1986-1992, the ESPRIT CNMA project worked on a communication architecture that is well adapted to the physical integration of industrial applications. This architecture, compatible with the Manufacturing Automation Protocol (General Motors, 1988), provides services for communication with manufacturing devices, for the transfer of files, for access to remote databases and for the administration of the communication infrastructure. This solved one of the most costly problems of industrial companies, namely the interconnection of heterogeneous equipment provided by different manufacturers.

Yet, other problems remained. A study made by the CCE-CNMA project in 1993 showed that the following problems were encountered by all users when attempting to integrate their applications:

- Existence of many application programming interfaces (APIs): a manufacturing application commonly accesses databases, uses one or more industrial messaging protocols and displays text on man-machine interfaces. Very different APIs are found for these tasks with different mechanisms to deal with the same problems such as memory management, error handling or event management.
- Complex application programming interfaces that require the initialisation of many parameters, some of which are never used.

- Non harmonized application programming interfaces: the same problems are solved in different ways by the different APIs, leading to increased costs for programmer training.
- Lack of tools for the development and debugging of new applications.
- Lack of mechanisms to ensure the consistency of data stored in the system, especially when these data are distributed over multiple machines of different suppliers.

An independent study presented in (Deregibus *et al.*, 1991) discusses a list of user requirements that is very close to those identified by the project.

3 THE CCE ARCHITECTURE

CCE (CIME Computing Environment) is an *open* platform for the integration of industrial applications (CCE-CNMA, 1994 and 1995a). It is designed to reduce the above mentioned problems. Its main properties are described in Section 4.

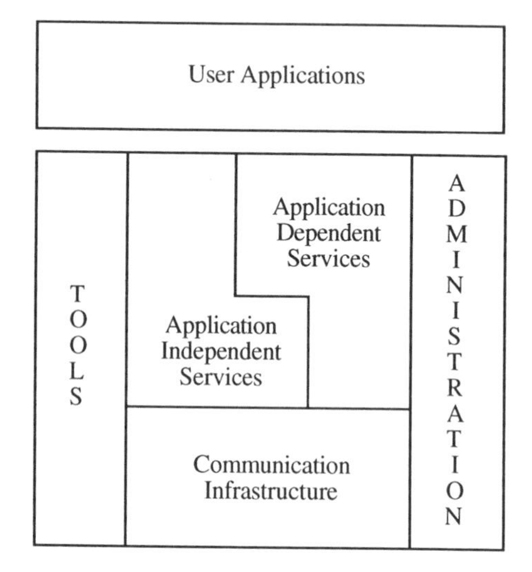

Figure1 The CCE architecture.

The CCE architecture (Figure 1) distinguishes the following components:

- a communication infrastructure
- application independent services
- application dependent services

- tools
- an administration

In the following sections, we briefly describe each of these components.

3.1. Communication infrastructure

At the lowest level of the architecture lies the communication infrastructure. The standard adopted by CCE is the CNMA architecture (CCE-CNMA, 1993), compatible with the Manufacturing Automation Protocol (General Motors, 1988). The services offered to the user by this communication profile are the following:

- MMS (Manufacturing Message Specification) for remote control and monitoring of industrial devices. With MMS, users can download and upload programs or data, start and stop programs, read or write variables (CCE-CNMA, 1995b).
- RDA (Remote Database Access) for access to relational databases using SQL. The protocol is independent of the database used, for example ORACLE, INGRES or INFORMIX.
- FTAM (File Transfer Access and Management) for the transfer and remote manipulation of files.
- X.500 Directory Service, to obtain information on users, applications or objects accessible in the network.

While CNMA is the recommended communication infrastructure, CCE is actually able to run on a variety of communication protocols, including the Internet Protocol Suite (TCP/IP) and proprietary protocols. CCE hides the protocols used by applications to communicate with one another.

3.2. Application independent services

A first layer above the communication infrastructure offers application independent services. At the upper interface of this layer, the user views a uniform environment consisting of CCE objects accessible through operations and able to send notifications when interesting events occur. The environment is composed of application independent objects such as variables, programs or domains.

One of the application independent services offered by CCE is transaction management, a service that offers the mechanisms needed to ensure consistency of objects stored in CCE. In this area, the project adopted the X/Open Transaction Demarcation Interface as the standard interface. Transaction management is available on CCE platforms that are based on an OLTP (On-Line Transaction Processing) execution environment (Gray and Reuter, 1993).

3.3. Application-dependent services

While application independent services can always be used to develop an application, there is often a need in industrial applications for higher level abstractions. Indeed, when designing its application, the user does not reason in terms of MMS variables, domains or programs but in terms of real objects such as pallets, tools, parts, etc. The application dependent interfaces offer services to manipulate objects of this kind with object-specific operations. CCE-CNMA defined such interfaces for mould management, electronic kanban and tool management. Additional interfaces have been considered for pallet and warehouse management.

As an example, let us consider tool management. In a manufacturing cell, numerical controllers of machine tools and machining centres have to be supplied with data that describe the dimensions and age of the tools used to manufacture parts. Depending on the manufacturer of the machines, these data are stored in very different ways: some data are stored in MMS structured variables, others in MMS domains. In CCE, the tool manager hides to the user the differences in data representations. The user can access the following services, among others, to manipulate tools:

- `CreateTool`: to create an object in CCE that stores data on a given tool.
- `DownloadToolData`: to transfer the tool data associated to a program and a given machine from CCE - where it is stored in a database - to the given manufacturing device. The user does not need to read the database, this work is done for him by CCE.
- `UploadToolData`: to transfer back the data once the machine is done with them. The data are updated in the database without user intervention.
- `GetToolData`: to read the current values of the tool data. CCE fetches the data where they are most up-to-date: in the CCE database or in the manufacturing device.

Application-dependent interfaces are one of the most important features of CCE that make it different from classical platforms like the OSF DCE (Distributed Computing Environment), Comandos (Cahill *et al.*, 1993) or ANSAware (Herbert, 1994). It is important to note that these platforms provide the means for the user to build distributed applications from scratch, and in this sense they can be said to be general-purpose. But a tremendous work must be done on these platforms to provide the user with the right abstractions (objects) that are used by industrial applications. In CCE, these abstractions are built-into the application independent and dependent interfaces, thus relieving the user from developing these abstractions and allowing him to concentrate on his application and not on meta problems.

3.4. Tools

A series of tools are available for the development of applications as well as for the configuration of the platform. In the following, we describe three of them.

The CCE Configuration Tool is an application designed by Alcatel-TITN-Answare that allows an administrator to create CCE objects, to monitor the operation of a running system and to modify the parameters of the platform or objects. Based on a graphical interface, it allows an easy access to the information needed by an administrator.

The SQL preprocessor allows the user to read or write data using the well-known SQL database manipulation language (Pleinevaux, 1992). The user views the system as a relational database and no longer as an object-based system. This enables the user to access CCE from offices and to reuse existing SQL applications. Object-oriented extensions based on the SQL 3 and ODMG-93 standards (Cattell, 1993) are being considered for CCE (Näf and Pleinevaux, 1995).

The Variable Generator allows a user to define simple C data structures and to automatically generate the data structures used by the application interfaces, for example MMSI, the standard interface to the MMS protocol. When the communication partner is a Programmable Logic Controller (PLC), the tool generates the corresponding PLC data structures as well.

4 CCE PROPERTIES

When using application dependent or independent interfaces, the user sees ASPI objects but does not know where these objects are stored, in which form and which protocol is used to access them. The platform has the following general properties:

- **CCE hides the nature of accessed data**: When accessing a CCE object, the user does not know whether this is an MMS variable stored in an industrial device, a database entry or an application variable.
- **CCE hides data location**: The place where the object is stored is unknown to the user. This may be in a manufacturing device, a database or a workstation.
- **CCE hides the communication protocols**: The protocol used to access the CCE object is hidden by the interface: it may be a standard protocol like MMS, a proprietary protocol like UNITE, an RPC protocol, the user does not know. This fundamental feature allows for example the migration from proprietary protocols to standard protocols without the user having to rewrite his application programs.
- **The platform is portable**: CCE has been designed and implemented in such a way that it can be easily and quickly ported on different operating systems. This is one of the major user requirements.
- **Object-orientation**: The functionality offered at the two application interfaces is presented using the object approach. The user accesses CCE objects through operations

and may receive notifications when significant events take place in the object. Objects belong to types that inherit attributes, operations and notifications from parent types. This style of interface allows a simple mapping after the analysis and design phases of the application.

- **The user can access all interfaces**: Three levels of interfaces are accessible to the user: the communication interfaces (MMS, RDA, FTAM), the application independent interfaces and the application dependent interfaces.
- **Integration of PCs under Windows 3.x**: The ASPI interface is available on Windows PC but in a DDE flavour. This allows a PC to access objects stored on other machines (UNIX, DCE, OLTP) with the same style of interface. The DDE style of interface enables the user to access CCE from any Microsoft application package supporting DDE, for example Access, Excel, etc.

5 CCE AS CIMOSA INTEGRATING INFRASTRUCTURE

The CIMOSA integrating infrastructure is a platform for the interpretation and execution of the different models that describe an enterprise. The requirements identified by CIMOSA (AMICE, 1992; Querenet, 1991) that must be met by a platform used as integration infrastructure are:

- interpret the behavior and execute the models produced by the CIMOSA methodology
- provide services common to all CIM systems
- manage resource availability
- manage data related to monitoring and control of resources
- ensure proper communication
- manage location, failure, access and performance transparencies
- handle heterogeneity of manufacturing devices
- use standards for the communication subsystem.

Conceptually, the integrating infrastructure offers four groups of services:

- *Business services* provide the functions necessary to control the execution of the models associated with the function and resource views.
- *Front-end services* provide the means to communicate with resources, either machines, humans or applications.
- *Information services* provide access in a unified way to all information of an enterprise. These services ensure consistency, integrity, and protection of an enterprise information.
- *Communication services* allow cooperation of the above services.

The service groups are themselves composed of integrating infrastructure services. For example, the front-end services are divided into machine front-end for integration of manufacturing devices, application front-end services for example to communicate with CAD/CAM and human front-end services for communication with operators.

5.1. Relationship between CIMOSA and CCE

CCE satisfies a large number of the requirements that are specified by CIMOSA for a platform used as integrating infrastructure:

- CCE offers application independent services for access to variables, transfer of bulk data, remote control of programs and event management.
- CCE offers adequate communication services for CIM systems, namely access to manufacturing devices and databases, file transfer, access to a name service.
- CCE is based on standard communication protocols defined by the International Standards Organisation (ISO).
- CCE offers access to the Manufacturing Message Specification (MMS) for communication with heterogeneous manufacturing devices. CCE integrates proprietary communication protocols as well, for example Unitelway, Modbus or SINEC.
- CCE provides location, access and performance transparencies.

As will be explained below, CCE offers services that comply with the above classification of CIMOSA integrating infrastructure services. CIMOSA considers two classes of information services: (1) system-wide data provides a unified access to data without concern for the location and structure of the data; (2) data management which provides services to store and retrieve in a unified way the data stored in the system.

In CCE, access to data is possible in three different ways: the CNMA application protocols, the ASPI interface and the SQL interface. When an application requires the use of a single application protocol to communicate with a remote application, the user can in this case directly call the application interface of this communication protocol.

When an application makes use of multiple application protocols to simultaneously access data in different stores, the user can still use the above approach or call the CCE Application Service Programming Interface (ASPI). This interface gives access to CCE objects that can be mapped on objects of all CNMA application protocols, in particular MMS objects, database entries and application specific objects. The main advantage of this approach is that the user does not know where the data are stored and which access method is used to read or write them. The ASPI offers in particular a filter mechanism which allows to access a group of objects having similar names.

For access to large collections of data, CCE offers the SQL interface (Pleinevaux, 1992). With SQL, the user is able to read or write data about objects stored either as MMS objects, Network Management objects, database entries or CCE objects. All these objects are viewed as tuples of relations on which the classical relational operations can be performed. An SQL pre-processor translates programs with embedded SQL statements into programs containing CCE interface calls.

CIMOSA consider two groups of communication services: system-wide exchange handles all intra-node data exchanges and forwards inter-node communications to communications management; the latter provides access to protocols such as the OSI protocols. CCE satisfies the requirements identified for the CIMOSA communications services. For intra-node communications, CCE offers the mechanisms of the underlying execution environment, namely DDE (Dynamic Data Exchange) for Windows or the DCE Remote Procedure Calls (RPCs) for the OSF DCE environment. For inter-node communications, CCE gives access to the CNMA communication architecture, the Internet Protocol Suite (with TCP/IP) or any proprietary protocol used by the application.

As can be seen from the above discussion, CCE covers the requirements of communication, front-end and information services. The only area in which significant work remains to be done is the business services, which are a distinctive aspect of CIMOSA. The idea is to allow execution of the models derived from the CIMOSA views. One of the benefits of this idea is that an evolution of the enterprise results in an update of the models which are themselves updated for use by the integrating infrastructure. In CCE, application knowledge is still mainly provided to the system in the form of programs, not in executable models. The object approach however, allows to store data and knowledge in entities - the objects - that can be dynamically created, deleted, updated or moved in the system. Addition of new object types and instances can also be made at any time while the system is up and running.

6 STATUS OF THE PROJECT

The project started in January 1993 and ended in May 1995. CCE has been specified, implemented and tested in 1993-1994. A validation and evaluation phase started in the summer of 1994 with four pilot installations: Aerospatiale, EFACEC, Magneti Marelli and Mercedes-Benz. The purpose of these pilots is to assess the validity of the design in real industrial environments, from car manufacturing to production of electrical transformers.

CCE is currently available on Lynx, AIX, SCO UNIX and Windows 3.1. On Windows, CCE gives access to MMS and the application independent interfaces through DDE (Dynamic Data Exchange), a protocol defined by Microsoft for communication between Windows applications. This approach allows the reuse of a large base of existing software packages available on Windows, for example In Touch or Excel.

CCE however is not completed. A good basis is available but complementary work needs to be done to address, among other issues, the following ones:

Evolution towards CORBA and OLE:
The heart of CCE is an entity that enables users to invoke operations on CCE objects. Two standards are emerging in this area. CORBA (OMG, 1991) from the Object Management Group, is a specification for an Object Request Broker (ORB), precisely this type of entity. The CORBA standard has been widely accepted by major vendors in the computer industry, the second version addressing the problem of ORB interoperability. OLE (Object Linking and Embedding) is the Microsoft standard for communication among Windows applications. OLE 2.0 replaces the DDE mechanism mentioned above by an RPC (Remote Procedure Call) mechanism. CCE will have to follow these evolutions, in both areas to allow for interoperability of CCE with office applications.

Definition by the user of new object types:
In the current version of CCE, the platform is delivered to the user with a limited number of object types. To introduce new types in its application, the user must implement special software components that provide the semantics of these types. In the present state of CCE, this operation requires a good knowledge of system programming. In the future, it is planned to have a tool allowing the user to define his object type, compile it and integrate it in the platform.

Definition of standard object types:
This area is of fundamental importance to the user. If an agreement can be found among users on a common definition of widely used manufacturing objects, then these definitions can be implemented by vendors in the application dependent interfaces, thus relieving the user of this task. The MMS Companion Standards (ISO 9506-3 and 4) are a first attempt in this direction but the results are not satisfactory.

Fault-tolerant infrastructure:
While the current version of CCE supports some degree of fault tolerance with persistent storage of CCE objects, there is a clear need for mechanisms that allow for quick recovery from failures. In this area, CCE could draw upon the work performed by the ESPRIT Delta-4 project (Powell, 1991) which specified, implemented and validated a communication architecture, based as far as possible on international standards, that is close to the CNMA architecture.

7 CONCLUSION

The ESPRIT CCE-CNMA project defined, implemented and validated an open platform for the integration of CIME applications. This platform is based on the CNMA communication architecture but is able to deal with proprietary communication protocols. The platform offers application dependent and independent services that allow the rapid creation of new industrial applications. Data nature, data location and communication protocols are hidden to the applications which can thus be reused in different contexts. Tools for the platform allow for automatic code generation, compiling programs containing embedded SQL statements and for configuration of the system.

CCE differs from other general purpose platforms like the DCE, Comandos or CORBA by the fact that it integrates a number of communication protocols like MMS or RDA and is available with predefined application dependent and independent interfaces, specifically designed for industrial applications.

CCE can be regarded as a realization of the CIMOSA integrating infrastructure. The provision of simple high level programming interfaces allows the reduction of application development time. Work remains to be done in the areas of CIMOSA business services, fault tolerance, standardisation of classes of manufacturing objects and tools for the integration of new object types in the platform.

REFERENCES

AMICE (1992), ESPRIT Consortium AMICE, *CIMOSA: Open System Architecture for CIM*. Springer-Verlag.

Cahill, V., Balter, R. and Harris, N.R. (1993) *The Comandos Distributed Application Platform*. Springer-Verlag.

Cattell, R.G. (1993), *The Object Database Standard: ODMG-93*. Morgan Kaufmann, San Matteo, California.

CCE-CNMA (1993), ESPRIT Project 7096, CNMA Implementation Guide, Revision 6.0.

CCE-CNMA (1994), ESPRIT Project 7096, Introduction to the CIME Computing Environment - A Platform for the Creation and Execution of Industrial Applications. Document 7096.94.08/D2.PD available on the CNMA information server at http://litwww.epfl.ch/~ppvx/cc_CCE-CNMA.html.

CCE-CNMA (1995a), ESPRIT CCE-CNMA Consortium, *CCE: An Integration Platform for Distributed Manufacturing Applications*. Springer-Verlag.

CCE-CNMA (1995b), ESPRIT CCE-CNMA Consortium, *MMS: A Communication Language for Manufacturing*. Springer-Verlag.

Deregibus, F., Bobbio, M., and Rusina, F. (1991) Open Systems and Manufacturing Software Integration Platforms, in *Proc. of the 7th CIM-Europe Annual Conference*, Turin, Italy, 33-43, Springer-Verlag.

General Motors (1988) *Manufacturing Automation Protocol*, Version 3.0.

Gray, J.N. and Reuter, A. (1993) *Transaction Processing: Concepts and Techniques*. Morgan Kaufmann, San Matteo, California.

Herbert, A. (1994) An ANSA Overview. *IEEE Network*. **8**(1), Jan./Feb. 1994, 18-23.

Näf, M. and Pleinevaux, P. (1995) On the use of SQL3 and ODMG to access CCE objects. Technical report 7096.95.03/F1.PD, available on the CNMA information server.

Object Management Group (1991) *The Common Object Request Broker: Architecture and Specification*.

Pleinevaux, P. (1992) An SQL Interface to CCE. EPFL-LIT Internal Report, December 1992.

Powell, D. (Editor) (1991) *Delta-4: A Generic Architecture for Dependable Distributed Computing*. Springer-Verlag.

Querenet, B. (1991)The CIMOSA integrating infrastructure. *Computing and Control Engineering Journal*, May 1991, 118-125.

17

A Software Engineering Paradigm as a basis for Enterprise Integration in (Multi-) Client/Server Environments

Dr. D. Solte
Research Institute for Applied Knowledge Processing (FAW) Ulm
Helmholtzstraße 16, 89081 Ulm, Germany, Tel: 0731/501-510,
Fax: 0731/501-111, solte@faw.uni-ulm.de

Abstract

The ability to build and execute enterprise models including data, service and process models is a topic of growing importance for industry. It addresses the problem to develop reasonable models of the enterprise but has to cope also broadly with implementation and execution issues in heterogeneous environments. With respect to implementation and execution, client/server architectures, request broker mechanisms and distributed data and applications are emerging as the future state-of-the-art. In this context, the existing heterogeneity of technological frameworks as well as coping with legacy systems is a crucial fact. Existing methodologies and tools are not overwhelming these problems. They often do not integrate aspects of enterprise- or process-modelling, CASE (Computer Aided Software Engineering), workflow management and client/server execution. As a consequence, new kinds of architectures are needed. This paper outlines an approach, developed at FAW for the described scenario. The main objective of this solution is to cope with heterogeneity by a neutralizing approach instead of standardization. The described software engineering paradigm supports a model-oriented development of distributed data, services and processes in a uniform way towards a neutralizing execution environment. The FAW software engineering paradigm complies with the specifications of CIMOSA and accomplishes the requirements of the CORBA architecture.

Keywords

Client/Server, Enterprise Engineering, Enterprise Integration, Execution Environment, Integrating Infrastructure, Operational Paradigm, Process Modelling, Software Engineering.

1 INTRODUCTION

Since ´87, the FAW (Research Institute for Applied Knowledge Processing, Ulm) has devoted continous efforts to develop a comprehensive framework for enterprise engineering and application development, which consists of an overall systems architecture and a pertaining software engineering paradigm (Holocher et al., 1993, Radermacher and Solte, 1994). The most important aspect when developing this overall strategy is the observation that none of the relevant standardization activities will really lead to a worldwide homogeneous standard for a wide body of services (Verall, 1991). As a matter of fact, heterogeneity will always prevail with respect to infrastructural components, but also in methodologies and other aspects of IT-technologies. Taking this into account, the main aim of the FAW software engineering paradigm is to define and implement a framework that actively supports enterprises in building an integrated and cooperative engineering and execution environment for distributed data, services and (business) processes. Related to the requirements described by (ESPRIT Consortium AMICE, 1993) (cf. Figure 1) this leads to the development of an integrating infrastructure named AMBAS (Adaptive Method Base Shell, c.f. Holocher and Solte, 1992) and an engineering environment named Ωmega (Operational Modelling Environment and Generator for AMBAS Applications).

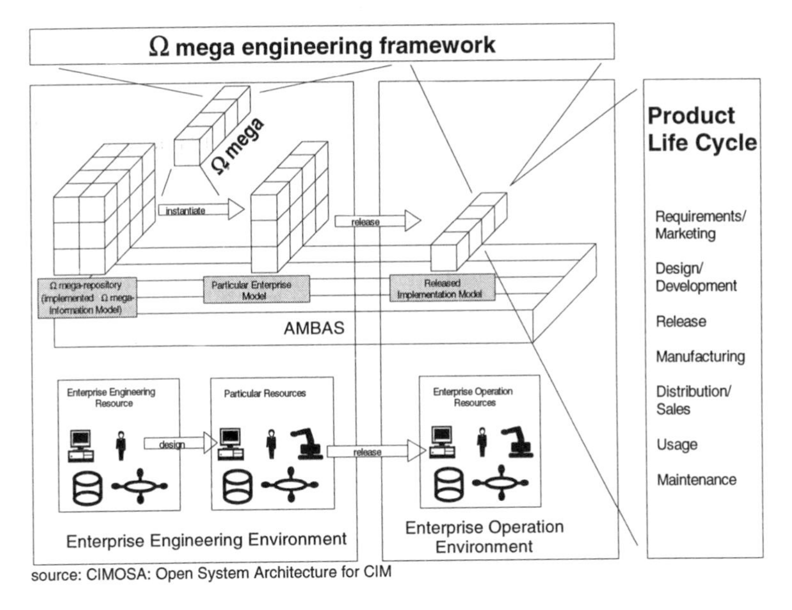

Figure 1 Positioning of Ωmega/AMBAS within CIMOSA.

An important issue for the design of the systems architecture was the consideration of migration needs of enterprises, since taking into account the existing infrastructural components and methodologies already used. This has lead to a so-called active federation strategy as a framework for a knowledge-based approach that tries to cover heterogeneity by neutralization instead of standardization. Neutralization is thereby performed by using models of the different components and translating the models using knowledge-based mechanisms as proposed in (Petrie (ed.), 1992).

Besides that more principal design aspect, the framework (c.f. Figure 2) consists of several parts which are combined in an overall architecture. The main parts are the information model, the process model (including methodological concepts) and the tool model. The tool model consists mainly of the repository, I/O editors, generators (Ωmega) and an execution environment (AMBAS).

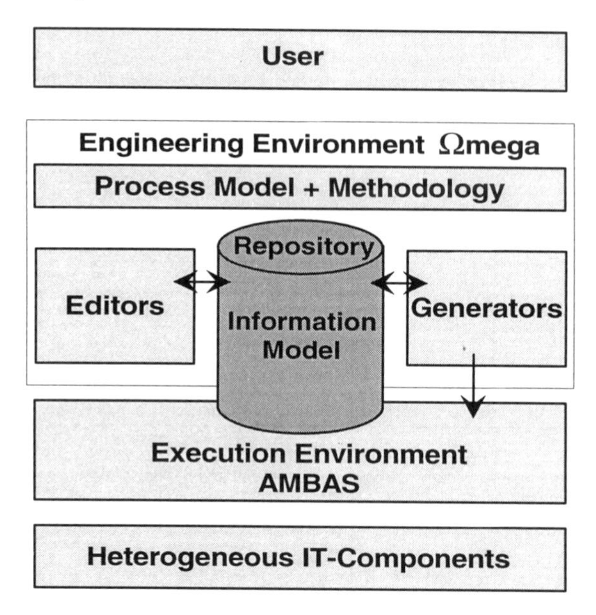

Figure 2 Enterprise engineering and execution framework.

The FAW software engineering paradigm has its roots in project TIMM (Technical Infrastructure and MultiMedia). TIMM established the first heterogeneous and distributed computing environment at FAW (c.f. Figure 3), ranging from personal systems over workstations to a mainframe (Solte and Heerklotz, 1991). Moreover, these systems were connected to special equipment such as PABX and state-of-the-art multimedia components by using several networks (ISDN, Ethernet, Token-Ring, Broadband, Sinec-H1).

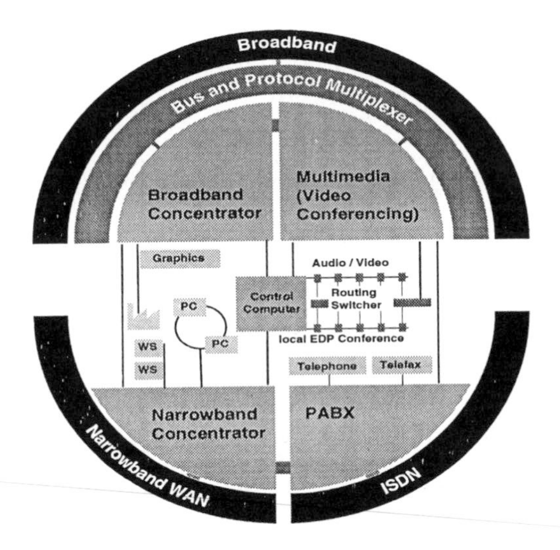

Figure 3 Heterogeneity at FAW.

The chief objective of the FAW software engineering paradigm is the definition of concepts and tool designs to support the cooperative development of distributed data and applications (including process-engineering and control) for this kind of heterogeneous platforms in particular consiering reliability aspects. The basic foundation has been laid by project MIDA (Model-Oriented Integration of Data and Algorithms) (Holocher et al., 1993) which continued the work of project SESAM (Decision Support for Job-Shop Scheduling) (Müller and Solte, 1994). Project KIWI 2000 (Communications Infrastructure 2000) further expanded the framework with special focus on the execution environment by incorporating several telecommunication devices and services. KIWI 2000, which has been finished in 1995, was the largest joint project of the State of Baden-Württemberg. It involved 14 industrial partners, 2 chambers of commerce and the German Telecom (Kopaczyk et al., 1993).

2 THE ENGINEERING PARADIGM

The FAW software engineering paradigm focuses on the development of distributed data, services and processes in (multi-)client/server architecture. By (multi-)client/server architecture we mean the possibility to distribute presentation, process logic and data management within heterogeneous environments, where these components are implemented as servers. One server can be executed by several clients, one client can have access to several servers and servers can act as clients.

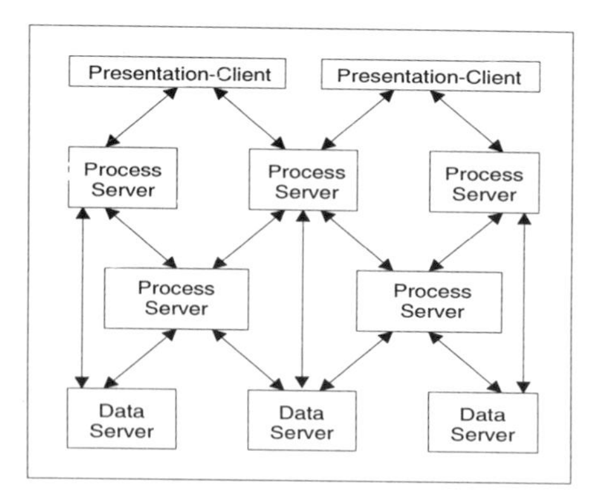

Figure 4 (Multi-)client/server architecture.

It is one important aspect that the binding of clients and servers should be possible during runtime, instead of compilation time. To support this kind of architecture, the execution environment contains mechanisms of a request broker (c.f. OMG, 1991) and the engineering environment consequently devides the applications logic from its implementation details.

Distributed data and methods (services, applications and processes) have to be modelled at an abstract model layer and will be translated automatically to corresponding (C++-) code after customizing the models due to implementation details. The compiled code can be executed by the neutralizing platform (AMBAS). In addition, one has to consider that information is needed to administrate productive implementations, and for systems management. To cover all these aspects, the information model comprises six representation layers, as depicted in Figure 5.

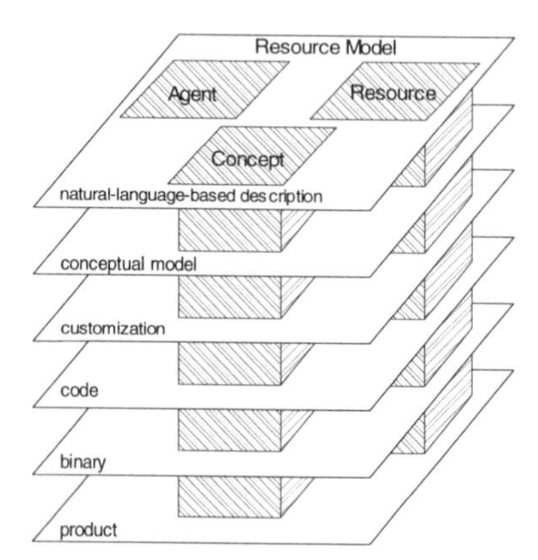

Figure 5 The Ωmega Information Model.

The first layer administrates natural language based descriptions. These are informal descriptions of all data and method components (services, applications and processes) of the enterprise (i.e. the ontology). The second layer represents the formal descriptions (models) of these components. At this layer - called the conceptual model - only the logical aspects of data, services, applications and processes are modelled. For reliability, reuse etc. it is important that no information about implementation details should be merged with the components logic. Instead, this information is modelled at a separate representation layer (customization). This includes technical and organizational aspects (communications infrastructure knowledge, e.g., GUI, CUI, file system, database management system, operating system, user model etc.) and incorporates the principle of definition moduls (conceptual model) and implementation models (customization) derived from modular programming concepts (c.f. Teufel, 1991) applied to models to the FAW software engineering paradigm. Using a conceptual model and a corresponding customization, a C++-representation is generated automatically and stored in the code-layer. The binary layer administrates the compiled components. The product layer captures information about installed (productive) data, services, applications and processes.

We have developed a specific strategy (the operational paradigm firstly introduced by Solte, 1987) to support the representation-framework (metameta-model) of this information model. We have chosen an approach with a modelling layer (analysis), with an orientation to rather classical concepts (data and functions) and with an implementation of data and functions in a strictly defined way as objects. Coming from a mathematical view functions are classified using their input and output signature leading to so-called function classes

$T\,(Y_{i,}\ Y_O) := \{T \mid T: Y_i \rightarrow Y_o\}$

where each concrete function represents an object of this class. Y_i specifies the functions input and Y_o the functions output. When implementing the function as an object, the function is encapsulated by a well-defined set of elementary access functions.

These are

create:	the creation of an object.
delete:	to delete an object.
assign:	the assignment of a value to the input of the function.
evaluate:	to activate asynchronously the evaluation of the function.
access:	to read the output of the function. If it is not yet evaluated, access also activates the method evaluate, otherwise access synchronizes the requesting process by waiting for the function to be terminated.
link:	to make the object only a reference to an already existing object.

When implementing data, we interpret them as special function classes $Y := \{id(y)$, e.g. $id: Y \rightarrow Y\}$. With this interpretation the same capsule, as it is defined for functions, can be implemented for data which leads to a uniform implementation and administration of data and functions for execution.

The operational paradigm overcomes problems using object-oriented principles "naively" for analysis (Holocher et al., 1994) but uses the full potential of object orientation for the implementation and execution of distributed applications. At FAW, we have developed a modelling language called OML (Operational Modelling Language) dedicated to support a reasonable combination of object orientation, semantic data modelling, a strict functional characteristic of the language (especially to support process modelling) and the capability to model constraints (but with a restricted constraint calculus). OML should mainly be perceived as the repository's metameta-model, it is not intended to be a solitaire modelling language for software engineering. We have already proved, that one could use different editors (supporting different languages) to build and modify models, e.g. STEP/Express (c.f. Anderl, 1993 and Grabowski, 1993) or GRAPES (c.f. Kaufmann, 1993).

The representation framework defined by the metameta-model OML allows the uniform knowledge representation of application concepts under different perspectives (Figure 6).

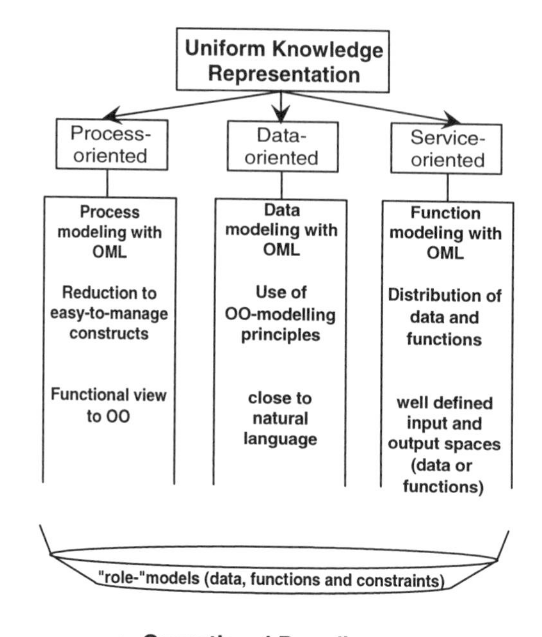

Figure 6 Uniform knowledge representation.

The strict functional characteristic allows the analysis and the requirements engineering in a process driven way but also on the basis of services (e.g., if specific technical servers like fax, telephone or others have to be modelled). On the other hand, the capabilities of semantic data modelling and object orientation allow the comfortable description of any kind of data. In addition, there is a possibility to collect data, functions and constraints in a context-oriented way by forming models, containing all these data, functions and constraints that fit to the specified context. This enables the integrated view of the different perspectives followed during modelling.

Part of the FAW software engineering paradigm is an analysis and design technique that forces the development of a model hierarchy, thereby allowing the different perspectives (e.g., process, service or data-oriented). At the bottom level, the models have to describe organizational or technical roles (by means of data that has to be provided; and functions the role is responsible for) of the enterprise. This leads to a role-based structure of the enterprise model (as depicted in Figure 7). A distribution logic could directly be derived from this structure. Role models are implemented as a whole and could be installed on all computers accessable by those organizations and agents that are capable of taking these roles. The structure within the information model, that supports the management of role-based enterprise models, is called resource model.

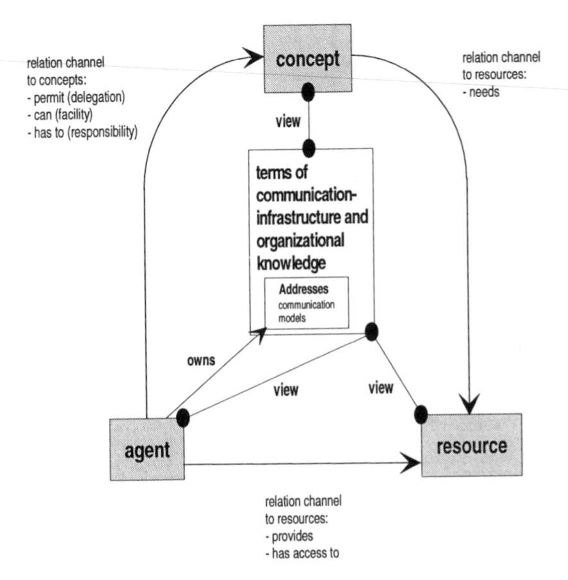

Figure 7 Construction of the Resource Model (Coarse Structure).

In the resource model all models are specializations of the term "concept" which has relations to terms "agent" and "resource". Agents are all acting objects (things that could take a role) of an enterprise which means technical systems like computers, printers, terminals and so on, but also humans or organizations. In addition, "concepts" need "resources" and an "agent" provides "resources". It is important that, based on these three top level terms of the enterprise´s ontology, one can define specific views forming the basis to model all kind of communication´s infrastructure and organizational knowledge. This is for example the definition of addresses as a possibility to describe communication models.

3 THE SYSTEMS ARCHITECTURE

The first operational prototype of the entire tool model for the FAW software engineering paradigm is now available. Ωmega (Operational Modelling Environment and Generator for AMBAS applications) contains specific components (in particular the repository/information model, I/O-Editors, generators and the neutralizing execution environment AMBAS) to develop models and to produce executable code. Available commercial products are integrated in Ωmega whereever appropriate.

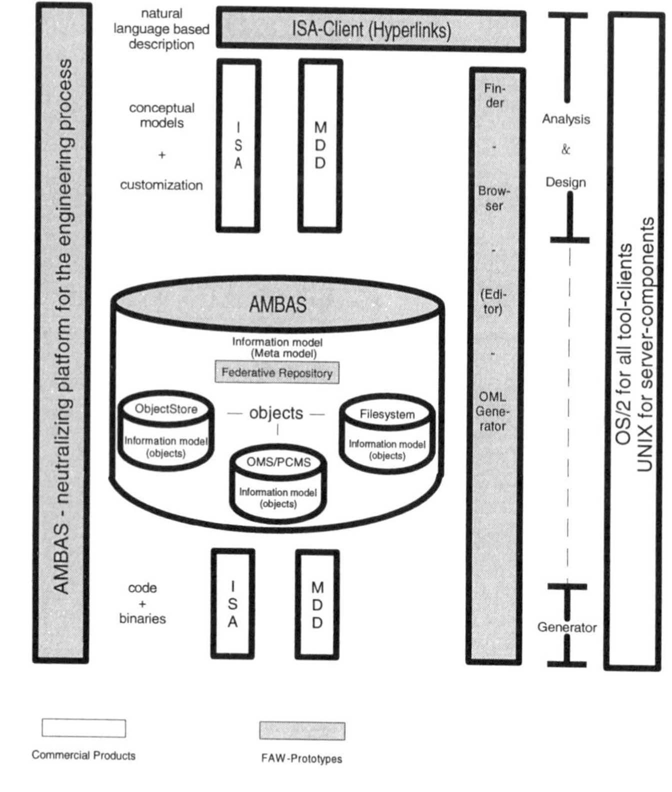

Figure 8 Components of Ωmega.

As indicated in Figure 8, Ωmega provides different components to edit the different representation layers of the knowledge representation framework defined by Ωmega´s information model. For the natural language based description, we use a generic editor that we build for any kind of AMBAS-objects. Part of this editor is a generator that maps the structure of implemented objects onto a page-layout. Besides the implementation of this editor with Ωmega (which means the use of the ISA dialogue manager for the GUI) we have developed a low-cost variant based on HTML. This allows to edit AMBAS-objects with any kind of HTML-editor (e.g. MOSAIC). Applying this to the natural language based description every page generated relates to an ontological component, defined as data, a service, an application or a process in the model. For the conceptual models, we have developed a graphics-oriented finder and browser which allows to navigate in the information model. In addition, textual editors can be used to edit models using the language OML. For the customization we have integrated the ISA dialogue manager for GUI components and the MAESTRO database designer (MDD) for relational database management systems. This means we are using the generator components of ISA and MDD in combination with our own generator for distributed data, services, applications and processes. All the client parts of Ωmega are available for OS/2, the processes logic could also be distributed on several UNIX-platforms.

As indicated, the FAW software engineering paradigm focuses on the development of distributed software for heterogeneous computing environments, covering a large variety of computer systems and communication facilities together with multimedia components and telecommunication components as well. A neutralizing execution environment for this kind of communication infrastructure comes as part of the systems architecture. Taking into account that several standardization proposals are competing worldwide and large companies are seeking proprietary interoperability solutions, the FAW software engineering paradigm favours a neutralizing approach instead of a standardization approach to bring companies into the position to actively integrate heterogeneous environments for their applications. By using a knowledge based approach (runtime repository including knowledge about the communications infrastructure and the available data and methods) this execution environment - AMBAS - has been implemented at FAW. AMBAS provides a CORBA-compliant execution environment (Common Object Request Broker Architecture from the Object Management Group (c.f. OMG, 1991)) which has been shown in (Eck et al., 1994) and meets the CIMOSA-specification of an integrating infrastructure (Heimann et al., 1994). However, it provides additional functionalities, especially for intelligent request brokerage and integration of existing even non-CORBA-compliant networking environments, e.g. DCE, TCP/IP, SNA, NetBIOS and DECnet. The AMBAS execution environment also supports aspects of a knowledge-based API (Application Programming Interface), a distributed operating system, a distributed data-base management system, workflow control and monitoring functionalities.

The knowledge representation framework (Ωmega information model) is implemented as an AMBAS-object which can alternatively be made persistent in file systems, the OMS/PCMS-Repository (part of MAESTRO II) or ObjectStore (an object-oriented database management system). The architecture of AMBAS is depicted in the following Figure 9.

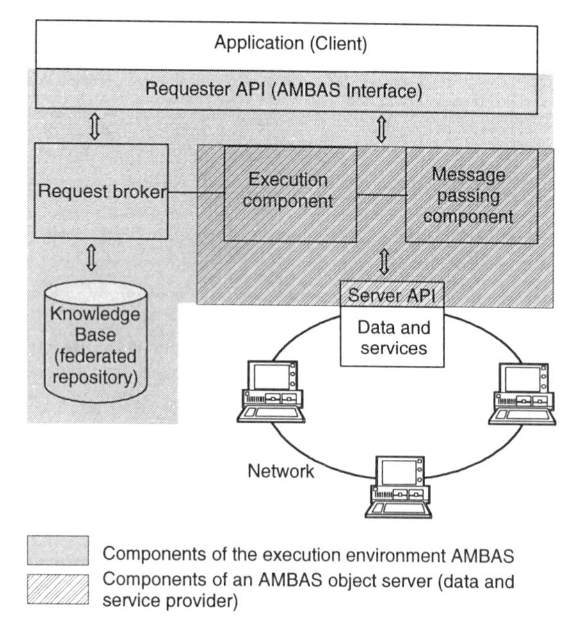

Figure 9 AMBAS - Architecture overview.

AMBAS provides its application programming interface (Requester API) in a problem-oriented fashion. Applications can issue problems to the system instead of specifying the name and location of the function to be executed by specifying the output requested and the input that should be used. The intelligent request broker facility of AMBAS searches for matching methods and presents the most appropriate methods with respect to the problem description based on preference elicitation and other search strategies for selection. Current work done on Eigenmodel-based systems (c.f. Bartusch et al., 1989) applies the algorithmic concepts of Job-Shop Scheduling to this matching process (Möhring et al., 1994).

Based on the data stored in the information model AMBAS supports the development of new kinds of intelligent decision support systems in heterogeneous distributed computing environments by employing AMBAS´ intelligent method base facilities. AMBAS allows the distribution of data and methods and eases the integration of new components (data, services, applications and processes) into this framework. This builds the fundamental basis for cooperative development of software including decision support systems with the possibility to transfer new algorithms directly into industrial use.

Within the FAW software engineering paradigm a generator has been implemented to produce C++ for AMBAS-executables directly from OML-models as depicted in the following Figure 10.

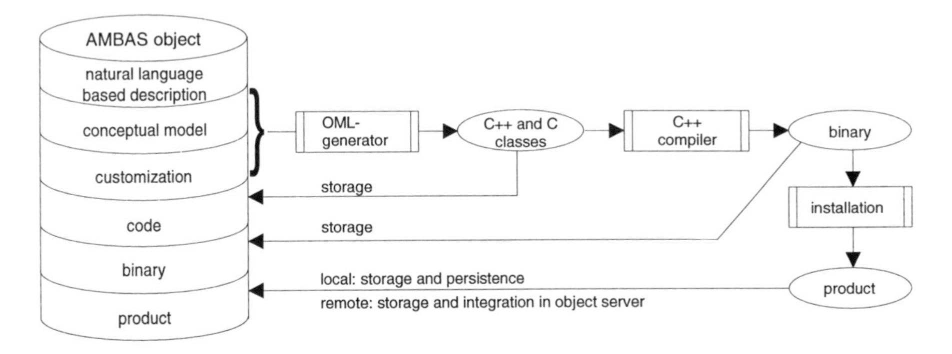

Figure 10 The Ωmega generator.

By using ObjectStore or a file system as the underlying technical framework for the knowledge administration, the generated C++-code can be made immediately persistent. Since data and methods (services, applications and processes) are implemented as objects the same way, AMBAS can be seen as a data and method base management system (extending the concepts of DBMS also to the administration of methods) with OML as its 4GL.

4 CONCLUSION

In this paper the main architectural components of the FAW Software engineering paradigm have been described. In addition, we gave an overview about the architecture of Ωmega and AMBAS, which implements this paradigm. This implementation of a neutralizing execution environment (AMBAS) and an engineering environment (Ωmega) has proven that an overall architecture for enterprise engineering, including process modelling and application development, can be built. Current work focuses on the enhancement of the prototypes. Several mechanisms (preference elicitation, graph-search et.) are added to the request-broker-component of AMBAS as well as monitoring-functions for intelligent load-balancing. Based on the complex modelling of the communication infrastructure knowledge, components of an autonomous and intelligent systems management especially for configuration, change & distribution are built to cover the needs to administrate large environments.

5 REFERENCES

Anderl, R. (1993) STEP-Grundlagen der Produktmodelltechnologie, in *Datenbanksysteme in Büro, Technik und Wissenschaft*, ed. W.A. Stucky.

Bartusch, M., Möhring, R. H., Radermacher, F. J. (1989) Design Aspects of an Advanced Model-Oriented DSS for Scheduling Problems in Civil Engineering; *Decision Support Systems*, Vol. 5, No. 4.

Eck, O., Heimann, M., Holocher, M. (1994) Abdeckung der OMG/CORBA-Spezifikation durch die FAW-Software Engineering Strategie, *FAW Technical Report*, FAW-TR-94015.

ESPRIT Consortium AMICE (Eds.) (1993) CIMOSA - Open System Architecture for CIM, *Research Reports ESPRIT*, Springer.

Grabowski, H. (1993) STEP als Integrationskern für die Produktdatengenerierung, *VDI-Zeitschrift* 135, Nr. 7.

Heimann, M., Holocher, M., Solte, D. (1994) AMBAS - A CIMOSA Compliant Execution Environment, Correspondence of the CIMOSA Integrating InfraStructure and the FAW Execution Environment AMBAS, *FAW Technical Report*, FAW-TR-94019.

Holocher, M., Michalski, R., Radermacher, F. J., Rapl, K., Solte, D. (1994) Gegenüberstellung von Konzepten der relationalen Modellierung der objektorientierten Modellierung und der vollen Modellierung, *FAW Technical Report*, FAW-TR-94002.

Holocher, M., Michalski, R., Solte, D., Vicuna, F. (1993) MIDA - An Open Systems Architecture for the Model-Oriented Integration of Data and Algorithms, *FAW Technical Report*, FAW-TR-93018.

Holocher, M., Solte, D. (1992) AMBAS - An Adaptive Method Base Shell; in *Enterprise Integration Modelling*, ed. C.J. Petrie, Jr., MIT Press.

Kaufmann, F. (1993) Erstellen von Modellen für Organisations- und DV-Lösungen, Entwurf und Spezifikation betrieblicher Objektsysteme mit der grafischen Entwurfssprache GRAPES, SNI-AG, Berlin, München.

Kopaczyk, A., Michalski, R., Rapl, K., Solte, D. (1993) Kommunikationsinfrastruktur 2000, *FAW Technical Report*, FAW-TR-93023.

Möhring, R. H., Müller, R., Radermacher, F. J. (1994) Advanced DSS for Scheduling: Software Engineering Aspects and the role of Eigenmodels; Proceedings *27th Annual Hawaii International Conference on System Sciences.*

Müller, R., Solte, D. (1994) *How to make OR-results available - a proposal for project scheduling*, will appear in special volume of annals of operations research.

OMG (1991) The common object request broker; architecture and specification, *OMG Document* No. 91.12.1.

Petrie, C.J. Jr. (Ed.) (1992) Enterprise Integration Modelling: Proceedings of the first international conference, MIT Press, Cambridge.

Radermacher, F. J., Solte, D. (1994) Die FAW-Software-Engineering Strategie für Multi-Client/Server-Umgebungen, Proceedings On-line '94.

Solte, D. (1987) Open Systems, Ein lernendes Verwaltungssystem für die rechnerunterstützte Methodenkonstruktion im Bereich des Operations Research, *VDI Reihe* 16, Nr. 38.

Solte, D., Heerklotz, K.-D. (1991) Knowledgebased Management of Distributed Ressources, *FAW Technical Report*, FAW-TR 91001.

Teufel, B. (1991) Organization of Programming Languages, Springer.

Verall, M. S. (1991) Unity Doesn't Imply Unification of Overcoming Heterogeneity Problems in Distributed Software Engineering Environments, *The Computer Journal*, Vol. 34, No. 6.

6 BIOGRAPHY

Dirk Solte received his doctoral degree after studying business engineering with focus on operations research and computer science at the University of Karlsruhe. Since 1988, he has been a senior scientist at FAW, heading the department "Communication Systems / Industrial Software Production" and co-heading the department "Enterprise Integration / Decision Support Systems". He is also responsible for the sophisticated technical infrastructure of FAW. This focuses his work in these domains to solutions in heterogeneous distributed environments. Dr. Solte has published several papers in these fields, directed a number of ambitious research and software development projects and consulted industry.

18

An Information Sharing Platform for Concurrent Engineering

T.I.A. Ellis, A. Molina, R.I.M. Young, R. Bell
Manufacturing Engineering Department, Loughborough University of Technology, Loughborough, Leicestershire, LE11 3TU
Email: r.i.young@lut.ac.uk
Telephone: +44 (1509) 263171 ext.2920
Fax: +44 (1509) 267725

Abstract

Research is currently being undertaken by several research laboratories into the computer aided support of simultaneous engineering. These support systems aim to assist organisations that have adopted the simultaneous engineering philosophy by providing consistent sources of product and manufacturing information, and by providing access to a diversity of applications that support decision making.

This paper reports on one such system, the Model Oriented Simultaneous Engineering System (MOSES), and the methodology used for its development. The structure of the system is described together with an outline of the novel research work that is contributing to particular elements of the system. The complexity of co–ordinating and structuring the design and development of such CAE systems necessitates the use of formal methods for their representation. A CAE Reference Model has been developed to guide the development of MOSES. The nature of the reference model is discussed.

Keywords

Concurrent Engineering, Product Model, Manufacturing Model, Reference Model

1 INTRODUCTION

The goals of Simultaneous Engineering (SE) are regularly espoused as being to produce products better, cheaper and faster. SE is seen as being largely dependent on increasing the consideration of life cycle activities in the early design stages [Nevins and Whitney 1989]. This has been achieved by the implementation and management of team working practices [Evans 1990]. The effect of such action is that design practicioners are faced with even greater volumes of product related data to digest and comprehend. As a consequence a requirement for the introduction of information technology systems has arisen [Kahaner and Lu 1993].

A dissatisfaction with the functionality of existing information technology (IT) tools has led to the current interest in SE support. However, the development of complete information technology systems for the support of simultaneous engineering is a relatively young research field concerned primarily with the creation of computer systems that enable the effective sharing of information [Jagannathan et al. 1991]. A complete specification for such systems has not yet been universally agreed. Industrially oriented systems have often focused on developing applications to support one of the many specialist tasks that come under the simultaneous engineering banner, whilst research systems have tended to embrace the wider information structure and software architecture issues. This has led to a number of diverse IT systems being proposed as SE support tools [Molina et al. 1994a].

The research reported on here is related to the development of one system, MOSES, that aims to support SE by providing two key information sources, a product and manufacturing model, together with an architecture that enables dedicated application environments to be added onto the system as required. In order to classify how the MOSES system compares with others it is essential to be able to document what kind of understanding of SE is reflected in MOSES, to what extent MOSES represents an organisational perspective, how MOSES has been developed and how MOSES has been implemented.

The similarities and differences between the goals of different system developers can be highlighted by exploring these issues. The focus of the second part of the paper is on a CAE Reference Model that has been developed to address these issues. It is the authors' opinion that the Reference Model could provide a common basis for comparison and a useful guide for the development of CAE systems that support simultaneous engineering.

2 THE MOSES CONCEPT

The MOSES project is a joint undertaking between Loughborough and Leeds Universities and is funded by the EPSRC. The specification of the research into MOSES focused on a computer based system that provides product and manufacturing information, enables decision support based on these information sources and is co–ordinated in a manner that makes it suitable for operation in a simultaneous engineering environment [ACME/SERC 1991]. The architecture of the system is shown in Figure 1 and consists of two data models (Product and Manufacturing Models) linked, by an integration environment, to a number of application environments. An application called the Engineering Moderator ensures that the evolving product design considers the different life cycle activities that are represented by the application environments.

The operation of MOSES is such that any number of application environments may be supported. Application environments are sometimes referred to as 'Design For X' applications. The configuration and functions of the selected application environments will depend on the needs of the host organisation e.g. a 'design for maintenance' application environment may be important to an automobile manufacture but not required by a disposable watch manufacturer. All product related information is stored within the product model as a design evolves. This is the sole consistent source of product information. Should an application be triggered, then it operates on product information from the product model and any product information that it generates is added to that model. The manufacturing model is the sole source for manufacturing information and hence all applications obtain their manufacturing information from it [Ellis et al. 1993a] . The user may trigger any of the application environments. Conversely they may also be triggered by the Engineering Moderator. The Moderator monitors the product model to ensure that conflicts in information requirements are identified.

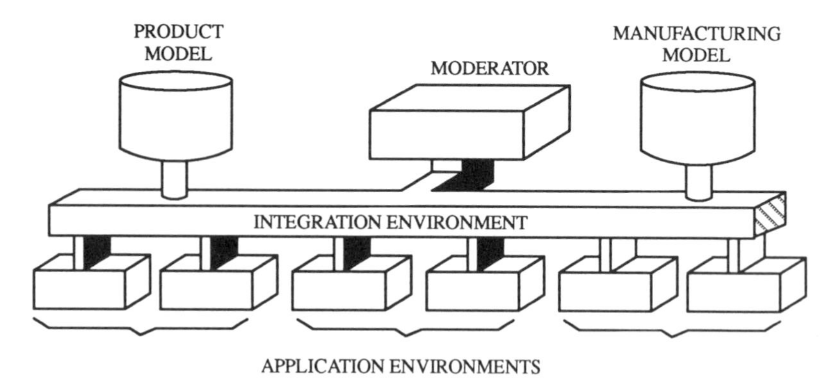

Figure 1 The Structure of the Model Oriented Simultaneous Engineering System (MOSES).

In developing the MOSES system it is intended that the information and procedures necessary to support that part of the simultaneous engineering process concerned with concurrent design for manufacture be identified, and that a prototype knowledge and software environment be used to demonstrate this. To this end an application environment for the support of design for manufacture is being developed. This makes extensive use of both product and manufacturing data and hence exercises and tests the validity of the data models. In order to ensure industrial relevance a total of eight collaborating organisations contribute to the project.

2.1 Product model

One of the data models key to the MOSES project is the Product Model (PM). The Product Model contains all data related to a products' life cycle and is based, wherever possible, on the evolving STEP standard [ISO CD 10303 – 1].

STEP ("Standard for the Exchange of Product Model Data") is the most significant effort towards the development of integrated product models is that currently being undertaken. This standard aims to define a neutral data format for the system independent representation and exchange of all product related life cycle data. The information scope and possible forms of implementation covered by STEP go far further than previous interchange formats such as IGES, SET and VDAFS which concentrated on representing geometric data [Bloor and Bowen 1991]. The aim is to define a standard that can be used as the basis for the development of application oriented software. The researchers on the project have contributed to the STEP standard and within the project several areas of product modelling that have yet to be addressed by the STEP community are being developed. In particular the representation of specifications and assembly relationships are being explored. The nature of the new work on specification, assembly and planned process representations is outlined below.

The aim of the specification modelling research is to build representations that allow specification information to be available for use in other stages of the product development life–cycle [Mckay & Erens 1994]. An IDEF0 activity model for the process of product specification has been established and a preliminary specification data model has been documented and tested through the use of case study data in a prototype implementation. The work to date has addressed different aspects of specification such as product variants and techniques that allow product families to be described in a conceptually efficient way. The work to date has concentrated on mechanical products and further research is being undertaken to establish the relationships between original and variant product specifications.

The product data framework available from previous research and STEP considered assemblies as lists of parts without reference to their functional or physical connectivity. This is not sufficient for applications such as tolerance analysis and design for assembly that require information on the nature of relationships between components [Henson et al 1993]. The research undertaken as part of MOSES has concentrated on combining the existing product data model framework with other data structures to support information about assemblies. The information requirements of the assembly data model have been defined and representations for the function and behaviour of components in an assembly are being researched.

If a product model is to support all life cycle activities then the information associated with the manufacture of the product must be represented. The nature of simultaneous engineering is such that as the product design evolves its manufacturability must be considered. This involves associating the product with both manufacturing processes and resources. The STEP community has developed representations for Integrated Generic Resources [ISO CD 10303–49] and an application protocol for the Numerical Control Process Plans for Machined Parts [ISO CD 10303–213]. The research in MOSES is looking to see how this can be extended to enable the information associated with design for manufacture to be represented. The work undertaken by STEP to date is aimed at representing the information that is in a complete process plan. Because by nature design for manufacture takes place on incomplete product designs then the information structures necessary are currently insufficient. To date a schema for planned processes has been developed that allows elements of a products description to be associated with particular manufacturing processes and activities. The resources used to perform each activity, and any manufacturing advice to be associated to any of these elements, can be modelled.

2.2 Manufacturing model

The second key data model within the MOSES system is the Manufacturing Model. It was found in previous projects that many applications require access to a consistent source of manufacturing information. Often this information is stored in different application systems and databases. In this scenario, data duplication can therefore be a problem. A goal of the MOSES project was to formulate the scope and logical structure of a Manufacturing Model that would integrate, and store without redundancy, the information contained in these separate sources.

The basis for the research into this type of data model came from pioneering work done during the IMPPACT project on a Factory Model [IMPPACT Esprit No. 2165]. Subsequent work on a Manufacturing Resource Model [Kimura 1991], Facility Model [Molina et al. 1992] and Manufacturing Model [Al–Ashaab and Young 1992] have helped shape the nature of the MOSES Manufacturing Model. This project collaborates with other standardisation bodies (e.g. Project 2 of the ISO/TC184/SC4/WG8) to develop an application oriented representation of Resource Usage Management information and function.

The Manufacturing Model describes and captures the information about the manufacturing situation of a company in terms of its manufacturing facility and capabilities at different levels of abstraction [Molina et al. 1994b]. Three entities can be regarded to be basic elements in the definition of any manufacturing environment: resources, processes and strategies. These entities are relevant and important for any type of manufacturing firm. The relations and interaction among them builds the manufacturing environment of a company. It is these three elements that the MOSES Manufacturing Model represents.

Manufacturing resources are all the physical elements within a facility that enable product manufacture e.g. production machinery, production tools, material handling equipment, storage systems etc. The resources are often organized into groups to create manufacturing facilities such as stations, cells or shops. Resources have to be represented in a function oriented manner in order to describe their role in supporting the design, manufacture and production activities. A description of the resources based on their physical properties and functional composition allows the capture of their capabilities. Being able to represent resource capability enables the support of design decisions (e.g. design for manufacture) and manufacturing functions (e.g. process planning). A complete schema for resources has been developed, however, only that part associated with chip forming machinery has been fully defined and implemented.

Manufacturing processes are those processes carried out in a facility in order to produce a product. There are in general two types of manufacturing processes, information and material processes. If a CNC machine is producing a component and the cutting tool breaks then the material processes terminates, however, the information process continues as the NC programme continues to be processed. To limit the scope of the model it was decided from the outset to consider only the machining and injection moulding processes. The Manufacturing Model contains associations between processes and resources. There are many alternative resource combinations that can be employed to perform a given process. It is the role of an external application to select from the alternatives based on a knowledge of the product. It is vital that the content of the Manufacturing Model remain non–product specific if its use is to be maximised.

The representation of structured resources and processes allows us to have a reliable representation of the manufacturing facilities and their capabilities in terms of process technology and equipment. In addition to this type of information, there is a need to represent the manufacturing strategies. Strategies are decisions made on the use and the organization of resources and processes. In effect they are the constraints imposed on the use of a certain type of resource or process that differentiate an organisations manufacturing operation from that of others. There are two types of decisions which make possible the formulation of manufacturing strategies: decisions made over time which define the structure, capacity and technology of the facilities, and the day to day decisions which determine how to use the facilities and related processes. In the Manufacturing Model, strategies refer to capacity, facility structure, technology (i.e resources and processes), production planning/material control and the organization of work.

The Manufacturing Model has been structured into four levels based on a defacto standard [BSI PD 6526:1990]:

1. Factory Level
2. Shop Level
3. Cell Level
4. Station Level.

These levels of abstraction provide manufacturing information for all hierarchical and functional activities within a manufacturing enterprise. For the data model to be of practical use it must reflect at any given instant the actual resources in the factory. A prototype Manufacturing Model has been developed that represents the manufacturing capability of a collaborators machine shop (Figure 2). The manufacturing model has been documented in EXPRESS [ISO CD 10303 – 11] to ensure consistency with the practices of other ISO standardization groups.

2.3 Engineering moderator

SE requires that a number of different tasks be inter–woven within a given time period. Each task has its own goals and a hierarchy is imposed that allows trade–offs between goals at different levels, in effect some tasks are given higher priority. The nature of the tasks is such that they can require pre and post actions. SE is therefore often considered to consist of a number of parallel tasks but inevitably there are some critical paths which impose a sequence on events.

Research into computer support for SE has often concentrated on trying to resolve this complex inter–relation of tasks. This has proved difficult because SE rarely involves straightforward 'individual' actions. Apparently discrete activities rely upon a complex network of factors and systems to date do not adequately deal with this interaction.

The MOSES approach to imposing a measure of concurrency on the operation of the system is to use an application called the Engineering Moderator. The aim of the Moderator is fourfold [Harding 1993] :

1. Promote co–operation and negotiation between SE team members
2. Identify significant problems within a design
3. Determine appropriate action in response to a problem
4. Maintain communication between interested agents until the problem is resolved

The Moderator bases its operation on assessing the product information currently available rather than on monitoring a sequence of tasks. As a design evolves the data related to the product is entered into the product model. The product model is therefore the most reliable indicator of the state of the current design. The moderator works by monitoring and analysing the proposed input to the product model in order to assess whether the user, or any of the application environments, should be advised of the design change. To achieve this the moderator has an inbuilt knowledge base which contains details of the types of product information change which are particularly sensitive to each application environment.

All the application environments access shared product information in the product data model. This leads to situations were tasks may be disrupted due to a mismatch of expectations. For example if a design is being analysed for design for manufacture then a design change may be recommended that conflicts with that of the design for function requirements. If undetected then inappropriate actions may be carried out. However, the moderator attempts to get the applications to realign their expectations. The moderator needs to ensure the timeliness of its intervention and the authors believe this is best achieved by being aware of changes to the product data model as they occur.

In effect the moderator enables the system to become self–organising because the way in which the system responds depends on the way in which the PM evolves. There is no fixed procedural task structure for the design process. Hopefully this improves the competence of the system and supports interaction and conversation because each application is basing its decision to act on the information available to it rather than on a false preconceived model of design tasks.

The sensitivity of the Moderaotor to product information change is critical because, if incorrect, it is possible that some applications could be continuously switched between tasks. A simplified prototype Engineering Moderator has been developed that successfully monitors product model changes and subsequently triggers the relevant application.

2.4 Application environments

SE systems are still often compartmentalised into predominantly problem centred DFX application groups. This is because it enables researchers and developers to deal with smaller information sets and also because research is often theoretically driven and involved in resolving highly specialised research topics. Rather than replace this view the authors believe it can prove an advantage if the information sources used by each application group are common, and that the working of each is co–ordinated. It is essential that future CAE systems be extensible. By modularising the application groups that use the services of the data models this becomes feasible.

The resources available to the project have necessitated that only one application environment, addressing design for manufacture, be fully researched [Ellis et al.1993b]. This shall be referred to as the Design For Manufacture Environment (DFME). An application environment concerned with design for function is being developed in research related to the project . Design for manufacture was selected because it is an application area that can exercise most elements of the MOSES system. The structure of information in the product and manufacturing models can be tested and the operation of the moderator investigated.

Design for manufacture (DFM) is one of many facets of simultaneous engineering and aims to support experienced designers by helping to ensure that certain key design rules are systematically addressed. The design rules help ensure that the product design can be manufactured with the processes and resources available to the organisation.

The DFME aims to provide reliable, and timely, enterprise specific design for manufacture advice to the user and must ultimately also support the generation of manufacturing information e.g. process planning. To perform design for manufacture on an evolving product design it is necessary to analyse the component in relation to process capability and available resources. This information is stored in the product model because it is of relevance to later activities such as producing process plans. Indeed the nature of post design manufacturing activities such as process planning and scheduling could be significantly changed by undertaking DFM using such a CAE support system.

Though the actions of the design for manufacture environment are determined by the amount of information available in the data models, it is useful to relate this to some traditional design model in order to show the type of activities undertaken. Elements of DFM are applied throughout the product design process. This is assumed to consist of conceptual, embodiment and detailed design phases [Pahl and Beitz 1984]. At the conceptual stage designers often assume a particular manufacturing method and are interested only in the general capability of processes e.g. material suitability, product size, estimated tolerance bands. At the embodiment stage the design team requires information on the capabilities of various processes e.g. form constraints, resource types, workholding methods. In the detailed design stage they will benefit from information related to the potential capabilities and availability of specific machines and tools and the way in which the combination of these effect geometric, surface and dimensional tolerances. We therefore propose that there are three query types that the DFME environment must be able to respond to:

1. Product suitability for manufacture by a given process
2. Product suitability for manufacture using a given resource type
3. Product suitability for manufacture using a specific resource combination.

Process suitability queries will prompt the DFME to determine which manufacturing processes are capable of manufacturing a given product, for a given volume, material and approximate shape. Resource Type suitability queries prompt the DFME to determine how suitable a component, or a feature of a component, is for manufacture by a given process and workstation type. Specific resource suitability queries prompt the DFME to determine how suitable a particular component, or component feature, is for being produced using a named resource, or resource set, that is available to the Enterprise. The nature of feedback to the user and product model is to list the feasible solutions and to comment on the implications of product design in terms of cost, lead time and quality. The DFM analysis considers the product specification, form and tolerances. It is assumed that the specification contains manufacturing due date, volumes and required batch sizes.

The DFME is partitioned in such a way that it has one governing application called an manufacturing strategist. The role of the strategist is such that it governs a hierarchy of application sub–groups. In the DFME a range of process groups are represented e.g. machining and injection moulding. In turn each of these has sub–experts e.g. machining would consist of experts for turning, milling etc. Each application group and expert models a set of constraints related to their particular subject area. The Strategist must determine which of these sub–groups to trigger based on the available product information. It may be that several sub–groups are activated. In this case the Strategist must evaluate between the proposed feasible alternatives returned by the sub–groups. The Strategist also overseas the requests from applications for information from data models and the users. The user may be asked to make selections in order to direct a search sequence or may have to provide additional product information in order to make a search feasible.

In essence the DFME contains mechanisms for making associations between product information and process/resource information. In order to demonstrate the principles involved in the research a much restricted implementation of the DFME is being developed that communicates with the product and manufacturing models. The role of the DFME is being documented using the Booch methodology [Booch 1991] and IDEF0 diagrams [Colquhoun et al. 1993].

3 THE CAE REFERENCE MODEL

The activity of co–ordinating and structuring complex CAE systems during design and development necessitates the use of formal methods. The MOSES research is being guided by the use of a CAE Reference Model (CAE–RM) to provide a framework for the integration of modelling methodologies and computer tools [Molina et al. 1994c].

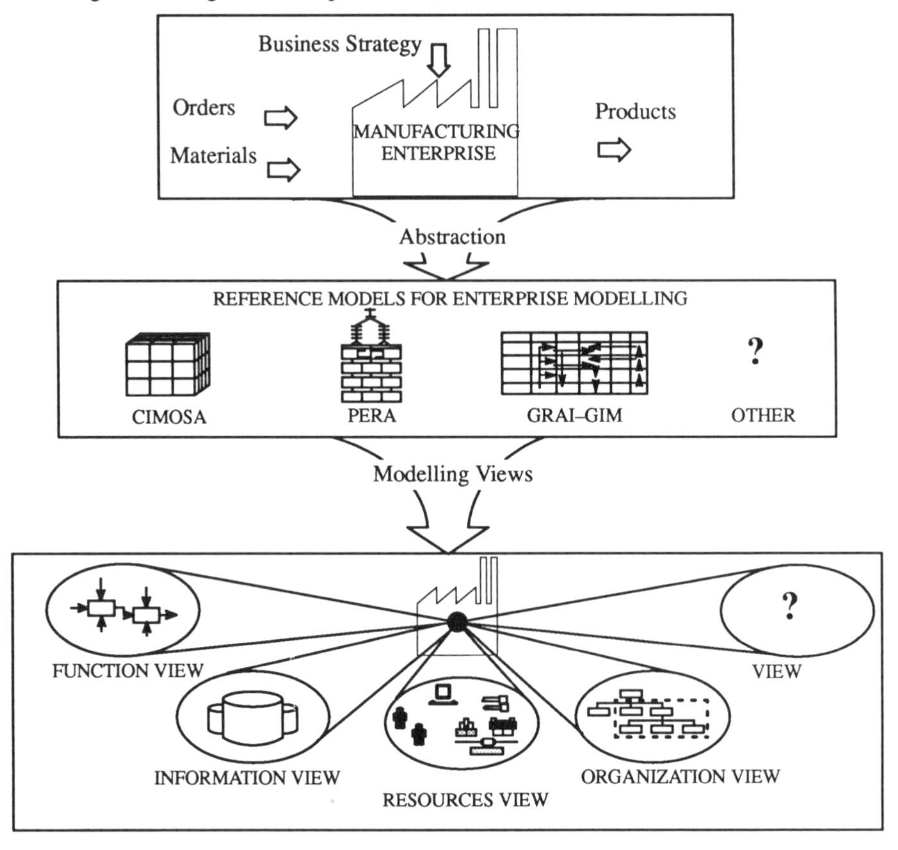

Figure 2. Reference Models for Enterprise Integration and related Modelling Views

The Booch Object Oriented Methodology [Booch 1991], IDEF0 [Colquhoun et al. 1993] and EXPRESS [ISO CD 10303 – 11] are methodologies that are specified within the CAE–RM to assist in the modelling of different aspects of a system. A number of supporting computer tools have been used to maximise the effectiveness of these methodologies. They include STEP

Programmer's Tool Kit (STEP Tools Inc. 1993) for compiling and refining EXPRESS schemas, and the RATIONAL ROSE (Rational 1993) software tool for Booch documentation.

Many current systems being developed to support SE are open in nature and distributed so as to allow remote operation and extensibility. The Reference Model for Open Distributed Processing or RM–ODP [SC21/WG7 N 755] is an emerging standard focused on open distributed systems. The CAE–RM that has been developed by the MOSES team is based on this reference model so as to enforce the generic and modular characteristic of the model.

The RM–ODP allows the description of an information system from different viewpoints: Enterprise, Information, Computation, Engineering and Technology. Each view represents a specific aspect of the information system. The views of the RM–ODP aim to allow a complex system to be described from a number of perspectives. The emphasis of each view is tailored to primarily represent either objectives, realisation or behavioural attributes. The research currently being undertaken is mainly involved with the definition of the first three viewpoints i.e. Enterprise, Information and Computation.

The Enterprise Viewpoint is associated with the specification of requirements for ODP systems. IDEF0 activity diagrams are employed to describe this view. The information viewpoint is particularly important to the development of MOSES since this sets the context for the development of the information models in an implementation independent form. The information flows, together with information structures, are represented at the Information Viewpoint. This view is documented using EXPRESS and IDEF0 models. The Computational Viewpoint focuses on the functional decomposition of the system into objects, the activities that occur within those objects and the interactions between the objects. Booch's diagrams are used to describe this view.

The last two viewpoints (Engineering and Technology) have not been fully developed by the MOSES team. The Engineering Viewpoint focuses on the infrastructure required to support distribution. This view enables the specification of the processing, storage and communication functions required to implement the system. The Technological Viewpoint focuses on the selection of the necessary technology to support the system. In our case we chose the object oriented database DEC Object/DB [Objectivity/DB 1991] and the object oriented programming language C++ [Stroustrup 1986].

A reference model should allow the scope of a system to be defined. SE systems are traditionally described in terms of their architectures and how to use them i.e. the use of formalised approaches to describe technology driven research. Whilst this is adequate for standalone systems, a vital element that is missing is how the system will support and be integrated in the organisation in a wider context.

A wide range of different reference models, frameworks and architectures exist for use in information system and manufacturing systems development. Among these reference models, the ones related to modelling and implementation of CIM systems can be used, in the opinion of the authors, to define a context for where future computer systems to support simultaneous engineering can be integrated within the enterprise. The reference models most often considered for the task of describing an integrated system, its life cycle and the methodology for its application are CIM–OSA [ESPRIT Project 688/5288], Purdue Enterprise Reference Architecture [Williams 1991] and GRAI–GIM [Doumeingts et al. 1992]. These reference models enable the creation of enterprise models that take into consideration different viewpoints (Figure 3). The authors believe that using these enterprise integration frameworks to set the context for the research into a Reference Model for Computer Aided Support of Simultaneous Engineering allows the creation of a more structured and flexible model. The MOSES CAE–RM achieves this, and ensures integration within a wider framework, by defining a mapping to the CIM–OSA architecture. The CIM–OSA architecture is such that a reference model for a technical support system, such as a CAE system, can be defined as a resource within the architecture. This is achieved by defining the inputs, outputs and required capabilities of the technical system. The information and function elements of the architecture set the context for the requirements and the use of that resource i.e. the CAE system (Figure 4).

4 DISCUSSION

The development of effective information structures for the representation of both product and manufacturing information will largely determine the success of CAE support for simultanoeus engineering. There is considerable research into product modelling and some of the work generated by the STEP community is being adopted by system vendors and hence slowly permeating into industry. The scope for development of the STEP standard is vast. Within the MOSES project the research into representing specification and assembly relationships should improve product model utility.

Research into manufacturing modelling has not received as much attention, however, the indicators are that this is a rapidly developing field. Several groups are focusing their effort on developing virtual factories and enterprise models. To support this concept the manufacturing models of the future will have to represent not only manufacturing capability and characteristics but also they will have to model the dynamic nature of a facility. This will enable them to support simulation and scheduling applications.

An important element of the MOSES Manufacturing Model is the ability to model strategies for the use of processes and resources. The way in which an enterprise operates is determined not only by the technology available but also by the rules and standards that have developed with the company. Modelling strategies is one way of ensuring that company differentiation is not lost by the implementation of IT support. It also potentially allows the effect of changing manufacturing strategies to be studied.

The element that distinguishes CAE support for simultaneous engineering from that of conventional CAE support is the management of the complex interaction of varied tasks. The MOSES project have chosen to use an application called the Engineering Moderator to enforce concurrency. This application bases its decision to act on the information content of the product model. Each addition to the product model is analysed and assessed to determine if either the user or application environments are likely to be interested in the change. Should this be the case then the relevant application or user is informed. Care is needed to ensure that the Moderator is not over–sensitive to change and hence constantly triggering actions. The advantages of this approach is that a preconceived design model is not followed and the evolving design is forced to consider all the life cycle activities represented by application environments.

The use of a CAE Reference Model to guide the design and development of the CAE system has proved valuable. The ability to represent different viewpoints of the same system is a major advantage. The RM–ODP provides a structured and well documented format for achieving this. The use of advanced computer tools which support modelling methods such as Rational Rose for Booch and STEP Tool Kit for EXPRESS accelerates the implementation of prototype software to prove system concepts.

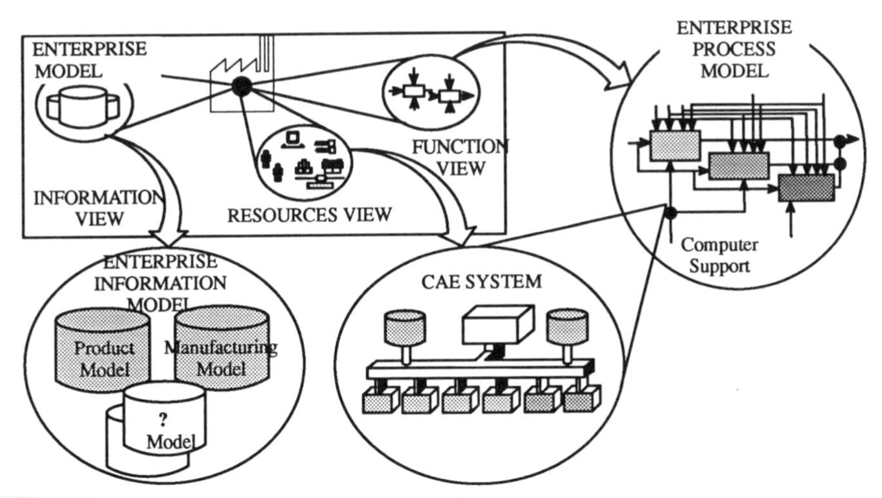

Figure 3. CIMOSA Enterprise Model as a Context for the MOSES CAE Reference Model and CAE System

The mapping of the CAE–RM to CIM–OSA has the advantage of setting IT support for SE in the context of an enterprise model. Potentially the influence of the system on existing working practices can then be modelled. The management of different perspectives in a company is often based on the strengths of the personalities involved and on the company culture [McCarthy 1994]. If IT support gives an equal status to each perspective then the nature of the design task in that company may be imbalanced and the design differentiation imposed by the company culture is lost. For this reason the system must be able to be customised to suit the organisation. It is only by studying this wider context that the impact of new technolgy can be predicted. Traditionally CADCAM has been a sequential development with technology considered before its effects on organisational and human aspects. The use of combined enterprise and CAE Reference Models should enable the selection of the most appropriate technology for the business. This process is lead by the need to achive complete enterprise integration.

5 CONCLUSIONS

The need to establish or sustain competitive advantage invariably imposes a requirement to introduce new information technology systems in virtually all contempory organisations at some stage in their development [Helliwell 1994]. Simultaneous engineering implementations have reached the level of maturity were organisations feel that they could benefit from more effective IT support.

The authors consider that the key issues in the CAE support of simultaneous engineering are information modelling, decision support and task co–ordination. The MOSES project has addressed these issues by:

- Developing Product and Manufacturing Information Models

- Creating an environment for decision support
- Investigating mechanisms for task co–ordination

The complexity of developing such systems imposes a requirement for an integrated and formal approach to system development. This has been achieved within the MOSES project by defining and using a CAE Reference Model. The authors' experiences demonstrate that the use of a CAE Reference Model is necessary for effective system development. The benefit of the CAE–RM has been enhanced by building a mapping to the CIM–OSA framework. This ensures that the CAE system meets the requirements imposed by the concept of an integrated enterprise.

6 ACKNOWLEDGEMENTS

The research reported on in this paper is that undertaken by the whole MOSES project team and not solely that of the authors. The research is part of an ACME funded project, 'Exploiting Product and Manufacturing Models in Simultaneous Engineering', pursued at Loughborough University of Technology and Leeds University, and supported by a group of industrial collaborators [SERC ref. GR/H 24273 and GR/H 24266 respectively].
The research undertaken by A. Molina is funded by the Mexican Government (CONACyT), Monterrey Institute of Technology (ITESM, Mexico) and the ORS Award Scheme (ORS/9226010).

7 REFERENCES

ACME/SERC Proposal, 1991, *Exploiting Product and Manufacturing Models in Simultaneous Engineering*, Department of Mechanical Engineering–University of Leeds, Department of Manufacturing Engineering–Loughborough University of Technology.

Al–Ashaab A., and Young R.I.M., 1992, "Information Models: An Aid to Concurrency in Injection Moulded Products Design", *Presented in the Winter Annual Meeting '92 of ASME*. Anaheim, California, November. 8–13, 1992.

Bloor M.S., and Owen J., 1991, "CAD/CAM product–data exchange: the next step", *Computer Aided Design*, Vol. 23, No. 4, May, pp. 237–242.

Booch G., 1991, *Object–oriented design with applications*, The Benjamin/Cummings Publishing Company, Inc..

BSI PD 6526:1990, CEN/CENELEC Report R–IT–01, Evaluation on CIM architectures, *British Standard Institution.*

Colquhoun, G., Baines, R.W., and Crossley R., 1993, "A state of the art review of IDEF0", *Int. J. Computer Integrated Manufacturing*, Vol. 6, No. 4, 252–264.

Corrigall M.J., Lee M.K., Young R.I.M., and Bell R., 1992, "Manufacturing Code Generation in a Product Modelling Environment", Proc. Instn Mech. Engrs., Vol. 206, pp. 165–175.

Doumeingts G., Vallespir B., Darricar D., and Chen D., 1992, *GIM, GRAI Integrated Methodology, A Methodology for Designing CIM Systems*, Version 1.0, Unnumbered Report, LAP/GRAI, University Bordeaux 1, Bordeaux, France, May.

Ellis, T.IA., Young R.I.M., and Bell R., (1993a). Modelling Manufacturing Process Information to Support Simultaneous Engineering. *Proceedings of the ICED 93*, Volume 2, N.F.M. Roozenburg (ed.), 9th International Conference of Engineering Design, August 17–19, 1993, The Hague, pp. 1081–1084.

Ellis, T.IA., Young R.I.M., and Bell R., (1993b). A Design For Manufacture Software Environment to Support Simultaneous Engineering. *Proceedings of the 9th National Conference on Manufacturing Research*, Alan Bramley and Tony Mileham, September 7–9, 1993, University of Bath, U.K. pp. 1081–1084.

ESPRIT Consortium AMICE, 1993, *CIMOSA: Open System Architecture for CIM, Project 688/5288*, AMICE, Volume 1, 2nd revised and extended edition, Springer–Verlag.

ESPRIT Project 688/5288, AMICE, Volume 1, *CIMOSA: Open System Architecture for CIM*, ESPRIT Consortium AMICE (Eds.), 2nd, revised and extended edition, Springer–Verlag, 1993.

Evans S., 1990, "Implementation framework for Integrated design teams", Journal of Engineering Design, 1(4), pp. 355–363.

Harding J.A. , 1993, The Engineering Moderator Within The MOSES Project, MOSES discussion document, moses–dd–21, 29/3/1993.

Helliwell G., 1994, Introducing IT into a Mature Production Related WOrk Environment: The Human Resource Factor, Journal of Information Technology, Vol.9 No.1, March 1994, pp. 39–50

Henson B.W. , Baxter J.E. , Juster N.P. , 1993, Assembly Representation Within a Product Data Framework, Proceedings of ASME Design Technical Conferences, USA, September 1993.

IMPPACT, Esprit No. 2165, *Proceedings of the Workshop Integrated Modelling of Products and Process using Advanced Computer Technologies*, Berlin, 26–27 February 1991.

ISO CD 10303 – 1, 1992, *Product Data Representation and Exchange – Part 1: Overview and Fundamental Principles.*

ISO TC184 SC4 WG8/N13, Project 2: "Resource Usage Management", Draft Scope presented at the London Meeting by B.R. Katzy, June, 1992.

ISO CD 10303–213, Numerical Control (NC) Process Plans for Machined Parts version 1.0, 1/5/1994.

ISO CD 10303–49, Integrated Generic Resources: Process Structure and Properties,

ISO CD 10303 – 11, 1993, *Industrial automation systems and integration – Product data representation and exchange – Part 11: Description methods: The EXPRESS language reference manual.*

Jagannathan V., Cleetus K.J., Kannan R., Matsumoto A.S., and Lewis J.W., 1991, "Computer Support for Concurrent Engineering", *CONCURRENT ENGINEERING Issues, Technology and Practice*, September, pp. 14–30.

Kahaner D., and Lu S., 1993, "First CIRP International Workshop on Concurrent Engineering for Product Realization", *Concurrent Engineering Research in Review*, Volume 5, pp. 6–14.

Kimura F., 1991, "Software for Product Realization", *International Symposium for International Trends in Manufacturing Towards the 21st Century*, October 18, Berlin, Germany.

McCarthy J., 1994, "The State–of–the–Art of CSCW: CSCW Systems, Cooperative Work and Organisation", Journal of Information Technology, Vol. 9 No. 2, June 1994, pp. 73–83

Mckay A. , Erens F., Relating Product Specifications and Product Definitions in a Product Model, Moses report series, moses–report–series–1, 14/1/1994.

Molina A., Mezgar I. , and Kovacs G., 1992, "Object Knowledge Representation Models for Concurrent Design of FMS", *Human Aspects in Computer Integrated Manufacturing*, G.J. Olling, F. Kimura (editors), Elsevier Science Publishers B.V., (North–Holland), IFIP, pp. 779–788.

Molina A., Al–Ashaab A.H., Ellis T.I.A, Young R.I.M, Bell R., 1994a, "A Review of Computer Aided Simultaneous Engineering Systems", Accepted for publication in the journal of *Research in Engineering Design*, June 1994.

Molina A., Ellis T.I.A, Young R.I.M, Bell R., 1994b, "Modelling Manufacturing Resources, Processes and Strategies to Support Concurrent Engineering", to be presented in the *First International Conference on Concurrent Engineering: Research and Applications*, August 29–31, 1994, Pittsburgh, PA, USA.

Molina A., Ellis T.IA., Young R.I.M., and Bell R., 1994c, "Methods and Tools for Modelling Manufacturing Information to Support Simultaneous Engineering", Preprints, *2nd IFAC/IFIP/IFORS Workshop Intelligent Manufacturing Systems – IMS'94*, (Ed. P. Kopacek), Vienna – Austria – June 13–15, 1994, pp. 87–93.

Nevins J.L. and Whitney D.E., 1989, *Concurrent Design of Product and Processes: A Strategy for the Next Generation in Manufacturing*, McGraw–Hill Publishing Company, N.Y.

Objectivity/DB, 1991, *Reference Manuals*, Objectivity Inc.

Pahl G., and Beitz W., 1984, Engineering Design: A Systematic Approach, Springer–Verlag.

Rational Object Oriented Software Engineering, 1993, RATIONAL ROSE ANALYSIS & DESIGN, SUN PLATFORM, Version 1.1.

SC21/WG7 N 755, 1993, *Draft Recommendation X.901: Basic Reference Model of Open Distributed Processing – Part 1: Overview and guide to use*, ISO/IEC JTC1/SC21/WG7.

Stroustrup B., 1986, *The C++ Programming Language*, Second Edition, Addison–Wesley.

STEP Tools Inc., 1993, *STEP Utilities Reference Manual*, Version 1.2 for UNIX Workstations.

Williams T.J., 1991, *The Purdue Enterprise Reference Architecture*, Report Number 154, Purdue Laboratory for Applied Industrial Control, December.

Young, R.I.M. and Bell R., 1992, Machine Operation Planning in Product Modelling Environment, International Journal of Production Research, Vol. 30, No. 11, pp. 2487–2513.

PART EIGHT

Pre-normative and Standardisation Issues

19

Development of GERAM, A Generic Enterprise Reference Architecture and Enterprise Integration Methodology

T. J. Williams
Chairman, IFAC/IFIP Task Force on Architectures for Enterprise Integration, Purdue University, 1293 Potter Engineering Center Room 308D, West Lafayette, Indiana, 47907-1293, USA, Phone: 317/494-7434, Fax: 317/494-2351, Email: tjwil@ecn.purdue.edu

Abstract
The IFAC/IFIP Task Force on Architectures for Enterprise Integration has developed an initial proposal for its **G**eneric **E**nterprise **R**eference **A**rchitecture and **M**ethodology (GERAM). The report briefly describes GERAM and the plans of the Task Force for its further development.

Keywords
Enterprise Integration, Reference Architecture, Integration Methodology, Modeling Tools, Modeling Languages, Glossary

1 THE TASK FORCE AND ITS WORK

The IFAC/IFIP Task Force on Architectures for Enterprise Integration was formed at the IFAC World Congress in Tallinn, Estonia, in August 1990. It's mission was to study the field of enterprise reference architectures for the purpose of picking a best one for future use from those already available. Failing this, the Task Force should recommend a method of development of a better one than those it had studied.

At the 1993 Congress in Sydney, Australia, the Task Force reported (Williams, et al, 1993) on its extensive analysis of the three major enterprise integration architectures available at that time: CIMOSA (AMICE Consortium, 1992), GRAI-GIM (Doumeingts, et al, 1992), and PERA (Williams, 1992). While each of these could be developed into a complete and satisfactory Reference Architecture and associated Integration Methodology, the group recommended that the synergy of the consolidation of the best features of each should be developed. This should give the best possible system for future enterprise integration programs.

Our first major proposal for such a consolidated architecture was authored by Peter Bernus and Laszlo Nemes, both Task Force Members, and presented at the Vienna Workshop Meeting of the Task Force in June 1994 (Bernus and Nemes, 1994). They applied the title GERAM (**G**eneric **E**nterprise **R**eference **A**rchitecture and **M**ethodology) to this more complete and usable architecture as suggested earlier by the Task Force.

2 WHAT IS GERAM?

As had been recommended in the previous report of the Task Force, the proposal by Bernus and Nemes builds upon the work carried out earlier by CIMOSA (AMICE Consortium, 1992), GRAI-GIM (Doumeingts, et al, 1992) and PERA (Williams, 1992) to organize the development and use of the modeling techniques, tools and methodologies needed by the potential user to consolidate and use the data required for integrating their own enterprise. The proposed system would also help the work of industrial consultants in aiding the user. It would help coordinate the products of tool and methodology developers to prepare suitable additional products to further enhance the developing system. Finally, it would outline the needs of further theory and concept development to be carried out by the researcher in this area.

As proposed by Bernus and Nemes, GERAM itself is a framework of six major components as follows:

1. Generic Enterprise Reference Architecture (GERA);
   This is the definition of enterprise related concepts, with the primary focus on the life-cycle of the enterprise. Since the life-cycle can be considered as a design process the architecture will also have to identify the intermediary results and components of this design process. It can be considered the model of the life cycle.
2. Generic Enterprise Engineering Methodology (GEEM);
   This is the description, on a generic level, of the processes of enterprise integration. In other words, the methodology is a detailed process-model, with instructions for each step of the integration project.
3. Generic Enterprise Modeling Tools and Languages (GEMT&L);
   The engineering of the integrated enterprise is a highly sophisticated, multidisciplinary management, design and implementation exercise during which various forms of descriptions and models of the target enterprise need to be created. To express these models more than one modeling language may be needed.
4. Generic Enterprise Models (GEMs);
   Generic enterprise models capture concepts which are common to all enterprises. Therefore the enterprise engineering process can use them as tested components for building any specific enterprise model.
5. Generic Enterprise Modules (GMs);
   Modules are **products**, which are standard implementations of components that are likely to be used in enterprise integration — either by the enterprise integration project or by the enterprise itself. Generic modules can be configured to form more complex modules for the use of an individual enterprise.
6. Generic Enterprise Theories (GTs);
   Theories which describe the most generic aspects of enterprise-related concepts. Generally called ontological theories. They may also be considered "meta models" because they consider facts and rules **about** the facts and rules of enterprise models.

The Bernus and Nemes Proposal also pointed out an interesting aspect of the generic integration process. This is that it is recursive in that the same Generalized Enterprise Reference Architecture and Methodology (GERAM) can be used to describe not only the manufacturing enterprise itself, but also three other recursively related enterprises as well. These are all as follows:

1. The design development and production of a product to be produced by a manufacturing enterprise. The product so produced will be labeled Entity 4.

2. The life history (concept, design, construction and operation) of the manufacturing enterprise itself. The manufacturing enterprise is Entity 3.
3. The life history of the engineering enterprise which developed the subject enterprise in question. The engineering design enterprise is Entity 2.
4. The strategic enterprise management process (life cycle) which developed the need and concept for the subject manufacturing enterprise in the first place as well as the engineering enterprise that produced this subject enterprise. The strategic enterprise is Entity 1.

That is: Entity 1 (The Strategic Enterprise) developed Entity 2 (The Engineering Enterprise), which further develops Entity 3 (the Manufacturing Enterprise) which then develops and produces the product (Entity 4) for the final customer user. This scenario with suitable adjustments would describe the relationship of any similar set of enterprises and their final products in any industrial interaction or indeed in any multi-enterprise endeavor of any type.

Bernus and Nemes proposed that this process of recursivity is limited to the above four stages and that there are no further stages in either direction from them.

Such a concept of universal usage of the principles of enterprise reference architecture applicability in all areas of human endeavor today (not just manufacturing as commonly assumed by many involved in our field) has major ramifications on the importance and potential future use of the results of the Task Force's work.

3 WHAT MUST GERAM ACCOMPLISH?

A generic enterprise reference architecture, such as GERAM will become, has many requirements that it should satisfy. Some of these are:

1. Give the best possible treatment of the scope of the enterprise from the systems point of view;
   It is necessary that all activities which are involved directly or indirectly in designing and operating or improving the enterprise should be covered by the architecture.
2. Provide a consistent modeling environment which will lead eventually to executable code for computer usage;
   The modeling views offered should cover a minimal necessary set but this set should be expandable. The ideal modeling environment should be modular so that alternative methodologies can be based upon it.
3. There should be a detailed methodology for use which development personnel of enterprises of all types can readily follow;
   This methodology must be technically correct as well as readily understandable and easy to use. It must be executable by real teams within acceptable cost, time and resource constraints.
4. It should promote good engineering practice for building reusable, tested and standard models;
   The apparent complexity must be kept low. Intricate details should be encapsulated in reusable engineering building blocks.
5. It should provide a unifying perspective for products, processes, management, enterprise development and strategic management;
   The architecture should tie and relate all aspects of enterprise integration and the related enterprise engineering to the rest of the activities of the enterprise.

4 A DESCRIPTION OF GERAM

GERAM as proposed (Bernus and Nemes, 1994) is based upon a so-called matrix graphical model of the life-cycle of an enterprise as developed by those authors and others for use in the Task Force Major Report (Williams, et al, 1993) as a basis for a comparison and evaluation of the capabilities of each of the candidate architectures studied there. This model has been structured to include a representation of the capabilities and strong points of each of these architectures.

Since our stated goal is to develop an architecture combining the best features of the earlier architectures to produce a synergistically superior product, the evaluation matrix also becomes a model of GERAM itself. The matrix is shown in Figure 1.

The other major contributions of the authors in describing GERAM beyond the earlier recommendations of the Task Force in its report are as follows:

1. The definition of the three major classes of users of GERAM;
   a) Those who would use GERAM as the basis for developing enterprise integration programs for their own companies or manufacturing plants. This would also include consultant engineering companies who help final user groups prepare and carry out these programs.
   b) Those entities, individuals, groups or companies, carrying out the development of new and improved tools and methodologies for carrying out the work of the personnel of Item 1a above.
   c) Researchers, academics, and others developing new modeling theories and ontologies to provide a theoretical underpinning for the work described for both groups above.

   It was noted by the authors that the matrix representation would probably be too complex for ready use by Group 1a above and maybe even 1b as well. Thus it was recommended by them that initially the PERA diagram of Figure 2 (Williams, 1992) should be used. The Matrix form of Figure 1 would be used by the research group and possibly the tool developers as well.
2. The definition of the six major components of GERAM as described in Section 2 above.
3. The discovery and promulgation of the recursivity of the Matrix diagram, and the PERA diagram as well, as also described in Section 2 above in the representation of all types of enterprise entities as noted there. This latter finding is a major factor in showing the correctness and probably the universal applicability of these architectural forms.
4. Following the Task Force recommendations, Bernus and Nemes also emphasized the need for a complete descriptive methodology to teach and guide users in preparing the enterprise integration programs. As the most complete at present, the Purdue Methodology (Williams, 1994) was presented as the initial choice.

5 CURRENT STATUS OF THE GERAM PROJECT

The Task Force Members have in general whole-heartedly welcomed the Bernus-Nemes Proposal as a solid step toward our eventual goal of a single, well-accepted, and standardized architecture with its associated tools and methodology. However, it is also very clear at this point that this is being considered only as a first step which will require much additional clarification, and development, and hopefully competition with other near future proposals which may emerge by others.

Bernus and Nemes thus proposed the following to develop the six components of GERAM:

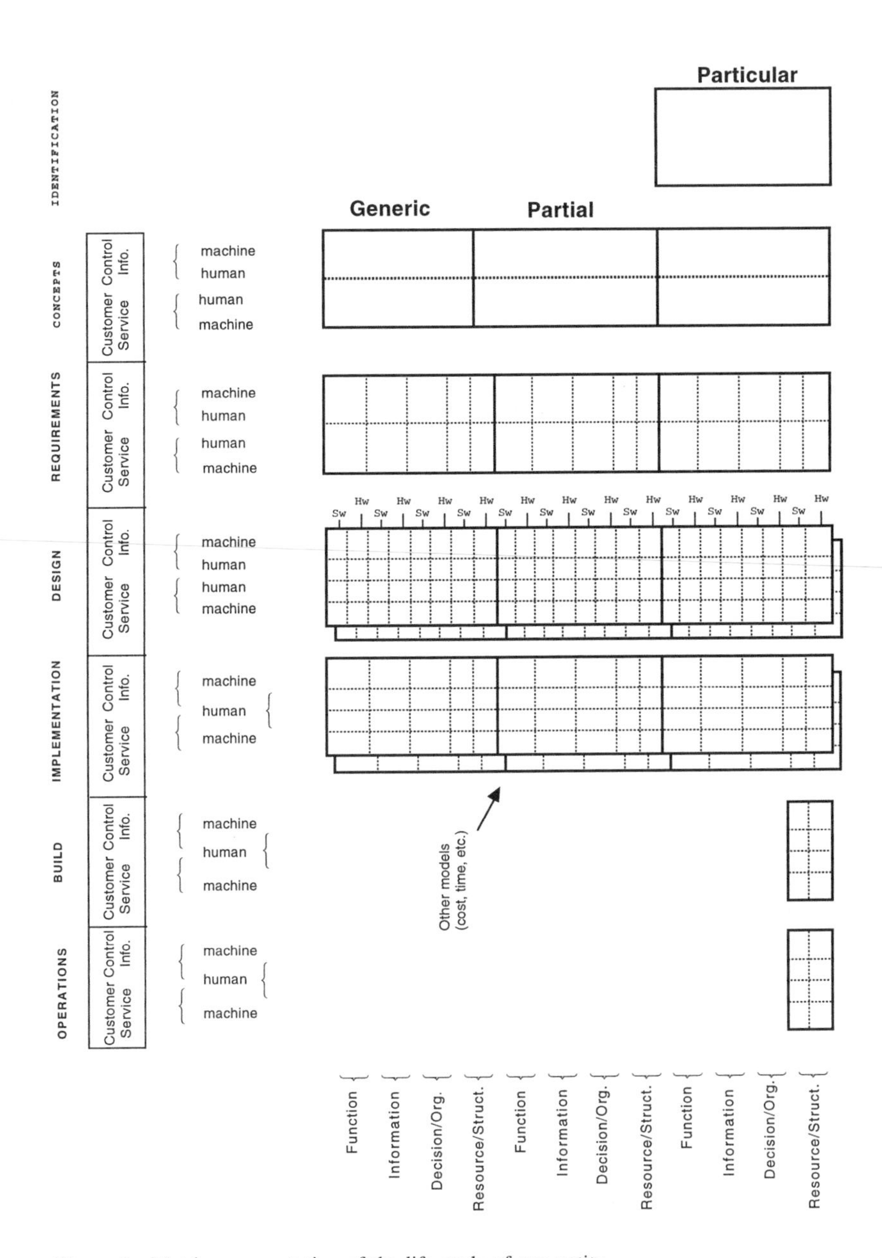

Figure 1 Matrix representation of the life-cycle of any entity.

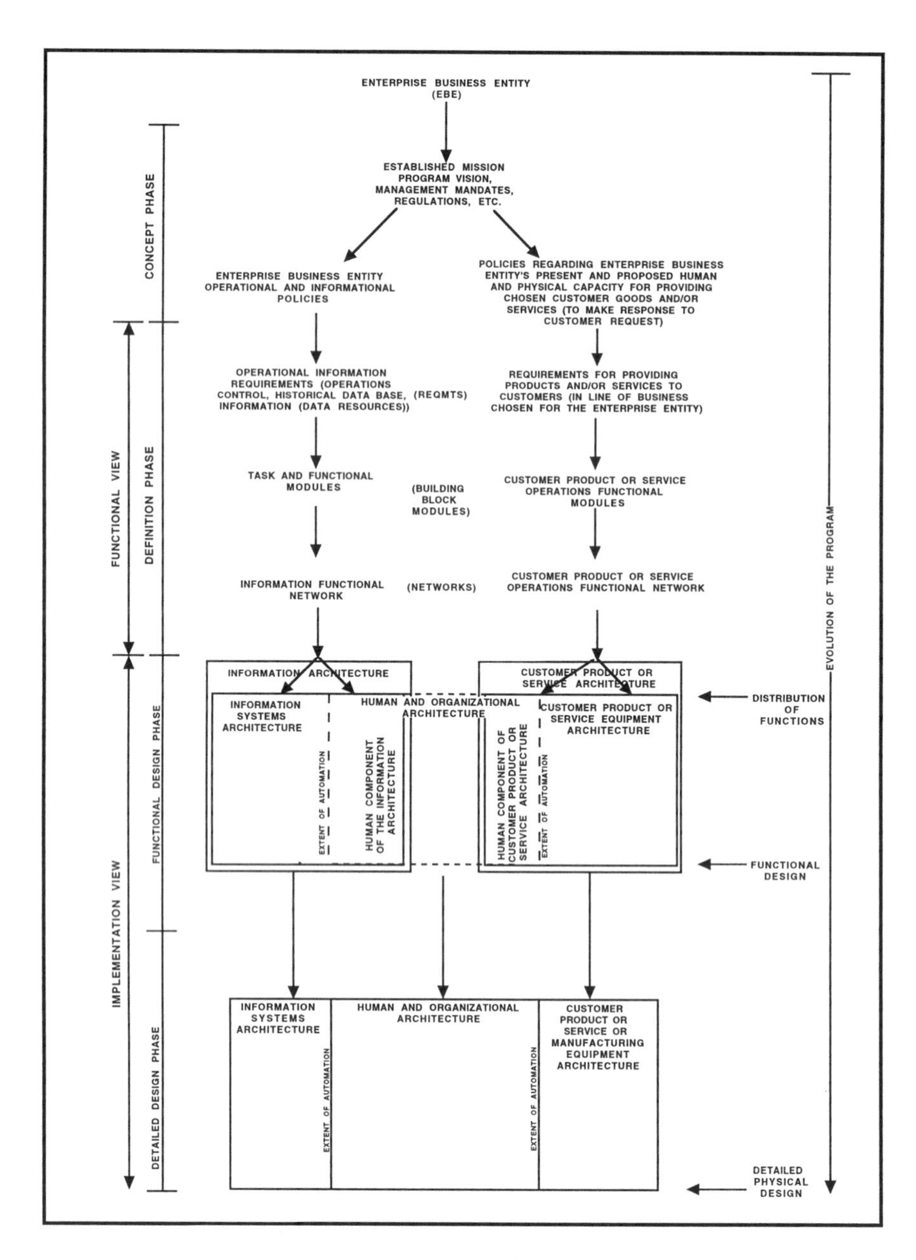

Figure 2 Detailing the development of an enterprise integration program.

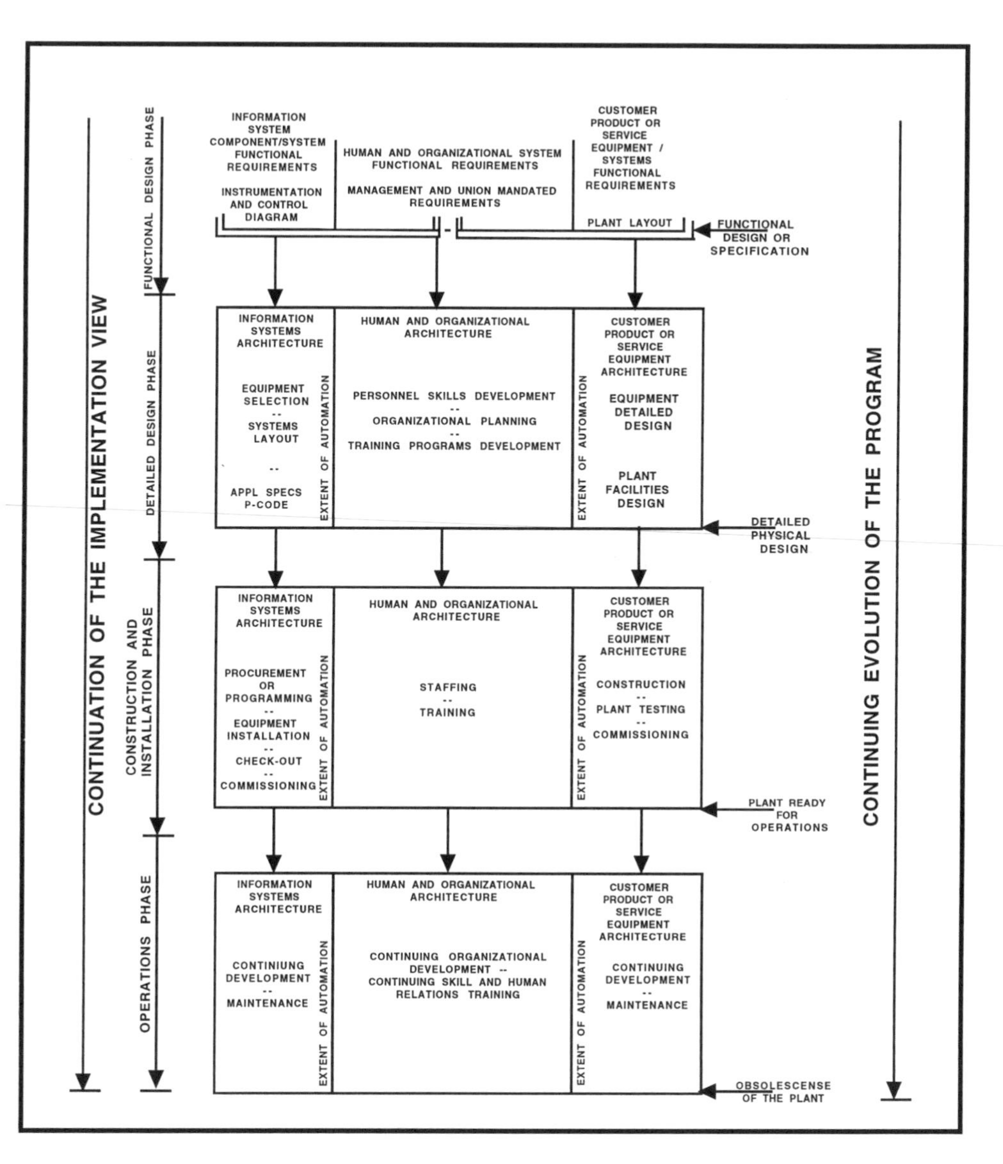

Figure 2 (cont.) The later phases in the detailing of the enterprise integration system evolution and its tasks.

1. Milestones to develop GERA:
   a) Take the Purdue Enterprise Reference Architecture, PERA, as the most complete in scope and add to it the areas discovered in the study of the four recursive matrices and their associated entities. This will include the development of the complete specification of GERA (i.e. including the concept, identification, and requirements and design of GERA). The specifications will also have to include the definitions of the other deliverables.
   b) Develop two presentation forms:
      1) Presentation of GERA for the user community:
         For the user community, preferably use the two tiered diagram format of PERA (Figure 2) (now presenting four such diagrams according to the four entities of Section 2). The two tiered diagram format will **group** together those "subject matter" areas which will belong together in the methodology. This presentation can be prepared on the basis of the PERA document by adding further chapters.
      2) Presentation of GERA for the developer and researcher (tools and theoretical aspects) community:
         The second presentation is for the methodology, modeling, research, and standardization community, preferably in the form of the four recursive matrices of Figure 1. The matrix presentation will be seen as a further development of the CIMOSA cube. The matrix presentation will use a refined subject matter presentation drawing on the one hand from the corresponding CIMOSA definitions and on the above subject matter groups (as developed for the user community).
2. Milestones to develop GEEM:
   a) Determine, given GERA, the areas which should be but are not yet covered by the Purdue Methodology (Williams, 1994).
   b) Extend the methodology regarding the specification and design of the decision system of the enterprise, based on GIM where these details are highly developed and tried.
   c) Based on the extensions to the CIMOSA modeling framework (GEMT) illustrate, with examples and specific instructions, the steps of the methodology.
3. Milestones to develop GEMT&L, GEM and GMs;
   Given the above uncertainties these milestones need to be re-evaluated in light of the feasibility of the first few steps.
   a) Determine which modeling tasks are not currently addressed by CIMOSA or GIM.
   b) Identify the complementing set of languages as needed. Draw on the IDEFX set of languages.
   c) Feed the result into the definition of the Generic Enterprise Engineering Methodology (GEEM) project. This ensures that the existing set of modeling constructs will be met by a compatible presentation of the Enterprise Engineering Methodology.
      Up to this point this is a feasible engineering project. The following tasks, however, need to be addressed by research.
4. Milestones to develop GTs;
   a) Integrate these languages (as described above) on the meta level — possibly using ontological theories such as developed in TOVIE (Fox, 1992) or IDEFX. Since the core CIMOSA languages are already based on one integrated meta-model this is not a project from scratch.
   b) Express the languages as views of an underlying ontology.

Comments by Task Force Members on this proposal as initially presented by Bernus and Nemes have concentrated on four major points:

1. The need on the part of final enterprise integration users for the simplest and most easily understandable and usable system of methodology, tools and illustrating architecture.
2. Means for allowing methodology and tool initiating and producing individuals and organizations to readily develop their offerings in this area and as equally readily incorporate them into the user's potential work package.
3. Means by which individuals and groups doing research and development in this field could easily contribute their findings to Item 2 above and eventually to Item 1.
4. The vital need for development and use by the Task Force of one agreed-upon Glossary. Even the most common terms in our field are used in different ways by each of the individual development groups in our Task Force, let alone others in the field.

Most prominent in the comments was the often expressed call for major help to potential users by developing a simple, easy-to-use, robust methodology with its associated tools and descriptive architecture. Excessive complexity was cited often as the principal impediment in the use of these technologies in industry today. In these times of drastically reduced engineering and research staffs in most companies, those companies simply do not have the personnel to devote the time necessary to learn and use the currently available systems in view of the present complexity and difficulty of understanding.

Bernus and Nemes had anticipated this message by proposing a two level system with the PERA two-tier diagram (Williams, 1992) (Figure 2) and the Purdue Methodology (Williams, 1994) as an initial recommendation for users applications. This would be supported by the second, or so-called matrix form of the architecture (Figure 1) to assist tool and methodology developers in their work. Finally they proposed the development of ontological models and theories as suitable and needed research tasks for academicians and other theoretical workers in the field.

The issue of a Glossary surfaced often during the Vienna meeting discussion (June 1994) and must be a priority topic in future meetings as even the most common terms have obviously different meanings when used by different individuals. An initial version of such a Glossary is included in the Purdue Methodology (Williams, 1994).

6 THE PATH AHEAD

The Task Force will thoroughly discuss the above results in its future meetings. The Task Force is committed to make a major report of its findings to date at the Triennial Congresses of its parent bodies: IFAC, the International Federation of Automatic Control and IFIP, the International Federation for Information Processing, both of which will be held in 1996.

The Task Force has committed itself to produce the following deliverables in its continuing program for the next two years of its current term:

1. A trade journal version of the GERAM definition article (Bernus and Nemes, 1994). This article will be produced in the near future to publicize the work of the Task Force.
2. A thorough review and early publication of the Purdue Methodology document (Williams, 1994) as the initial version of GEEM.
3. An investigation of the potential for developing a Hypertext version of Item 2 above.
4. Investigation of the need for specializing the Methodology of Items 2 and 3 for use by specific industries.
5. Development of a report and instructional guide for the use of CIMOSA Constructs with the PERA diagram and Methodology.
6. A teachable book for university and industrial classes on the CIMOSA architecture, tools, models and languages.

7 REFERENCES

AMICE Consortium (1992) AMICE: CIMOSA, ESPRIT Project 5288, Milestone M-2, AD2.0 **Volume 2**, Architecture Description, Document RO443/1, Consortium AMICE, Brussels, Belgium.

Bernus, P., and Nemes, L. (1994) "A Framework to Define a Generic Enterprise Reference Architecture and Methodology." Paper submitted for the Third International Conference on Automation, Robotics and Computer Vision (ICARCV '94), Singapore.

Doumeingts, G., Vallespir, B., Zanettin, M., and Chen, D. (May 1992) *GIM, GRAI Integrated Methodology, A Methodology for Designing CIM Systems*, Version 1.0, Unnumbered Report, LAP/GRAI, University Bordeaux 1, Bordeaux, France.

Fox, M.S. (1992) "The TOVIE Project, Towards a Common Sense Model of the Enterprise," in *Enterprise Integration,* Ch. Petrie (Ed.), MIT Press, Cambridge, Massachusetts, USA.

Williams, T.J. (1992) *The Purdue Enterprise Reference Architecture*, Instrument Society of America, Research Triangle Park, North Carolina.

Williams, T.J., et al. (1993) IFAC/IFIP Task Force on Architectures for Integrating Manufacturing Activities and Enterprises, *Architectures for Integrating Manufacturing Activities and Enterprises Technical Report,* Theodore J. Williams, Editor, Purdue University, West Lafayette, Indiana, USA.

Williams, T.J. (Ed.) (1994) *A Guide to Master Planning and Implementation for Enterprise Integration Programs, Report 157,* Purdue Laboratory for Applied Industrial Control, Purdue University, West Lafayette, Indiana, USA.

8 BIOGRAPHY

Professor Theodore J. Williams is Professor Emeritus of Engineering and Director Emeritus of the Purdue Laboratory for Applied Industrial Control having retired in December 1994 after thirty years of service at Purdue University. He has been Chairman of the IFAC/IFIP Task Force since 1990. He is a Past President of the Instrument Society of America, of the American Automatic Control Council, and of the American Federation of Information Processing Societies.

20

CEN activities on enterprise modelling and enterprise model execution and integration services

D.N. Shorter
Convenor of CEN TC310 WG1
IT Focus, Folia, Flowers Hill, Pangbourne, Berks RG8 7BD, UK
Telephone: 01734 843949, Fax: 01734 842493,
Email: david@itfocus.demon.co.uk

Abstract

Today's manufacturing enterprise needs both increased integration of its processes and resources, while at the same time retaining the ability to respond rapidly to an ever changing market requirements. Enterprise modelling aims to meet these potentially conflicting objectives. The paper describes CEN's work to establish the standards that are needed to support Enterprise Modelling for manufacturing, in particular on a framework of concepts, an evaluation of and preliminary standards for modelling constructs, and a statement of requirements for the environment in which model constructs are to be developed and executed (EMEIS).

1 OBJECTIVES FOR THE WORK

There are two potentially conflicting business drivers which are of particular importance for today's manufacturing enterprise. These are:

- Developing, maintaining and improving the integration of manufacturing processes and resources.
- Redesigning manufacturing and business processes to meet changing market conditions.

Integration of data and integration of communications are prerequisites and other standards work is addressing these requirements. This paper describes a third and complementary approach – integration through an enterprise model.

Management techniques such as Just in Time, Business Process Re-engineering, the Virtual Enterprise etc. all place new demands on the ability to redeploy manufacturing resources flexibly, and to do this in such a way as to retain the degree of integration necessary to reduce work in progress and associated lead times. Enterprise modelling is intended to be both to be supportive of manufacturing integration, and to facilitate organisational and process redesign.

2 THE ENTERPRISE MODELLING APPROACH

Enterprise Modelling for manufacturing is a technique which sets out to meet these objectives by explicitly modelling a manufacturing enterprise.

For a particular enterprise, the model needs to represent all relevant aspects of the agents, processes and resources which are involved; to do this in a way which ensures consistency and completeness; and to provide the possibility of executing these models to direct enterprise operations (possibly after a translation process).

CEN TC310 WG1 (previously CEN/CENELEC WG/ARC) is a standardization committee which has been working to develop the standards necessary for such models in five work areas:

- Producing a Framework for Enterprise Modelling which sets out the general requirements and terminology for such enterprise models.
- Evaluating technologies which might form general components or 'building blocks' for models.
- Developing a Statement of Requirements for the IT services (EMEIS) that are required for the execution and integration of model components, following an evaluation of existing approaches.
- Producing a vocabulary for CIM systems architecture for guidance to those working in the field, and as an organising document for its own work.
- Working to produce a PreStandard for building blocks, known as 'Constructs for Views'. This is the main item of on-going work.

How the WG is organised will be discussed after describing these work areas in somewhat more detail.

3 THE ENV 40 003

In 1991 CEN published the Framework for Enterprise Models (ENV 40 003) which defines '... a framework for ... computer-based modelling of enterprises, focusing on Discrete Parts Manufacturing. Models generated using this framework will ultimately be computer executable and possibly enable the daily operations of an enterprise to be run, monitored and controlled by such models.' This Framework was developed from a substantial contribution from the AMICE consortium which produced the CIMOSA concept (Open System Architecture for CIM), with further inputs from industry and academia.

In particular, the modelling concepts of the Framework support:

- ***Genericity*** – the ability to specialise from general models to partial models used by a specific industry sector, and even by a particular enterprise.
- ***Model*** lifecycle issues, e.g. the ability to encompass the three main phases of requirements specification, design and implementation

- The selective perception provided by ***Views***. Each View concentrates on some aspect of the model for a particular purpose and hide irrelevant aspects to reduce complexity. Whether there is to be just one set of views is still a matter of some discussion, but the ENV (and the current work of the WG1) has adopted for the moment the four views of Function, Information, Resource and Organisation.

The first two of these Views, Function and Information, correspond to 'classical' functional and information analysis. The third, Resource, supports the modelling of the use, consumption and production of (largely) material resources. The last, Organisation, is more novel – it allows for the specification of which part of the business has the responsibility for initiating which manufacturing activity and under what conditions. (It is somewhat similar to the use of explicit modelling of workflow in office environments.)

The explicit representation of enterprise entities according to the Views of the ENV provides the following benefits:

- Function View: Improved flexibility and ability to adapt through more flexible coupling between processes.
- Information View: Capture of 'organisational memory' and improved information re-use; a prerequisite for inter-enterprise trading, and for a manageable and robust IT structure.
- Resource View: Again flexibility and adaptability; reduction of work in progress.
- Organisation View: Increased ability to adapt and respond to changing market conditions and appropriate manufacturing logistics.

These Framework concepts have, of course, a close correspondence to the modelling framework of CIMOSA on which they were based.

How the model is implemented, e.g. as computer-interpretable data, is not the main issue for WG1. For example, while object-oriented approaches are very powerful in terms of their ability to handle complexity, they are not the only approach to implementation. All object-oriented methods contain some key technical concepts which can help in managing the representation of complex requirements and designs (encapsulation, inheritance, polymorphism). But until recently, the tools which allow an OO analysis of a manufacturing situation and the design of appropriate software structures did not easily lead to the development of executable code. Generally at some stage the developer is required to develop fragments of code to supply the required functionality. This situation is changing however as new tools become available.

4 EVALUATION REPORT ON CONSTRUCTS

After the development of the ENV 40 003, WG1 set out to clarify requirements for generic model components (components which could be specialised to meet particular enterprise needs). These components needed to be in accord with the ENV, and also capable of computer representation and checking.

A number of detailed requirements were identified from ENV 40 003. These were then consolidated into a set of detailed questions that should be asked about particular modelling methods.

Following a call for input, the WG received 11 sets of material to review. Each was summarised in terms of its source, documentation available, status, description, evaluation summary and the implications for the WG in its task of producing standardized constructs. Tables contained the detailed evaluations for each method.

At the time that the evaluation was carried out (1992), the main conclusions were that:

- The different methods covered to a varying extent some part of what was needed – but no one method was sufficiently comprehensive for the WG's objectives.
- The more comprehensive approaches were research projects that had not yet been sufficiently demonstrated to be practicable.
- Generic model components could be realised using both function oriented and object oriented approaches and that any standard needed to be open to both.
- Generic model components particularly designed to address one View of the ENV 40 003 could not be studied in isolation from components designed for the needs of other Views.

This report was published by CEN as R-IT-06 in April 1993, and has been republished in late '94 in a special issue of 'Computers in Industry' on CIM Architectures (Ed: Prof. H Wortmann, Sept. 1994, Elsevier Science B.V.) by kind permission of the Nederlands Normalisatie-instituut.

5 REQUIREMENTS FOR ENTERPRISE MODEL EXECUTION AND INTEGRATION SERVICES (EMEIS)

5.1 Background

The work item 'Enterprise Model Execution and Integration Services' is aiming to provide a specification of the standards required for the (possibly distributed) computing environment that is needed to execute an enterprise model. WG1 has now produced a statement of requirements which is outlined in this section and as assessment of available technology described in Section 7. This statement will be developed into another ENV after comments and further inputs have been received from interested parties.

5.2 The 'basic reference model' for EMEIS

In order to put the different kinds of requirement in an overall context, the WG developed a simple model of the key concepts and their relationships. This had three main components:

- Model development services – the collection of services which in total provide the environment within which model components are ***developed and tested*** before release to the EMEIS for use as an executable model of an enterprise, or as part of such a model.
- Model execution services – the collection of services which (i) ***embed*** a model component into EMEIS, so converting it into a runnable entity and (ii) provide all the ***operations*** services that are particular to the execution of such runnable entities, or for the provision of other CIM-specific services, over and above the IT Base services .
- IT Base services – the ***general IT services*** which are not CIM specific. These are concerned with systems qualities such as portability of applications, interworking between open systems and distribution transparency in distributed platforms.

The relationships between these are illustrated in Figure 1.

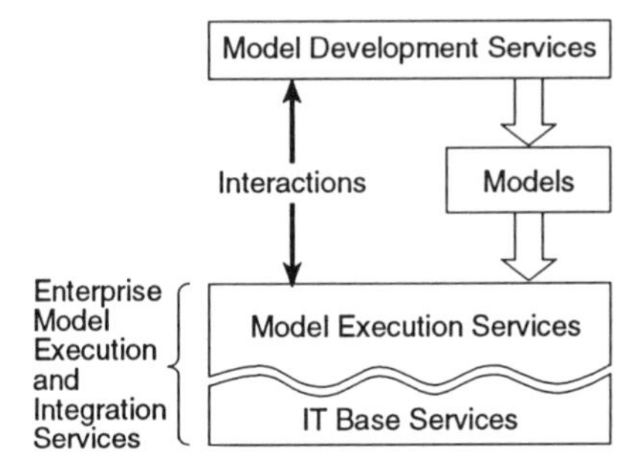

Figure 1 The reference model for EMEIS

5.3 Guiding principles for EMEIS requirements

In developing the statement of requirements, the WG articulated and used a number of guiding principles. These were:

- Conformance to existing standardization, including ENV 40 003.
- Openness to different modelling approaches – because enterprise models 'compliant' with ENV 40 003 are difficult to find.
- Description of EMEIS in, at least, two levels of abstraction (functional representation and IT representation).
- The use of services and service requesters (clients) to achieve a degree of technological independence.
- Services to be encapsulated (describe the 'what' not the 'how').
- System structuring according to system theory.
- Use of an Applications Program Interface to provide transparency of service provision, application portability etc.
- Use of protocols to control data exchange.
- EMEIS as a requirements framework should contain abstractions for the mechanisms which support the EMEIS requirements, not the mechanisms themselves.

The principle embedded in ENV 40 003 of enterprise integration through the use of an explicit model led the WG to make some particular assumptions about the relationships between:

- The environment and services which are needed to develop the model or model component (MDS),
- The services required to embed and operate the model (MXS),
- The General IT Services which provide access to enterprise data.

5.4 CIM-specific requirements

Led by consideration of particular requirements and design rules for CIM, including that of the essentially physical nature of the manufacturing process, the following requirements were identified (not all of which are unique to CIM but which have been given a particular interpretation):

- What is to be provided by MDS, including the necessary scope of an enterprise model and the stages in its development.
- The required granularity of model, including guidance on when to stop modelling.
- The need to support three different types of model component (compiled, interpretable and parameter-driven).
- The ability to compose models, possibly expressed in different languages corresponding to different Views, into one integrated model; the need for a mechanism to support that process; and the need to address life-cycle issues for model components.
- What is involved in embedding a model to make it an executable entity.
- Where model bindings should be established, corresponding to the rate of change (from organisational change to resource management to immediate 'make' instructions).
- The services that are needed to provide the resources necessary for a model component to execute.
- The need for binding mechanisms in both MXS and MDS (both early and late binding).
- The characteristics of the information that needs to pass between the MDS and MXS.
- Incremental development of model components.
- End user visibility of model components and the operation of those components (or assemblages of these).

Based on these requirements, and especially the principle of services, the report proposes a way of identifying what future EMEIS standards might be needed. An initial set of possibilities is identified as:

- A common mechanism for describing model components in different languages and provided by different suppliers.
- A common mechanism for interactions (messages) between MDS and MXS.
- A standardized way of describing model behaviour.
- A common set of semantics for the states of model components and the signalling of those states.
- A common procedure or process for declaring, registering and withdrawing model components.

It should be noted that there are many similarities here with the requirements for Integrated CASE tools in software engineering.

5.5 General requirements for EMEIS

Services are classified into three groups (although the boundary between EMEIS and IT Services is not clear cut):

- Model Development Services, MDS
- Model Execution Services, MXS, in turn comprising
  - Model Embedding Services
  - Model Operation Services
- General IT Services

Often services can be decomposed into other services, but from the point of view of specifying EMEIS requirements, decomposition is required only where it is desired to constrain EMEIS services and service components in some way.

The report sets out a list of IT-related service requirements for MDS and MXS, covering such issues as lifecycle requirements, simultaneous development, namespace management, recovery procedures, etc.

General IT services have to provide support for model communication with the real world, including access to the current state of the enterprise, the presentation of information to human operators and the acceptance of human intervention etc. These services are to provided using possibly heterogeneous and distributed IT resources. General IT services should be decoupled as far as possible from the underlying technology, make maximum use of services which are the subject of available standards, and use existing services and protocols for agent protocols. Again, a list contains more detailed requirements.

6 EVALUATION OF EMEIS CONTRIBUTIONS

In preparing the review of relevant project initiatives, CEN TC310/WG1 prepared a checklist against which the contributions were reviewed. Experience in this evaluation contributed to the Statement of Requirements described above. A necessarily condensed version of the checklist follows:

Q1 What visibility is there of concepts of model, models, model components?

Q2 How does the model represent derived functionalities? Embedded processes (include control, timing and behaviour)? Necessary information and other resources? What else is represented explicitly in this approach (e.g. obligations as in Eiffel)?

Q3 What is the level of integration, from (lowest) communication between running model processes, through shared data or services, to meta-models, shared semantics and semantic unification (highest)?

Q4 What binding paradigm is used (concept, interpretation)? Where does binding take place etc.?

Q5 What modelling language is used to represent the model(s) or model components and what attributes of the model does this language capture?

Q6 Does the model support predictability of run-time behaviour (including performance)? Is the use of resources and methods predetermined or opportunistic?

Q7 What is the paradigm used for invocation of executable model component? (Eager evaluation, lazy evaluation, controlled evaluation, context-driven application protocol)?

Q8 What CIM-specific semantics or application-specific semantics are visible during model execution? (e.g. CIM ontology, class library or CIM-specific notions such as consumable resources)

Q9 Is the question of lifecycle (of model(s), of model component) addressed and, if so, how? Is the approach used linked to the binding mechanism used, e.g. by maintenance of a binding trail allowing components to be withdrawn or replaces and the consequences managed?

Q10 What is the definition of service in the contribution being evaluated? The ISO TC184 SC5 WG1's definition was used as a reference concept here.

Q11 How are the 'General IT Services' and other services accessed? Five sub questions asked about: the assumed execution environment; properties of the execution environment; visibility of IT service components; when and how such services are invoked; and how Enterprise Model Execution and Integration Services are characterised in the approach.

Q12 How is the 'Level' between the model and its executing environment (the one in which the model is executed, its execution environment) described?

These questions were used as a set of concerns to assess six contributions, as summarised in Table 1. Each proposal was summarised based on the contributions supplied. The WG wishes to express its thanks to the sources of these contributions, and also for the assistance giving in reviewing the assessments and clarifying various points.

Table 1 Contributors to the EMEIS evaluation

CIMOSA	N198, N199, N245 [N198 (CIMOSA, Group A) 'Framework for Integrating Infrastructure', CIMOSA, 15 June 1992; N199 (CIMOSA, Group A) 'Formal Reference Base for IIS', CIMOSA, 15 June 1993; N245 (CIMOSA, Group A) 'Relation between CIMOSA model and IIS, Kurt Kosanke, 11/5/93; pages 2-3 (N245).
Flatau	N219 'Framework for an Integrating Infrastructure' Ulrich Flatau and Dieter Wüsterfeld, October 1992. This proposal is essentially that contained in ISO TC184 SC5 WG1's N300 as of late 1992.
MIDA	246, MIDA: An Open Systems Architecture for the Model-Oriented Integration of Data and Algorithms; M. H. R. Michalski, D. Solte, F. Vicuna; FAW Ulm, 11/5/93; pages 1...25.
PISA	N238, Framework for IIS, W Gielingh, 5/3/93.
TOVE	N239, A common-sense model of the enterprise, Mark S Fox, J Chionglo, Fadi G Fadel, 19/2/93.
ISA	A framework for information systems architecture, J A Zachman, IBM Systems Journal, Vol 26, No 3, © 1987 IBM; Extending and formalising the framework for information systems architecture, J F Sowa and J A Zachman, IBM Systems Journal, Vol 31, No 3, © 1992 IBM.

Following the summary of each project, an assessment was made using the questions above as a common frame of reference. These assessments were later sent to the projects and revised where necessary after clarification.

The main conclusions of the evaluation report were that:

- Carrying out the assessments made a major contribution to WG1's understanding of requirements, and greatly contributed to the Statement of Requirements described earlier. It would have been desirable to align the questions above with that Statement, and then to repeat the assessment exercise but this was not possible in the time available.
- Several of the initiatives contain requirements statements and design proposals covering at least some part of EMEIS requirements – but none encompasses all the reference concepts or functionality seen as necessary by WG1. At the time of the evaluation (mid-late 1993) most (all?) of the contributions were research or development projects, and industrial experience was lacking, especially for MXS.

7 FUTURE WORK ON EMEIS

The reports on EMEIS Statement of Requirements and the EMEIS evaluations are being published by CEN to elicit wider feedback. It is then likely that the work on EMEIS requirements will be reworked and submitted for ballot as a European PreStandard (ENV) on EMEIS Requirements.

At a suitable time, probably as the work on Constructs for Views nears completion, the WG intends to issue a call for new input, based on a somewhat revised Statement of EMEIS Requirements, or against an initial draft of the intended ENV.

During these developments, the WG is particularly concerned to maintain appropriate liaisons and further develop these where necessary, to ensure that new standards are developed only where the requirements cannot be fulfilled by existing or developing standards. Liaisons with ISO TC184 SC5 WG1 and WG4 will continue through cross membership and exchange of documents. Other liaisons that have been identified as important are with the IFAC/IFIP Task Force on Architectures for Enterprise Integration, and with a number of activities in ISO/IEC JTC1 SC21.

At the risk of introducing unmanageable complexity, the WG would welcome expressions of interest in liaison from other standardization groups. Section 11 describes how experts wishing to contribute more directly can be nominated to participate more directly in the work of WG1.

8 VOCABULARY FOR CIM SYSTEMS ARCHITECTURE

During the course of its work, WG1 identified the need for 'common and documented understanding of terms and definitions in the field of CIM Systems Architecture.' This is particularly a problem because many of the words used are already used with somewhat different meanings in other fields, e.g. in Systems Engineering, and in IT generally.

The terms and definitions of the vocabulary were produced taking into account already established terms in ISO (in particular the terms used by ISO TC184 SC5 WG1) and elsewhere. The group is maintaining links with others active in this field and is exchanging documents.

The draft N26 was been produced in 2/3/94 and recommended for circulation by CEN. Depending on that response, the document may progress to a standards document. Alternatively it could remain as a very necessary aid to WG1's future work.

9 PRESTANDARD ON CONSTRUCTS FOR VIEWS – WORK IN PROGRESS

WG1 is now working on the definition and representation of the further concepts that are required. In particular, if model components are to be procured from multiple sources, or indeed later re-used elsewhere within an enterprise, then there needs to be an adequate way of describing the significant properties of a model in terms of these concepts (a standardized modelling language), and of specifying precisely what each concept signifies (e.g., in a formal specification in a language such as EXPRESS).

A further desirable objective is for a common graphical representation to allow easier understanding of model components and the relationships between these – for both model-user and model-developer.

This work item is setting out to meet these objectives by creating a European Standard (initially an ENV) as an input to and a precursor of international work in ISO. Substantial input has been received from two European initiatives and WG1 also maintains links with international developments. However, the work is not just standardizing the status quo – it is open to new proposals for constructs in accordance with ENV 40 003.

10 CEN TC310 WG1, SYSTEMS ARCHITECTURE FOR MANUFACTURING

CEN TC310 WG1 is a working group of CEN, the European Committee for Standardization located in Brussels, and in particular of Technical Committee 310, Advanced Manufacturing Technologies. The mission of TC310 is to make sure that the standards required for Advanced Manufacturing Technologies are available to and known by European industry.

Priority is given to using international standards wherever this is possible and to undertake work in Europe only where a need has been identified which is not being met by developments internationally. For example, CEN TC310 WG1 maintains an active liaison with ISO TC184 SC5 WG1, Modelling and Architecture, and several members are active participants in both committees. The work programme and strategy of TC310 is documented in a CEN/CENELEC document known as M-IT-04, which provides an organised classification of work items, their time scales and the strategy for their development..

The current work of TC310 can be grouped into four main areas:

- Product Data Exchange, concentrating on the specific European needs for STEP developments,
- CIM Systems Architecture, the subject of this overview,
- Standards Parts Libraries – standards for representation of standard parts to be used by CAE systems,
- Ergonomics, where future directions for Europe are being set by a workshop,

The principal objective for WG1, CIM Systems Architecture, is to ensure that the requirements of European industry are met, so that maximum advantage can be taken of standardization for enterprise modelling and the use of development environments that will influence the industrial organisation, management and manufacturing approach to improving efficiency. WG1 has noted that enterprise modelling is potentially a powerful enabler in planning and implementing organisational and structural change.

A secondary objective is to ensure that existing expertise is deployed in the preparation of the standards, and that the necessary information is available to interested parties in the form of publicly available standards.

Enterprise integration via enterprise modelling is a new technology and requires R&D input for progress to be made. This standards work therefore has a strong dependence on the results of European research initiatives such as ESPRIT, and is acting as an important technology transfer mechanism in making those results available to European industry.

11 WHERE NEXT

Having completed a Framework for Enterprise Modelling as ENV 40 003, WG1 is now working to standardise requirements for constructs in accordance with that framework, and to standardize requirements for the environment within which models will be executed. When complete, this work will form the basis for the longer term development of standardized constructs and services within the executing environment.

As constructs become standardized, they will be more widely accepted and be able to deliver similar benefits to a wider community of users. At the same time, greater acceptance of these concepts will encourage further tool development, the establishment of libraries of useful particular models and increased availability of skilled expertise to apply and teach the technology.

Standardization offers particular benefits in promoting exchangeability and re-use of model components. It should also encourage the development of model-related products and services by providing a recognised and stable basis for these developments (e.g., a common and non-proprietary form of representation, common usage of terms etc.).

By participating in this work, companies, research organisations and consultants can influence its direction and gain insights into how these developments can be applied in their own business, products or services. In particular, participation can allow the merits of a particular technical approach to be presented to an informed group of experts, and if accepted, to become a constituent part of an emerging European standard. (Participation in the work is open to experts nominated by their National Standards Organisation, who must be members of CEN, and occasionally to invited experts.)

BIOGRAPHY

David Shorter is an independent IT consultant trading as IT Focus, offering strategic consultancy in IT-related matters based on his experience as Director of IKBS for the UK Alvey programme, as manager of the external research and technical coordination programmes for SD-Scicon and on his work in the development of Frameworks and Reference Models for enterprise modelling, and the representation of these using object-oriented techniques. He is also convenor of the CEN Working Group on Enterprise Modelling for CIM (TC310 WG1).

INDEX OF CONTRIBUTORS

KEYWORD INDEX